OR/MS AT WORK

INTERNATIONAL SERIES IN OPERATIONS RESEARCH & MANAGEMENT SCIENCE

Frederick S. Hillier, Series Editor
Stanford University

Saigal, R. / *LINEAR PROGRAMMING: A Modern Integrated Analysis*
Nagurney, A. & Zhang, D. / *PROJECTED DYNAMICAL SYSTEMS AND VARIATIONAL INEQUALITIES WITH APPLICATIONS*
Padberg, M. & Rijal, M. / *LOCATION, SCHEDULING, DESIGN AND INTEGER PROGRAMMING*
Vanderbei, R. / *LINEAR PROGRAMMING: Foundations and Extensions*
Jaiswal, N.K. / *MILITARY OPERATIONS RESEARCH: Quantitative Decision Making*
Gal, T. & Greenberg, H. / *ADVANCES IN SENSITIVITY ANALYSIS AND PARAMETRIC PROGRAMMING*
Prabhu, N.U. / *FOUNDATIONS OF QUEUEING THEORY*
Fang, S.-C., Rajasekera, J.R. & Tsao, H.-S.J. / *ENTROPY OPTIMIZATION AND MATHEMATICAL PROGRAMMING*
Yu, G. / *OPERATIONS RESEARCH IN THE AIRLINE INDUSTRY*
Ho, T.-H. & Tang, C. S. / *PRODUCT VARIETY MANAGEMENT*
El-Taha, M. & Stidham , S. / *SAMPLE-PATH ANALYSIS OF QUEUEING SYSTEMS*
Miettinen, K. M. / *NONLINEAR MULTIOBJECTIVE OPTIMIZATION*
Chao, H. & Huntington, H. G. / *DESIGNING COMPETITIVE ELECTRICITY MARKETS*
Weglarz, J. / *PROJECT SCHEDULING: Recent Models, Algorithms & Applications*
Sahin, I. & Polatoglu, H. / *QUALITY, WARRANTY AND PREVENTIVE MAINTENANCE*
Tavares, L. V. / *ADVANCED MODELS FOR PROJECT MANAGEMENT*
Tayur, S., Ganeshan, R. & Magazine, M. / *QUANTITATIVE MODELING FOR SUPPLY CHAIN MANAGEMENT*
Weyant, J./ *ENERGY AND ENVIRONMENTAL POLICY MODELING*
Shanthikumar, J.G. & Sumita, U./*APPLIED PROBABILITY AND STOCHASTIC PROCESSES*
Liu, B. & Esogbue, A.O. / *DECISION CRITERIA AND OPTIMAL INVENTORY PROCESSES*
Gal, T., Stewart, T.J., Hanne, T./ *MULTICRITERIA DECISION MAKING: Advances in MCDM Models, Algorithms, Theory, and Applications*
Fox, B. L./ *STRATEGIES FOR QUASI-MONTE CARLO*
Hall, R.W. / *HANDBOOK OF TRANSPORTATION SCIENCE*
Grassman, W.K./ *COMPUTATIONAL PROBABILITY*
Pomerol, J-C. & Barba-Romero, S. / *MULTICRITERION DECISION IN MANAGEMENT*
Axsäter, S. / *INVENTORY CONTROL*
Wolkowicz, H., Saigal, R., Vandenberghe, L./ *HANDBOOK OF SEMI-DEFINITE PROGRAMMING: Theory, Algorithms, and Applications*
Hobbs, B. F. & Meier, P. / *ENERGY DECISIONS AND THE ENVIRONMENT: A Guide to the Use of Multicriteria Methods*
Dar-El, E./ *HUMAN LEARNING: From Learning Curves to Learning Organizations*
Armstrong, J. S./ *PRINCIPLES OF FORECASTING: A Handbook for Researchers and Practitioners*
Balsamo, S., Personé, V., Onvural, R./ *ANALYSIS OF QUEUEING NETWORKS WITH BLOCKING*
Bouyssou, D. et al/ *EVALUATION AND DECISION MODELS: A Critical Perspective*
Hanne, T./ *INTELLIGENT STRATEGIES FOR META MULTIPLE CRITERIA DECISION MAKING*
Saaty, T. & Vargas, L./ *MODELS, METHODS, CONCEPTS & APPLICATIONS OF THE ANALYTIC HIERARCHY PROCESS*
Chatterjee, K. & Samuelson, W./ *GAME THEORY AND BUSINESS APPLICATIONS*
Hobbs, B. et al/ *THE NEXT GENERATION OF ELECTRIC POWER UNIT COMMITMENT MODELS*
Vanderbei, R.J./ *LINEAR PROGRAMMING: Foundations and Extensions, 2nd Ed.*
Kimms, A. /*MATHEMATICAL PROGRAMMING AND FINANCIAL OBJECTIVES FOR SCHEDULING PROJECTS*
Baptiste, P., Le Pape, C. & Nuijten, W./ *CONSTRAINT-BASED SCHEDULING*
Feinberg, E. & Shwartz, A./ *HANDBOOK OF MARKOV DECISION PROCESSES: Methods and Applications*
Ramík, J. & Vlach, M. / *GENERALIZED CONCAVITY IN FUZZY OPTIMIZATION AND DECISION ANALYSIS*
Song, J. & Yao, D. / *SUPPLY CHAIN STRUCTURES: Coordination, Information and Optimization*

OPERATIONS RESEARCH/ MANAGEMENT SCIENCE AT WORK

edited by

Erhan Kozan
and
Azuma Ohuchi

Springer Science+Business Media, LLC

Library of Congress Cataloging-in-Publication Data

Kozan, Erhan.
Operations research, management science at work / edited by Erhan Kozan and Azuma Ohuchi.
p. cm. – (International series in operations research & management science; 43)
Includes bibliographical references and index.
ISBN 978-0-7923-7588-3 ISBN 978-1-4615-0819-9 (eBook)
DOI 10.1007/978-1-4615-0819-9
1. Operations research. 2. Management science. I. Kozan, Erhan. II. Ohuchi, Azuma. III. Title. IV. Series.

T57.6 .O64545 2002
658.4'034--dc21

2002016068

Originally published by Kluwer Academic Publishers in 2002

Printed on acid-free paper.

Table of Contents

Recent Algorithmic Advances for Arc Routing Problems 1
Gianpaolo Ghiani, Alain Hertz and Gilbert Laporte

Sequencing and Scheduling for Non-Serial Permutation Flowshops 21
Robert Burdett and Erhan Kozan

Improved Multiple Ant Colonies System for Traveling Salesman Problems 41
Kawamura Hidenori, Masahito Yamamato and Azuma Ohuchi

The Video Placement and Routing Problem 53
Iradj Ouveysi, L. Sesana and Andrew Wirth

A Markov Chain Tabu Search Approach to the Evolutionary Allocation Problem 71
Hussein A. Abbass, Michael Towsey, Erhan Kozan and Julius Van der Werf

A Game of Corruption in a Stylized South-East Asian Timber Industry: A Genetic Algorithm Approach 93
Ryan McAllister and Michael Bulmer

A General Framework for Constructive Meta-Heuristics 111
Marcus Randall

The Immune Distributed Competitive Problem Solver using Major Histocompatibility Complex and Immune Network 129
Naruaki Toma, Satoshi Endo, Koji Yamada and Hayao Miyagi

Search Algorithms for Nurse Scheduling with Genetic Algorithms 149
Ahmad Jan, Masahito Yamamoto and Azuma Ohuchi

Parallel Agent Passing Tabu Search Algorithm for Graph Partitioning Problem 163
Taichi Kaji

Heuristic Techniques for Scheduling Railway Track Maintenance 177
Maree Lake, Luis Ferreira and Erhan Kozan

Optimising a Horizontally Integrated Push/Pull Hybrid Production System in a Foundry 189
Paul Corry and Erhan Kozan

Design of Adaptive Self Navigated Airship in Simulated Environment 205
Keiko Motoyama, Keiji Suzuki, Hidenori Kawamura, Masahito Yamamoto and Azuma Ohuchi

On the use of a Structural Modelling for Collaborative Learning Using the Concept Map 215
Hajime Saito, Azuma Ohuchi and Takashi Maeda

Global Convergence Property of Error Back-Propagation Method for Recurrent Neural Networks 227
Keiji Tatsumi, Tetsuzo Tanino and Masao Fukushima

A Material Planning Model for Mixed Model Assembly Lines 243
Erhan Kozan and Peter Preston

A Fuzzy Programming Model for a Cross-Border Logistics Problem under an Uncertain Environment in Hong Kong 261
Stephen Leung and K. K. Lai

The DSS Architecture Based on Non-Mathematical Problems Specification and Model/Solver Independence 281
Tsuyoshi Mukohara and Yasuki Sekiguchi

Hyper Sensitivity Analysis of Productivity Measurement Problems 299
Leonid Churilov and Moshe Sniedovich

Reduced Game Property of Linear Values with Equal Treatment Property 317
Tsuneyuki Namekata and Theo S.H. Driessen

Alternative Transformations and Duality of Linear Fractional Programming 333
G. C. O'Brien and Lifen Wu

Max-Min Optimization of the Multiple Knapsack Problem: An Implicit Enumeration Approach 351
Takeo Yamada

A Planning System for Precision Farming Based on an Autonomous Tractor 363
Keiji Suzuki, Kazuki Takamatsu, Taku Okuno, Azuma Ohuchi and Yukinori Kakazu

Spot Price Forecasting Models for Risk Aversion in Victoria's Electricity Market 375
Patrick Tobin and Alan Brown

Refining Diffusion Models for State-Dependent Queues 395
Toshikazu Kimura

A System Dynamics Model for Helicopter Operations 405
Yi Yue

INDEX 423

PREFACE

The principle aim of this book, entitled "Operations Research/Management Science at Work", is to provide a summary snapshot of recent research in and applications of Operations Research (OR)/ Management Science (MS) in the Asia-pacific region. The book emphasises research having potential industry interest, covering a wide range of topics from major fields of OR/MS in a systematic and coherent fashion and shows the elegance of their implementations. The book is intended to serve the needs of applied researchers who are interested in applications of OR/MS algorithms. The book includes descriptions of many real-world problems together with their solutions; we hope the reader will appreciate their applicability.

The Asia-pacific region has embraced business applications of decision support systems in recent years. Given that many of these applications are unaffected by legacy models or systems it has meant that state of the art OR/MS techniques have been embedded in them. Moreover, the increased use of OR/MS techniques in this region provides opportunities for identifying methodological advances that are taking place as a result of the unique nature of the applications. These also provide opportunities for exploring synergies and interfaces that exist between OR/MS, from the point of view of applications as well as theoretical advances.

We invited some eminent OR/MS researchers and practitioners in the Asia-pacific region to submit papers for publication in this book. We also invited some of the best papers from the Second Australian Society for Operations Research (ASOR) Queensland Branch and Operations Research Society of Japan (ORSJ) Hokkaido Chapter joint workshop.

Editing a book that covers a wide range of OR/MS applications is a very difficult and time-consuming task. The book could not have been successfully accomplished without support and constructive feedback from all the contributors. We would like to thank them for their efforts, for reviewing each other's work, and for providing feedback on their own chapters.

All chapters were carefully selected, peer reviewed and subject to a thorough and stringent refereeing process, in accordance with the usual high standards of the International Series in Operations Research and Management Science, Kluwer Academic Publishers. In general, papers were each reviewed by at least three referees who made valuable contributions, thus improving the quality of the chapters. We would like to thank the referees and all others who have contributed in so many different ways to the success of this book.

It is an honour for us on behalf of the ASOR and ORSJ to present to the community of OR /MS a book which will give a unique opportunity to keep up-to-date with OR/MS in Australia, Japan and other countries in the Asia-pacific region.

Also, we express our gratitude to the editor of The International Series in Operations Research and Management Science, Professor Frederick Hillier and to Kluwer Academic Publishers Gary Folven for their help and encouragement throughout the book.

Editors

Erhan Kozan and Azuma Ohuchi

RECENT ALGORITHMIC ADVANCES FOR ARC ROUTING PROBLEMS

Gianpaolo Ghiani[1], Alain Hertz[2] and Gilbert Laporte[3]

[1] *Dipartimento di Informatica e Sistemistica, Universitá di Napoli "Federico II", via Claudio 21, 80125 Napoli, Italy.*
ghiani@univa.it

[2] *Département de mathématiques, École Polytechnique Fédérale de Lausanne, CH-1015, Lausanne, Switzerland.*
Alain.Hertz@epfl.ch

[3] *GERAD and École des Hautes Études Commerciales, 3000 chemin de la Côte-Sainte-Catherine, Montréal, Canada H3T 2A7.*
gilbert@crt.umontreal.ca

Abstract This article reports on some recent algorithmic development for the *Rural Postman Problem* (CPP) and for the *Capacitated Arc Routing Problem* (CARP). Heuristics are described for the RPP and for the CARP. A branch-and-cut algorithm is described for the RPP.

Keywords: Arc routing, Rural Postman Problem, Tabu Search, Variable Neighbourhood Search, Branch-and-cut.

1. INTRODUCTION

In *Arc Routing Problems* (ARPs), the aim is to determine a least cost traversal of all edges or arcs of a graph, subject to some side constraints. Such problems arise naturally in several applications related to garbage collection, mail delivery, snow clearing, meter reading, school bus routing, etc. (Eiselt, Gendreau and Laporte, 1995b; Assad and Golden, 1995). In addition, a number of industrial applications such as laser beam plotting (Ghiani and Improta, 2001) and task sequencing (Anily, Gendreau and Laporte, 2000) have recently been described. Billions of dollars are at stake in arc routing activities and there exists a sizeable potential for savings. Yet, ARPs have received far less attention than so-called node or vertex routing problems common in collection and delivery operations.

The modern theory of arc routing is rooted in the work of Leonhard Euler (1736) who was called upon to solve the celebrated Köningsberg bridges problem. In the early eighteenth century, the inhabitants of Köningsberg (now Kaliningrad, Russia) debated whether there existed a closed walk using each of the seven bridges of the Pregel river exactly once (Figure 1a). Using a graph representation (Figure 1b), Euler showed there was none and put forward necessary and sufficient conditions for the *unicursality* of an undirected graph, i.e., for the existence of a closed traversal using each of its edges exactly once: the graph must be connected and all its vertices must have an even degree. Necessary and sufficient conditions for the universality of a directed or of a mixed graph were found later and are stated in Ford and Fulkerson (1962). When a graph G is Eulerian, it is relatively easy to determine a least cost closed traversal of all its edges or arcs. Such methods have been proposed by Hierholzer (1873) and van Aardenne-Ehrenfest and de Bruijn (1951), for example. For a survey of different approaches, see Fleischner (1990).

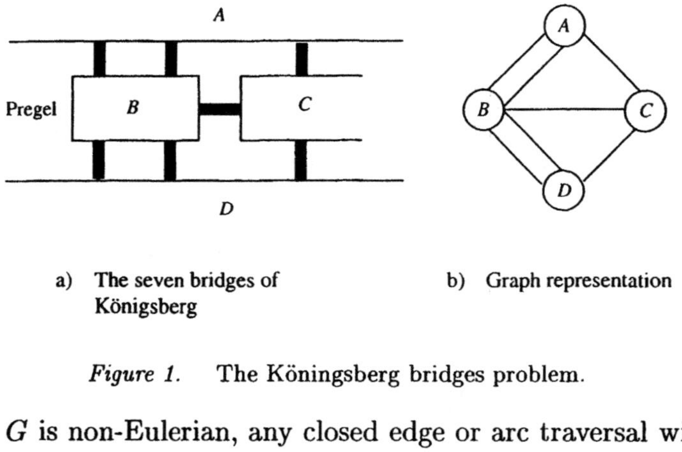

a) The seven bridges of Königsberg

b) Graph representation

Figure 1. The Köningsberg bridges problem.

If G is non-Eulerian, any closed edge or arc traversal will necessarily contain repeated edges or arcs. The problem is then to determine a least cost traversal of the graph in which each edge or arc is traversed *at least* once. The problem, known as the *Chinese Postman Problem* (CPP) was first posed by Guan (1962) for undirected graphs. The standard solution methodology, proposed by Edmonds and Johnson (1973), consists of first determining a least cost augmentation of G in order to make it Eulerian, and of then determining a Eulerian cycle on the augmented graph. If G is undirected, the augmentation problem is solved by first identifying all odd-degree vertices (there is always an even number of such vertices), and of then computing a least cost matching of these vertices, where the matching cost between two vertices is equal to the cost of a shortest chain linking them in G. Formally, the augmentation problem of an undirected graph $G = (V, E)$ can be cast as an integer linear program as follows. Let $x_{ij}(i < j)$ be an integer variable equal to the number of copies of edge $(v_i, v_j) \in E$ that must be added to G in order to make it Eulerian, and let c_{ij} denote the cost of a least cost chain between v_i and v_j. A non-empty subset S of V is called *odd* if it contains an odd number of odd degree vertices. Given $S \subseteq V$, let $\delta(S) = \{(v_i, v_j) \in E : v_i \in S, v_j \in V \setminus S \text{ or } v_i \in V \setminus S, v_j \in S\}$. The problem is then:

$$\text{(UCPP)} \quad \text{Minimize} \sum_{(v_i,v_j)\in E} c_{ij}x_{ij} \tag{1}$$

subject to

$$\sum_{(v_i,v_j)\in\delta(S)} x_{ij} \geq 1 \quad (S \subset V, S \text{ is odd}) \tag{2}$$

$$x_{ij} \geq 0 \quad ((v_i, v_j) \in E) \tag{3}$$

$$x_{ij} \text{ integer} \quad ((v_i, v_j) \in E). \tag{4}$$

Edmonds and Johnson show that the polyhedron of solutions to (2) and (3) is equal to the convex hull of feasible solutions to (2)–(4). They also show how (1)–(3) can be solved efficiently by computing a minimum least cost matching problem on the odd degree vertices of V.

A similar approach was suggested to solve the augmentation problem for a directed graph (Edmonds and Johnson, 1973; Orloff, 1974; Beltrami and Bodin, 1974). Let I denote the set of vertices v_i for which the number of incoming arcs exceeds the number of outgoing arcs by s_i, and J, the set of vertices v_j for which the number of outgoing arcs exceeds the number of incoming arcs by d_j. In other words, s_i can be seen as a supply, and d_j as a demand. Define c_{ij} as the length of a shortest path from v_i to v_j. Again, the x_{ij} values correspond to the number of extra copies of arc (v_i, v_j) (deadheaded arcs) that must be added to G in order to make it Eulerian. Then the augmentation problem can be cast as a *Transportation Problem*:

$$\text{(DCPP)} \quad \text{Minimize} \sum_{v_i\in I} \sum_{v_j\in J} c_{ij}x_{ij} \tag{5}$$

subject to

$$\sum_{v_j \in J} x_{ij} = s_i \qquad (v_i \in I) \tag{6}$$

$$\sum_{v_i \in I} x_{ij} = d_j \qquad (v_j \in J) \tag{7}$$

$$x_{ij} \geq 0 \qquad (v_i \in I,\ v_j \in J). \tag{8}$$

If G is a mixed graph, the augmentation problem is NP-hard and can be solved by means of a branch-and-cut approach (Nobert and Picard, 1996). Then a Eulerian circuit can be obtained by using the Ford and Fulkerson (1962) procedure. Details are provided in Eiselt, Gendreau and Laporte (1995a).

In many contexts, it is not necessary to *traverse* all edges of E, but to *service* only a subset R of *required edges*, traversing if necessary some edges of $E \backslash R$. Determining a least cost traversal of all edges of R is known as the *Rural Postman Problem* (RPP), a problem introduced by Orloff (1974) and shown to be NP-hard by Lenstra and Rinnooy Kan (1976). Another related problem with several practical applications is the *Capacitated Arc Routing Problem* (CARP), first studied by Golden and Wong (1981). Here, edges (v_i, v_j) have a *weight* $q_{ij} \geq 0$, and each edge with a positive weight must be traversed by exactly one of several vehicles starting and ending their trip at a *depot* located at a vertex v_0, subject to the constraint that total weight of all edges serviced by a vehicle cannot exceed its capacity Q. The CARP is also NP-hard since the RPP reduces to it whenever $q_{ij} = 1$ for $(v_i, v_j) \in R$, $q_{ij} = 0$ for $(v_i, v_j) \in E \backslash R$, and $Q = |R|$.

The purpose of this article is to report on some recent algorithmic developments for the undirected RPP and CARP. In the following two sections, we describe new heuristics for the RPP and the CARP, respectively. Then in Section 4, we present a recent branch-and-cut algorithm for the undirected RPP. Conclusions and directions for future research are provided in Section 5.

2. HEURISTICS FOR THE UNDIRECTED RURAL POSTMAN PROBLEM

As is common in combinatorial optimization, the RPP can be solved by applying a constructive procedure followed by an improvement phase. One of the best known constructive heuristics is due to Frederickson (1979). Improvement procedures were developed more recently by Hertz, Laporte and Nanchen-Hugo (1999). Our exposition is

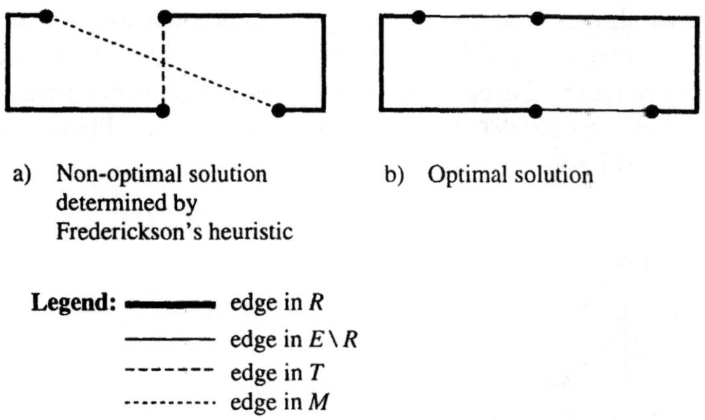

Figure 2. Example for which Frederickson's heuristic does not yield an optimal solution.

limited to undirected graphs. Extensions of some of the following results to directed graphs can be found in Mittaz (1999).

2.1 Frederickson's Constructive Heuristic

Frederickson's heuristic works along the lines of Christofides's (1976) heuristic for the undirected *Traveling Salesman Problem* (TSP).

Step 1. Construct a shortest spanning tree over an auxiliary graph H having one vertex for each connected component of required edges in G, and where the cost between two vertices is equal to the length of a shortest chain between the correspondent components in G; denote by T the set of edges in the shortest spanning tree.

Step 2. Solve a minimum cost matching problem (with respect to shortest chain costs) on the odd-degree vertices of the graph induced by $T \cup R$, and let M be the edge set induced by the matching.

Step 3. Determine a Eulerian cycle in the graph induced by $T \cup R \cup M$.

If R is connected, this algorithm yields an optimal RPP solution. Otherwise, it has a worst-case performance ratio of 3/2 provided the costs satisfy the triangle inequality. Figure 2 illustrates a case where Frederickson's heuristic fails to produce an optimal solution.

2.2 Improvement Heuristics

While several postoptimization heuristics are widely known for vertex routing problems such as the TSP, fewer such procedures are available for the RPP. In a recent article, Hertz, Laporte and Nanchen-Hugo

(1999) have shown that standard operations such as vertex insertions and edge exchanges can become quite intricate when implemented in an arc routing context. Three of the postoptimization procedures developed by the authors are SHORTEN, DROP-ADD and 2-OPT. These are sketched and illustrated here.

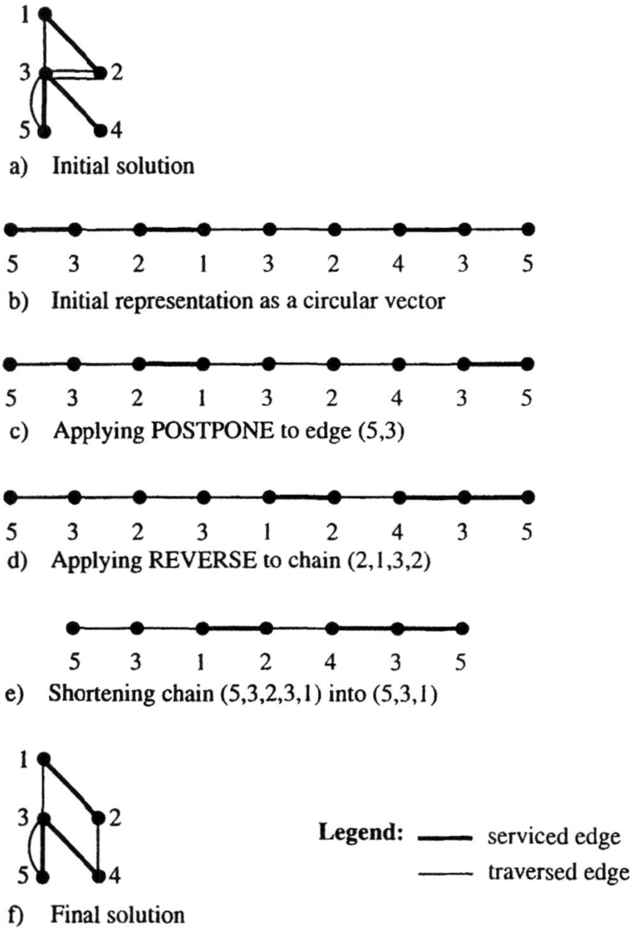

Figure 3. Illustration of procedure SHORTEN.

Procedure SHORTEN

An RPP solution S can be written as a circular vector $S = (v_{i_1}, \ldots, v_{i_r}, \ldots, v_{i_t} = v_{i_1})$. Using in turn each vertex v_{i_r} as a starting point for the circular vector, SHORTEN first attempts to create a route $S' = (v_{j_1}, \ldots, v_{j_s}, \ldots, v_{j_1})$ through the same edge sequence as S

but with a different starting vertex, in which the chain $(v_{j_1}, \ldots, v_{j_s})$ contains no required edge. It then replaces this chain by a shortest chain between v_{j_1} and v_{j_s}. Two operations are used to transform S into S': POSTPONE and REVERSE. Operation POSTPONE can be applied whenever an edge (v_i, v_j) appears in S first as a serviced, and later as traversed without being serviced. POSTPONE makes the edge first traversed and then serviced. Operation REVERSE simply reverses a chain starting with serviced edge (v_i, v_j) and ending with a non-serviced edge (v_k, v_i). Procedure SHORTEN is illustrated on the graph depicted in Figure 3a, where vertex 5 is used as a starting point. The RPP solution can be written as the circular vector (5,3,2,1,3,2,4,3,5) shown in Figure 3b. Applying POSTPONE makes the first edge (5,3) traversed and the last edge (3,5) serviced (Figure 3c). Applying REVERSE to the chain (2,1,3,2) yields the solution shown in Figure 3d, at which point the chain (5,3,2,3,1) is shortened into (5,3,1) (Figure 3e). The postoptimized RPP solution is illustrated in Figure 3f.

Procedure DROP-ADD

Procedure DROP-ADD consists of successively removing each edge from an RPP solution (by calling DROP) and reinserting it (by calling ADD). To remove an edge, DROP first makes it non-required and then calls SHORTEN. To insert an edge (v_i, v_j) into an RPP solution, ADD simply links its two extremities to the current solution by means of shortest chains and calls SHORTEN. DROP and ADD are illustrated in Figures 4 and 5, respectively.

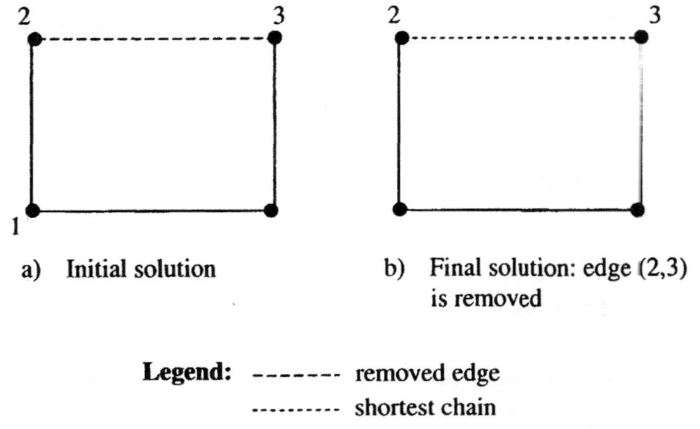

Figure 4. Illustration of procedure DROP.

Procedure 2-OPT

Procedure 2-OPT is analogous to the 2-opt heuristic introduced by Croes (1958) for the TSP. It successively removes two edges from a feasible RPP solution and reconnects differently the two remaining segments of the tour by means of shortest chains (Figure 6). SHORTEN is then called. If any required edge is missing from the solution (this may occur if one of the removed edges was required), these are reintroduced by means of ADD.

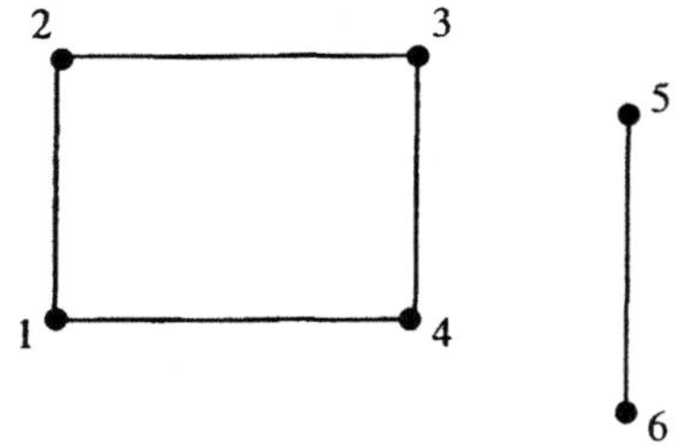

a) Initial solution

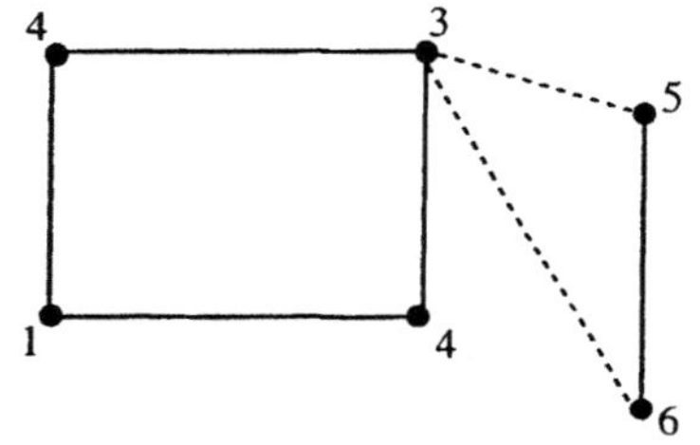

b) Final solution before SHORTEN: edge (5,6) is added

Legend: --------- shortest chain

Figure 5. Illustration of procedure ADD.

2.3 Computational Results

Extensive computational tests were conducted by Hertz, Laporte and Nanchen-Hugo (1999) on several classes of graphs. The main results of these tests can be summarized as follows. Frederickson's heuristic is very fast but generally produces suboptimal solutions, up to 10% worse than the best known solutions. Procedure 2-OPT reduces this gap considerably, often to nothing, within relatively short computing times. As a rule DROP-ADD is much faster then 2-OPT (sometimes by two orders of

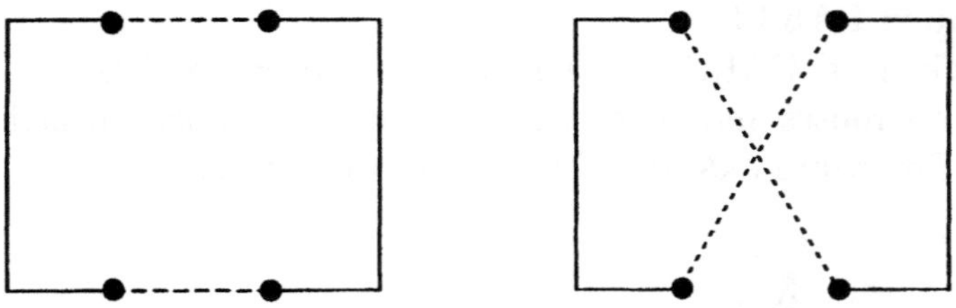

a) Solution before edge exchange

b) Solution after edge exchange

Legend: ------- removed edge

--------- shortest chain

Figure 6. Illustration of procedure 2-OPT.

magnitude), but it is not quite as effective. If solution quality really matters, the best option is to apply 2-OPT after Frederickson's heuristic. If computing time is critical, then 2-OPT should be replaced by DROP-ADD.

3. HEURISTICS FOR THE UNDIRECTED CAPACITED ARC ROUTING PROBLEM

Early construction and improvement heuristics for the undirected CARP are path-scanning (Golden, DeArmon and Baker, 1983), construct-strike (Christofides, 1973), modified construct-strike and modified path-scanning (Pearn, 1989). Recently Greistorfer (1994) and Hertz, Laporte and Mittaz (2000) have proposed more efficient algorithms based partly on tabu search. In what follows, we summarize the heuristics developed by the latter authors for the undirected CARP. Again, extensions to the directed case were obtained by Mittaz (1999). We first present four procedures developed by Hertz, Laporte and Mittaz: PASTE, CUT, SWITCH, and POSTOPT. We then show how these can be combined into constructive and improvement heuristics for the undirected CARP, and integrated with two highly efficient local search heuristics: a tabu search algorithm called CARPET and a variable neighbourhood search algorithm.

3.1 Basic Procedures

The following procedures can be used as subroutines in more elaborate heuristics.

Procedure PASTE

Given a CARP solution made up of m vehicle routes, PASTE merges all routes into a single RPP tour, possibly infeasible for the CARP. Procedure PASTE is illustrated in Figure 7.

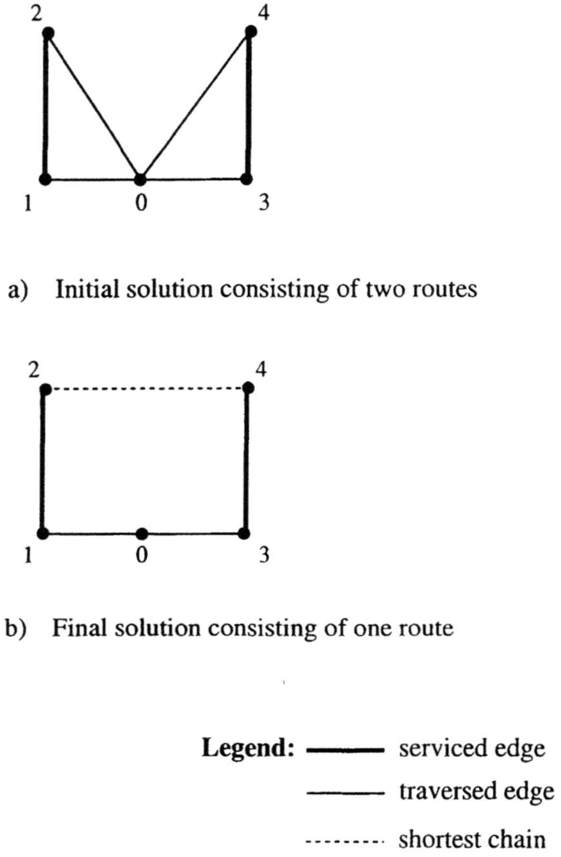

Figure 7. Illustration of procedure PASTE.

Procedure CUT

This procedure is the reverse of PASTE. Given an RPP solution, CUT transforms it into a feasible CARP solution. Starting at the depot v_0, it determines a vertex v incident to a serviced edge on the RPP tour such that the total weight of the first chain (from v_0 to v) does not exceed Q and the weight of the second chain (from v to v_0) does not exceed $Q(\lceil d/Q \rceil - 1)$, where d is the total weight of the RPP solution. When several choices are possible for v, the selection is made so as to minimize $L(v)$, the sum of the shortest chain lengths between v and v_0, and between v_0 and v', the first vertex of the first required edge after v on the RPP tour. Vertex v is then linked to v_0 by means of a shortest chain and some of the edges of this shortest chain may be serviced as long as the vehicle capacity is not exceeded. Vertex v_0 is then linked to v' by means of a shortest chain, and the same procedure is recursively applied to the remaining part of the tour until a feasible CARP solution has been reached. Procedure CUT is illustrated in the example depicted in Figure 8, where the depot is located at vertex 0. The numbers in square boxes are edge weights (for required edges only). The remaining numbers on the dashed lines or on the edges are shortest chain lengths. In this example, $Q = 11$ and $d = 24$. The procedure first computes $Q(\lceil d/Q \rceil - 1) = 22$, which means the first route must include edge (0,1) (then the remaining weight is 20, which does not exceed 22), and can include up to three required edges without having a weight exceeding Q. Three choices are therefore possible for v: 1, 3 and 4, yielding $L(1) = 3 + 6 = 9$, $L(3) = 4 + 4 = 8$ and $L(4) = 6 + 6 = 12$, meaning that vertex 3 is selected. The first route includes the chain (0,1,2,3) followed by a shortest chain to the depot. The procedure is then reapplied starting from vertex 3.

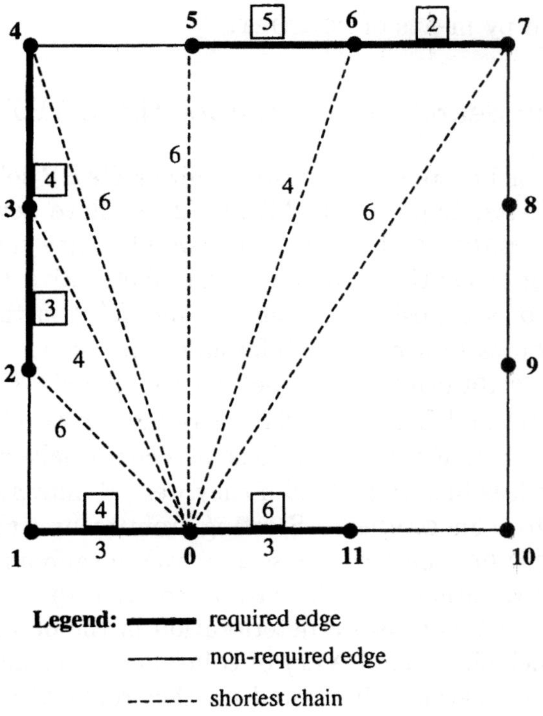

Figure 8. Illustration of procedure CUT.

Procedure SWITCH

Given a route in which a vertex v appears several times, this procedure reverses all minimal subroutes starting and ending at v as long as these do not include the depot. The role of this procedure is to create a better mix of solutions within a search procedure. For example, if PASTE and CUT were repeatedly applied to a given CARP solution, then this solution may repeat itself indefinitely. The chances of this happening are reduced if SWITCH is periodically applied.

Procedure POSTOPT

Procedure POSTOPT can be applied to postoptimize any CARP solution. It successively calls PASTE, SWITCH, CUT and it then applies SHORTEN to each route until a termination criterion has been met, such as a given number of iterations.

3.2 A Simple Consturctive Algorithm for the CARP

A feasible CARP solution can be obtained as follows. First generate an RPP solution by means of Frederickson's algorithm, and then a CARP solution by means of CUT. Improvements to the feasible solution can be obtained by means of POSTOPT.

3.3 A Tabu Search Algorithm for the CARP

A more elaborate way to postoptimize a CARP solution is to use tabu search. Hertz, Laporte and Mittaz (2000) have developed such a heuristic, called CARPET. It works with two objective functions: $F(x)$, the solution cost of solution x, and a penalized objective $F'(x) = F(x) + \beta E(x)$, where β is a positive parameter and $E(x)$ is the total excess weight of all routes in a possibly infeasible solution x. As in the tabu search heuristic TABUROUTE developed for the *Vehicle Routing Problem* (Gendreau, Hertz and Laporte, 1994), parameter β is self-adjusting. It is initially equal to 1. Every λ iterations, it is either halved if all previous λ solution were feasible, or doubled if they were all infeasible. The value of λ is equal to 5 in CARPET. Starting from an initial solution (see Section 3.2), CARPET performs a search over neighbour solutions, by moving at each iteration from the current solution to its best non-tabu neighbour, even if this causes a deterioration in the objective function. To define the neighbourhood $N(x)$ of solution x, consider a route S of x, an edge (v_j, v_k) serviced in S, and another route S' containing only the depot or an edge of R with an extremity distant from v_j or v_k by at most α, where α is a parameter. A neighbour of x is then obtained by dropping (v_j, v_k) from S adding it to S', using procedures DROP and ADD described in Section 2.2. Whenever an edge is removed from route S at a given iteration, it cannot be reintroduced into this route for θ iterations, where θ is randomly selected in [5,10]. The algorithm terminates after a set number of iterations without improvement in F or F', or after a preset time, or whenever a solution having the same value as that of a known lower bound is encountered.

3.4 A Variable Neighbourhood Search Algorithm for the CARP

Variable neighbourhood search (VNS) is a technique put forward by Mladenović and Hansen (1997) which can be embedded within a local search algorithm such as tabu search. The basic idea is to define several neighbourhoods to perform the search, thus reducing the risk of becoming trapped on a local optimum. The VNS algorithm developed by Mittaz (1999) for the CARP uses the same framework as CARPET, except that feasibility if preserved at all times (only $F(x)$ is used). A lower bound $\underline{m} = \left\lceil \sum_{(v_i,v_j)\in R} q_{ij}/Q \right\rceil$ on the number of routes in a feasible solution is first computed, and given a solution x and a value of k in $[2, \underline{m}]$. A neighbourhood $N_k(x)$ is defined by applying a procedure similar to POSTOPT to k of the routes. This procedure is called PASTE, SHORTEN, SWITCH and CUT. Procedure SHORTEN is then applied to each route. Since there are usually several ways to select k routes from a CARP solution and also several ways to choose the vertex v in order to reverse subroutes, only a limited number of choices are considered. Also, another neighbourhood $N_1(x)$ is obtained as in CARPET, but by forcing the solution to remain feasible at all times. For $k = 1, \ldots, \underline{m}$, the neighbourhoods $N_k(x)$ grow from fine to coarse in the sense that a small value of k will tend to yield a neighbour solution close to x, while a large value of k will produce a widely different solution. The idea is to use the various neighbourhoods in succession. Mittaz tested three strategies: 1) from $\underline{m}$ to 1, 2) from 1 to $\underline{m}$; 3) in random order. The first strategy turned out to be the best.

3.5 Computational Results

CARPET was run on a wide variety of test problems. On a first set of small scale instances ($7 \leq |V| \leq 27; 11 \leq |E| \leq 55; R = E$) used by DeArmon (1981), CARPET outperforms all previous heuristics mentioned at the beginning of Section 3, even if run for only one second on a Silicon Indigo 1 machine (195 MHz, IP 28 processor). With this running time, the average deviation to a lower bound is 3.4%. This can be compared with previous heuristics which produced average deviations ranging from 3.9% to 13.9%. When no time limit is imposed, CARPET achieves solution values exceeding that of the lower bound by an average of 0.17%, in times ranging from 0 to 330.6 seconds. CARPET was also applied to 25 instances proposed by Benavent (1997) with $24 \leq |V| \leq 50$ and $34 \leq |E| \leq 97$ where it produced solutions having an average deviation of 1.1% from a lower bound. Finally, CARPET was tested on 270 larger and denser instances containing 20, 40 or 60 vertices, with edge densities in [0.1, 0.3], [0.4, 0.6] or [0.7, 0.9] and $|R|/|E|$ in [0.1, 0.3], [0.4, 0.6] and [0.8, 1.0]. Average deviations from a lower bound were then equal to 5.5% when CARPET was run for up to 60 seconds and to 1.6% when it was run for up to 600 seconds. VNS produced results similar to those of CARPET on the DeArmon instances (the deviation from optimality was identical and computation times were about 20% shorter), but it yielded much better results on the larger randomly generated instances. When no time limits were imposed, the average deviation from optimality and computing times on the 270 instances were 0.71% and 349.81 seconds for CARPET, and 0.54% and 42.50 seconds for VNS. In other words, on larger instances VNS is better than CARPET in terms of solution quality, but also faster.

4. AN EXACT ALGORITHM FOR THE UNDIRECTED RURAL POSTMAN PROBLEM

We are aware of four exact algorithms for the undirected RPP. The first, due to Christofides, Campos, Corberán and Mota (1981) uses branch-and-bound combined with Lagrangean relaxation. Corberán and Sanchis (1994) described an integer linear programming formulation solved by means of a branch-and-cut algorithm in which the separation problems are solved visually. Letchford (1996) added so-called path-bridge inequalities to the Corberán and Sanchis formulation. Finally, Ghiani and Laporte (2000) introduced a new and more compact formulation which, when solved by branch-and-cut, yields excellent results on test problems. In what follows, we summarize some of the Ghiani and Laporte results.

4.1 Dominance Relationships

Let $C_k(k = 1, \ldots, p)$ be the kth connected component of the subgraph of $G = (V, E)$ induced by the set R of required edges. Let V_R be the set of vertices v_i such that an edge (v_i, v_j) exists in R, and $V_k \subseteq V_R(i = 1, \ldots, p)$ the vertex set of C_k. Denote by c_e the cost of edge $e \in R$. A vertex $v_i \in V_R$ is R-odd (R-even) if and only if an odd (even) number of edges of R are incident to v_i. Christofides *et al.* (1981) proposed the following graph reduction:

Step 1. Add to $G_R = (V_R, R)$ an edge between every pair of vertices of V_R having a cost equal to that of the corresponding shortest chain on G.

Step 2. Delete one of two parallel edges if they have the same cost, and all edges $(v_i, v_j) \notin R$ such that $c_{ij} = c_{ik} + c_{kj}$ for some v_k.

We now recall the Corberán and Sanchis (1994) formulation. Given $S \subset V$, let $\delta(S)$ be the set of edges of E with one extremity in S and one in $V \backslash S$. If $S = \{v\}$, then we write $\delta(v)$ instead of $\delta(S)$. Let $x_e = x_{ij}$ represent the number of additional (deadheaded) copies of edge $e = (v_i, v_j)$ that must be added to G to make it Eulerian. The formulation is then:

$$\text{(URPP1)} \quad \text{Minimize} \sum_{e \in E} c_e x_e \tag{9}$$

subject to

$$\sum_{e \in \delta(v)} x_e = 0 \bmod (2) \qquad (\text{if } v \in V_R \text{ is } R\text{-even}) \tag{10}$$

$$\sum_{e \in \delta(v)} x_e = 1 \bmod (2) \qquad (\text{if } v \in V_R \text{ is } R\text{-odd}) \tag{11}$$

$$\sum_{e \in \delta(S)} x_e \geq 2 \qquad (S = \bigcup_{k \in P} V_k, P \subset \{1, \ldots, p\}, p \neq \emptyset) \tag{12}$$

$$x_e \geq 0 \text{ and integer} \qquad (e \in E). \tag{13}$$

In this formulation, constraints (10) and (11) force each vertex to have an even degree, while constraints (12) ensure connectivity. In what follows, we recall some dominance relations that will enable a reformulation of the problem without the non-linear constraints (10) and (11).

Dominance relation 1 . (Christofides *et al.*, 1981).
Every optimal RPP solution satisfies the relations

$$x_e \leq 1 \qquad (\text{if } e \in R) \tag{14}$$

$$x_e \leq 2 \qquad (\text{if } e \in E \backslash R) \tag{15}$$

Dominance relation 2 . (Corberán and Sanchis, 1994).
Every optimal solution satisfies

$$x_e \leq 1 \qquad (\text{if } e = (v_i, v_j), v_i, v_j \text{ belong to the same component } C_k). \tag{16}$$

Dominance relation 3 . (Ghiani and Laporte, 2000).
Let $x(e^{(1)}), x(e^{(2)}), \ldots, x(e^{(\ell)})$ be the variables associated with the edges $e^{(1)}, e^{(2)}, \ldots, e^{(\ell)}$ having exactly one end vertex in a given component C_k and exactly one end vertex in another given component C_h. Then, in an optimal solution, only the variables $x(e^{(r)})$ having a cost $c(e^{(r)}) = \min\{c(e^{(1)}), c(e^{(2)}), \ldots, c(e^{(\ell)})\}$ can be equal to 2.

Now define a 0/1/2 edge as an edge e for which x_e can be at most equal to 2 in (URPP1), and a 0/1 edge as an edge e for which x_e can be at most equal to 1. Denote by E_{012} and E_{01} the corresponding edge sets.

Dominance relation 4 . (Ghiani and Laporte, 2000).
Let $G_C^* = (V_C, E_C)$ be an auxiliary graph having a vertex v_i' for each component C_i and, for each pair of components C_i and C_j, an edge (v_i', v_j') corresponding to a least cost edge between C_i and C_j. Then, in any optimal (URPP1) solution, the only 0/1/2 edges belong to a *Minimum Spanning Tree* on G_C^* (denoted by MST_C).

4.2 A New Binary Formulation Using Only Edge Variables

Using Dominance relation 4, formulation (URPP1) can now be rewritten by replacing each 0/1/2 edge e belonging to a given MST_C by two parallel 0/1 edges e' and e''. Denote by $E'(E'')$ the sets of edges $e'(e'')$, and let $\overline{E} = E_{01} \cup E' \cup E''$. In formulation (URPP1), constraints (13) are simply replaced by

$$x_e = 0 \text{ or } 1 \qquad (e \in \overline{E}). \tag{17}$$

Ghiani and Laporte (2000) also replace the modulo relations (10) and (11) by the following constraints called cocircuit inequalities by Barahona and Grötschel (1986):

$$\sum_{e \in \delta(v) \setminus F} x_e \geq \sum_{e \in F} x_e - |F| + 1 (v \in V, F \subseteq \delta(v), |F| \text{ is odd if } v \text{ is } R\text{-even},$$
$$|F| \text{ is even if } v \text{ is } R\text{-odd}). \quad (18)$$

Thus the new undirected RPP formulation, called (URPP2) and defined by (9), (18), (12) and (17), is linear in the 0/1 x_e variables. Constraints (18) can be generalized to any non-empty subset S of V:

$$\sum_{e \in \delta(S) \setminus F} x_e \geq \sum_{e \in F} x_e - |F| + 1 (F \subseteq \delta(S), |F| \text{ is odd if } S \text{ is } R\text{-even},$$
$$|F| \text{ is even if } S \text{ is } R\text{-odd}), \quad (19)$$

which are valid inequalities for (URPP2). If S is R-odd and $F = \emptyset$, constraints (19) reduce to the known R-odd inequalities (Corberán and Sanchis, 1994)

$$\sum_{e \in \delta(S)} x_e \geq 1. \quad (20)$$

If S is R-even and $F = \{e_b\}$, they reduce to the R-even inequalities introduced by Ghiani and Laporte (2000):

$$\sum_{e \in \delta(S) \setminus \{e_b\}} x_e \geq x_{e_b}. \quad (21)$$

Ghiani and Laporte (2000) have shown that constraints (19) are facet inducing for (URPP2). They have also developed a branch-and-cut algorithm in which connectivity constraints (12) and generalized cocircuit inequalities (19) are dynamically generated. In practice, it is sufficient to generate constraints of type (18), (20) and (21) to identify a feasible RPP solution.

4.3 Computational Results

The branch-and-cut algorithm developed by Ghiani and Laporte (2000) was tested on several sets on randomly generated instances generated in the same manner as those of Hertz, Laporte and Nanchen-Hugo (1999). Instances defined on random planar graphs with $|V| = 50, 100, \ldots, 350$ were solved to optimality with very few nodes in the branch-and-cut tree and within reasonable computing times. For example, the 350 vertex instances required an average of 22.4 nodes and 332.5 seconds on a PC with a Pentium processor at 90 MHz with 16

Mbytes RAM. At the root of the search tree, the average ratio of the lower bound over the optimum almost always exceeded 0.997. R-even inequalities played a key role in the problem resolution. These results outperform those reported by Christofides *et al.* (1981), Corberán and Sanchis (1994) and Letchford (1996) who solved much smaller randomly generated instances ($|V| \leq 84$) with far more branching.

5. CONCLUSIONS AND FUTURE DEVELOPMENTS

Arc routing problems lie at the heart of several real-life applications and their resolution by good heuristics or exact algorithms can translate into substantial savings. Over the past few years, there has been a revived interest in the study of these problems. In the area of heuristics, specialized procedures such as SHORTEN, DROP-ADD and 2-OPT (Hertz, Laporte and Nanchen-Hugo, 1999) have been proposed for the undirected RPP. These can also be used for the solution of constrained versions of the RPP such as the CARP (Hertz, Laporte and Mittaz, 2000) or adapted to the directed case (Mittaz, 1999). In addition, powerful local search heuristics have been developed for the CARP. Two approaches stand out: tabu search and variable neighbourhood search. In the field of exact algorithms, branch-and-cut appears to be the most promising approach. In the past two decades this method has known a formidable growth and considerable success on related problems such as the TSP (see Padberg and Rinaldi, 1991; Grötschel and Holland, 1991; Jünger, Reinelt and Rinaldi, 1995). Recent advances made by Corberán and Sanchis (1994), Letchford (1996) and Ghiani and Laporte (2000) indicate that this method also holds much potential for arc routing problems.

In coming years, we expect to see the development of similar heuristics and branch-and-cut algorithms to more intricate and realistic arc routing problems, including those defined on directed and mixed graphs, and problems incorporating a wider variety of practical constraints. Extensions to the area of locating-routing problems in arc routing context (see Ghiani and Laporte, 1999) are also to be expected.

6. ACKNOWLEDGEMENTS

This research was partly supported by the Ministero dell' Universitá e della Ricerca Scientifica e Tecnologia (MURST), Italy, and by the Canadian Natural Sciences and Engineering Council (NSERC) under grant OGP0039682. This support is gratefully acknowledged. Thanks are due to two referees for their valuable comments.

References

Anily, S., Gendreau, M. and Laporte, G. Optimal sequencing of tasks on a tree shaped structure. Ricerca Operativa 2000; 29:3-14

Assad, A.A. and Golden, B.L. "Arc routing methods and applications". In *M.O. Ball, T.L. Magnanti, C.L. Monma and G.L. Nemhauser, Network Routing*, Handbooks in Operations Research and Management Science, Amsterdam: North-Holland, 1995.

Barahona, F. and Grötschel, M. On the cycle polytope of a binary matroid. Journal of Combinatorial Theory, Series B 1986; 40:40-62

Beltrami, E.L. and Bodin, L.D. Networks and vehicle routing for municipal waste collection. *Networks* 1974; 4:65-94

Benavent, E. `ftp://indurain.estadi.uv.es/pub/CARP` . 1997

Christofides, N. The optimum traversal of a graph. Omega 1973; 1:719-732

Christofides, N. Worst-case analysis of a new heuristic for the traveling salesman problem. Report No 388, Graduate School of Industrial Administration, Pittsburgh, PA, Carnegie Mellon University, 1976.

Christofides, N., Campos, V., Corberán, A. and Mota, E. An algorithm for the rural postman problem. Imperial College Report I C.O.R. 81.5, London, 1981.

Corberán, A. and Sanchis, J.M. A polyhedral approach to the rural postman problem. European Journal of Operational Research 1994; 79:95-114

Croes, G.A. A method for solving traveling-salesman problems. Operations Research 1958; 6:791-812

DeArmon, J.S. A comparison of heuristics for the capacitated Chinese postman problem. Master's Thesis, University of Maryland, College Park, MD, 1981.

Edmonds, J. and Johnson, E.L. Matching, Euler tours and the Chinese postman problem. Mathematical Programming 1973; 5:88-124

Eiselt, H.A., Gendreau, M. and Laporte, G. Arc routing problems, part I: The Chinese postman problem. Operations Research 1995a; 43:231-242

Eiselt, H.A., Gendreau, M. and Laporte, G. Arc routing problems, part II: The rural postman problem. Operations Research 1995b; 43:399-414

Euler, L. Solutio problematis ad geometriam situs pertinentis. Commentarii academiae scientarum Petropolitanae 1736; 8:128-140

Fleischner, H. Eulerian Graphs and Related Topics (Part I, Volume I), Annals of Discrete Mathematics 45, Amsterdam: North-Holland, 1990.

Ford, L.R. and Fulkerson, D.R. *Flows in Networks.* Princeton, NJ.: Princeton University Press, 1962.

Frederickson, G.N. Approximation algorithms for some postman problems. SIAM Journal on Computing 1979; 7:179-193

Gendreau, M., Hertz, A. and Laporte, G. A tabu search heuristic for the vehicle routing problem. Management Science 1994; 40:1276-1290

Ghiani, G. and Improta, G. Optimizing laser-plotter beam movement. Journal of the Operational Research Society 2001, forthcoming.

Ghiani, G. and Laporte, G. Eulerian location problems. Networks 1999; 34:291-302

Ghiani, G. and Laporte, G. A branch-and-cut algorithm for the undirected rural postman problem. Mathematical Programming 2000; 87:467-481

Golden, B.L., DeArmon, J.S. and Baker, E.K. Computational experiments with algorithms for a class of routing problems. Computers & Operations Research 1983; 10:47-59

Greistorfer, P. Computational experiments with heuristics for a capacitated arc routing problem. Working Paper 32, Department of Business, University of Graz, Austria, 1994.

Grötschel, M. and Holland, O. Solution of large-scale symmetric traveling salesman problems. Mathematical Programming 1991; 51:141-202

Guan, M. Graphic programming using odd and even points. Chinese Mathematics 1962; 1:273-277

Hertz, A., Laporte, G. and Mittaz, M. A tabu search heuristic for the capacitated arc routing problem. Operations Research 2000; 48:129-135

Hertz, A., Laporte, G. and Nanchen-Hugo, P. Improvement procedures for the undirected rural postman problem. INFORMS Journal on Computing 1999; 11:53-62

Hierholzer Uber die Möglichkeit einen Linienzug ohne Widerholung und ohne Unterbrechung zu umfahren. Mathematische Annalen 1873; VI:30-32

Jünger, M., Reinelt, G. and Rinaldi, G. The traveling salesman problem. In *Network Models*, Handbooks in Operations Research and Management Science, M.O. Ball, T.L. Magnanti, C.L. Monma and G.L. Nemhauser (eds), Amsterdam: North-Holland, 1995.

Lenstra, J.K. and Rinnooy Kan, A.H.G. On general routing problems. Networks 1976; 6:273-280

Letchford, A.N. Polyhedral results for some constrained arc-routing problems, Ph.D. Thesis, Department of Management Science, Lancaster University, United Kingdom, 1996

Mittaz, M. Problèmes de cheminements optimaux dans des réseaux avec contraintes associées aux arcs, Ph.D. Thesis, Department of Mathematics, École Polytechnique Fédérale de Lausanne, Switzerland, 1999.

Mladenović, N. and Hansen, P. Variable neighbourhood search. Computers & Operations Research 1997; 34:1097-1100

Nobert, Y. and Picard, J.-C. An optimal algorithm for the mixed Chinese postman problem. Networks 1996; 27:95-108

Orloff, C.S. A fundamental problem in vehicle routing. Networks 1974; 4:35-64

Padberg, M.W. and Rinaldi, G. A branch and cut algorithm for the resolution of large-scale symmetric traveling salesman problem. SIAM Review 1991; 33:60-100

Pearn, W.-L. Approximate solutions for the capacitated arc routing problem. Computers & Operations Research 1989; 16:589-600

van Aardenne-Ehrenfest, T. and de Bruijn, N.G. Circuits and trees in oriented linear graphs. Simon Stevin 1951; 28:203-217

SEQUENCING AND SCHEDULING FOR NON-SERIAL PERMUTATION FLOWSHOPS

Robert Burdett and Erhan Kozan

School of Mathematical Sciences
Queensland University of Technology
GPO Box 2434 Brisbane 4001Australia.
{burdett, kozan}@.qut.edu.au

Abstract Most flowshop problems that have been researched have dealt with serial precedence requirements such that the flowshop has become synonymous with the serial assembly line. However, non-serial precedence requirements can also occur for which the resulting problem is still a type of flowshop. This research addresses in particular, non-serial versions of the three standard flowshop problems, namely the intermediate storage, no-intermediate storage and no-wait flowshop scenarios. The outcomes of this research include a number of new and alternative mathematical formulations for finding the optimal sequence of the products to be processed through the non-serial line. Secondly recurrence equations for scheduling purposes have also been developed.

Keywords: Sequencing, scheduling, flowshop, mixed-model assembly.

1. INTRODUCTION

The problem of how best to sequence a number of different jobs (products), with varying processing times, on a number of fixed resources (stations or machines) in a set order or routing (sequence), is known as the flowshop sequencing and scheduling problem. This is a well-known scheduling problem, which because of its complexity has had much attention by researchers both past and present. For example, see Cambell et al (1970), Koulamas (1998), Murata et al (1996), Nawaz et al (1983), Ogbu and Smith (1990), Osman and Potts (1989), Reeves (1995), Taillard (1990), (1993), and Widmer and Hertz (1989).

A major application of this theory is to the assembly of items in a mixed-model assembly process. Mixed-model assembly line (MMAL) is a special type of assembly process that is commonly used to produce highly customised and complicated equipment with many components and sub-assemblies. The term mixed-model refers to the production of more than two model variants, which are intermixed and processed simultaneously on the assembly line.

Various other mathematical models and solution techniques have also been proposed to strictly solve mixed-model assembly line sequencing problems, however much of the emphasis behind these approaches has been to incorporate just-in-time (JIT) principles into the assembly line. A substantial body of literature can be found on such sequencing issues and good reviews of the sequencing literature from before the 1990's can be found in Groeflin et al (1989), Sumichrast and Russell (1990), and Bard et al (1992). Work that is more recent includes Buzacott (1990), Bolet et al (1994), Bolat (1997), Duplaga et al (1998).

Much of the sequencing and scheduling theory for flowshops/MMAL has dealt primarily with serial processes. Serial production structures arise most commonly in assembly line processes in which items are processed on one station after the other, until total completion. Each station along the way performs specific tasks, which must be performed before proceeding tasks on the successor station. When a serial assembly line may have a number of sub-assembly operations that are prerequisites, or alternatively when several serial lines may merge and then continue as one, the process can be called non-serial (NS). Many industrial processes are non-serial processes like this. However, sometimes for simplicity only separate serial portions are considered when sequencing and scheduling. It should be noted however that when a non-serial assembly line is split into a serial one the makespan may not always be calculated correctly as the total time through the final assembly line (FAL). A correct makespan value may only be calculated with total accuracy if the whole process is considered. Hence, an alternative approach would be to treat the whole NS problem at once. The result is an extension of the general serial flowshop problem for which very little research appears to exist.

Since the flowshop problem is a special case of the jobshop, for which non-serial precedence requirements are nothing new, some recent and relevant research includes Lalsare and Sen (1995), Agrawal et al (1996), and Anwar and Nagi (1997). This research in summary considered the job shop or assembly shop type environment in which each finished product to be scheduled has a unique bill or material (BOM) and routing from which a precedence network can be obtained. This is also known as a multi-level, multi-product environment. The objective in particular was the minimisation of the makespan. In some of this research, setup times, lot sizes and due dates were also incorporated. Forward and backward iterative sequencing heuristics, which used dispatching rules or critical path methods were the type of solution technique developed in that research. It should be noted that in the type of jobshop addressed in that research, non-permutation sequences occur. In the flowshop situation that we address, non-permutation sequences do not occur, however this is not to say that they do not in other circumstances. This factor makes these solution techniques unsuitable for the problem in this paper. Alternative solution techniques based upon well-known meta-heuristics have also been applied to this problem by Kim and Kim (1996), and Roach and Nagi (1996).

Dauzere-Peres et al (1998) addressed three extensions of the standard scheduling problem that can be found in practice. In particular these were multiple resources, resource flexibility, and non-linear routing. Multiple resources imply that operations may need several resources before they can be performed, and for each resource required, a set of candidate resources to choose from is also available. This is referred to as resource flexibility. Lastly non-linear or non-serial precedence requirements may occur, which imply that an operation may have more than one predecessor or successor on the routing. In this problem the duration of an operation is resource dependent but which are known in advance. The problem is then to assign resources to operations and to sequence operations. A disjunctive graph representation was used, and a connected neighbourhood structure was proposed. A local search algorithm based upon TS was proposed and a numerical comparison was performed. This research considered a general shop scheduling problem and is also a type of resource constrained project scheduling problem (RCPSP) for which Patterson et al (1990), and Ozdamar and Ulusoy (1996) have also performed some recent work.

The research in this paper consists of the following. Firstly in Section 2 the variables used in this research are listed. In Section 3 the main mathematical formulations are developed while in Section 4, scheduling theory including recurrence equations and alternative mathematical formulations are developed for operations scheduling. A numerical investigation is presented in Section 5 and lastly a summary of the outcomes and conclusions of this research are given in Section 6.

2. NOMENCLATURE

A list of the variables are as follows:

M Number of final assembly line stations

N Number of different model (product or job) types that can be sequenced.

L Number of products in sequencing period to be sequenced.

i *models* $i = 1,..,N$

j *stations* $j = 1,..,M$

k stages k=1,.., L

$s(k)$ **The product in the *k*th position of the sequence**
$S = \{s(1), s(2), \ldots\ldots, s(L)\}$ **where** $s(k) \in (1, N)$. **The product in the *k*th position is also known as the *k*th stage.**

$X_{i,k}$ Binary variable that equals 1 if model *i* is sequenced in the *k*th position and zero otherwise

d_i Demand for model *i* for the current sequencing period

$C_{j,k}$ Completion time of stage *k* on station *j*

$D_{j,k}$ Departure time of stage *k* from station *j*

$I_{j,k}$ Idle time on work station *j* between stage *k*-1 and stage *k*

$Wait_{j,k}$ The time that the product in the *k*th position of the sequence waits in an intermediate storage area between stations *j* and *j*+1 after processing on station *j*.

$W_{j,k}$ The time that the product in the *k*th position of the sequence waits on the next station *j*+1 after leaving station *j*. Note that the successor station is used as a pseudo intermediate storage area.

$P_{i,j}$ Processing time for model type *i* on station *j*

$B_{j,k}$ The amount of time that stage *k* must wait on station *j* because its path is blocked downstream.

$B^{after}_{j,k}$ This is the post processing blocking which is equivalent to $B_{j,k}$.

$B^{before}_{j,k}$ Pre-processing blocking. The time that station *j* is blocked prior to processing stage *k* due to the use of station *j* as an intermediate storage area for predecessor operations.

$DUR_{j,k}$ The actual time spent on station *j* by the *k*th stage. The duration is calculated as follows, $DUR_{j,k} = \sum_{i=1}^{N} P_{i,j} X_{i,k} + B_{j,k}$.

SPN An activity on node (AON) directed graph representing the station precedence requirements, that is the station precedence network. If there is an edge $(j, j') \in SPN$, then the operation on station *j* is a direct precedence requirement for the operation on station j'.

$succ_j$ Graph operator that gives the successor of station *j*.

$$succ_j \in \{j' : (j, j') \in SPN\}$$

$pred_j$ Graph operator that gives the set of predecessors to station *j*.

$pred_j \in \{j' : (j', j) \in SPN\}$. Note that the total number of predecessor operations of *j* is given by the following operator, $\# pred_j$.

3. MATHEMATICAL FORMULATIONS

3.1 General Aspects of Sequencing Models

The aim of the problem is to find the sequence that gives the minimum makespan in a non-serial process with convergent operations and either intermediate storage, no-intermediate storage, or no-waiting time. Regardless of which flowshop scenario has occurred the following sequencing equations are required:

$$\sum_{i=1}^{N} X_{i,k} = 1 \qquad \forall k \tag{1}$$

$$\sum_{k=1}^{L} X_{i,k} = d_i \qquad \forall i \tag{2}$$

These equations ensure that each job is sequenced, and only one job is sequenced in each position. The makespan equation, which is the objective of each of the models, is also the same for each flowshop. However, the formulation of such an expression initially poses some difficulties. Since the process is non-serial, the makespan cannot be calculated with the same ease as for a serial line. For example, in a serial flowshop the makespan is calculated in one of two ways. The makespan is either the total time spent processing items on the first station plus the time that it takes the last item in the sequence to be processed on the rest of the stations. Alternatively, it is the time it takes the first item in the sequence to be processed on each of the stations plus the remaining time it takes all of the other items in the sequence to be processed on the last station (including idle time). It should be noted that all critical paths pass through each of the M stations for the serial flowshop.

However, for the non-serial scenario there are many more critical paths which pass through different combinations of the M stations depending on the station precedence network (SPN) and the processing times. An exact expression for the makespan is therefore somewhat difficult to write algebraically. One possible form is the following, which is based upon recursive equations.

$$C\max = \phi_M \tag{3}$$

$$\phi_j = \sum_{k=1}^{L} DUR_{j,k} + \max_{\forall j' \in pred_j} \{\phi_{j'}\} \qquad \forall j : \#pred_j = 0 \tag{4}$$

$$\phi_j = DUR_{j,L} + \max_{\forall j' \in pred_j} \{\phi_{j'}\} \qquad \forall j : \#pred_j \neq 0 \tag{5}$$

Essentially ϕ_j defines the maximum time of any critical path to station j. The makespan is hence just the maximum critical path to the last station M. It should be noted that the recurrence equations (4) and (5) are also based upon the first way in which the makespan is calculated for the serial case. Due to the complexity of these equations and the extra constraints and variables required it may be more beneficial in practice to generate a specific makespan expression for each particular problem occurrence. A third and more elegant alternative however is possible which allows the makespan and the subsequent mathematical model to be drastically simplified.

In previous flowshop models the idle time before the first stage is generally not required and is set as zero for simplicity. By removing this constraint and replacing it with another equation, the correct value for the idle time before the first stage is processed, can be obtained. Therefore, because the process is convergent, and the idle time before the first stage is processed can be calculated, the makespan equation can be simplified to equation (6).

$$C\max = \sum_{k=1}^{L}\left(\sum_{i=1}^{N} P_{i,M} X_{i,k} + I_{M,k}\right) \tag{6}$$

This expression is just the time before the first stage is processed on the last station plus the total time taken to process all stages on the last station and the idle time between. Hence, no recursive functions are needed to check the length of each critical path. The rest of the constraints that are required for each particular flowshop scenario will now be discussed.

3.2 Intermediate Storage (IS) Scenario

The following constraints are also required for the IS flowshop.

$$I_{j,k} + \sum_{i=1}^{N} P_{i,j} X_{i,k} + Wait_{j,k} = Wait_{j,k-1} + \sum_{i=1}^{N} P_{i,j'} X_{i,k-1} + I_{j',k}$$
$$\forall (j, j'): succ_j = j';\ k=2,..,L; \tag{7}$$

$$I_{j,1} + \sum_{i=1}^{N} P_{i,j} X_{i,1} + Wait_{j,1} = I_{j',1} \qquad \forall j, j' : succ_j = j' \tag{8}$$

$$I_{j,k} = 0 \qquad \forall j :\# pred_j = 0;\ k=1,..,L \tag{9}$$

$$Wait_{M,k} = 0 \qquad k=1,..,L \tag{10}$$

$$Wait_{j,1} = 0 \qquad \forall j : \# pred_{succ_j} = 1 \tag{11}$$

This model is similar to the original serial IS flowshop model that was first proposed by Wagner (see French 1982). However, it differs from this formulation in a number of ways. One such way is constraint (8), that allows the correct pre-first job (stage) idle time to be calculated. The following expression is a by-product of equation (8) that is not necessarily required in the model.

$$I_{j,1} + \sum_{i=1}^{N} P_{i,j} X_{i,1} + Wait_{j,1} = I_{j',1} + \sum_{i=1}^{N} P_{i,j'} X_{i,1} + Wait_{j',1}$$

$$\forall j, j' : succ_j = succ_{j'} \qquad (12)$$

It should be noted that equation (8) with the inclusion of $I_{j,1}$ terms allows the makespan to be expressed so simply as previously mentioned. Another difference in the model is the balance equation (7) that has been modified for the non-serial case. The modification here is simply to change the j+1 subscript to j' which is the successor of station j.

The other constraints in this model include (9), which enforces zero idle time on any station with no predecessors. Constraints (10) and (11) respectively enforce no waiting time on the last station or for any first stage operation performed on a station that is the only predecessor of another.

3.3 No-intermediate Storage (NIS) Scenario

3.3.1 Post processing blocking scenario

The first NS-NIS flowshop scenario has at its core the main assumption that all operations that have a common successor also have the same departure time. The following constraints are required in a model for this scenario.

$$I_{j,k} + \sum_{i=1}^{N} P_{i,j} X_{i,k} + B_{j,k} = \sum_{i=1}^{N} P_{i,j'} X_{i,k-1} + B_{j',k-1} + I_{j',k}$$

$$\forall (j, j') : succ_j = j' ;\ k=2,..,L; \qquad (13)$$

$$I_{j,1} + \sum_{i=1}^{N} P_{i,j} X_{i,1} + B_{j,1} = I_{j',1} \quad \forall j, j' : succ_j = j' \qquad (14)$$

$$I_{j,k} = 0 \qquad \forall j : pred_j = \{\ \};\ k=1,..,L \qquad (15)$$

$$B_{M,k} = 0 \quad k=1,..,L \qquad (16)$$

The correct value for the idle time before the first stage is processed is obtained by equation (14) which is similar to (8) which was for the IS case. The NIS flowshop is actually a no-wait (NWT) flowshop in duration and hence the standard precedence relationship equation can be modified for the non-serial case as shown in equation (13). The modification here was to change the original j+1 subscript to j' which stands for the successor of j. Therefore, for any pair of stations (j, j') such that $j \rightarrow j'$, a constraint is required.

Equation (15) ensures that there is no idle time on any station that has no predecessors while equation (16) ensures that there is no blocking on the last station. In general there is no blocking on any station that has no successors (ie $\#succ_j = 0$); however, since the process is assumed to be convergent, this only occurs for the last station. It is also true that there is no blocking time for the first stage (ie item) in the sequence on any station *j* if the $\#pred_j = 0$ and $\#pred_{succ_j} = 1$. This constraint is ensured by equation (14) in the model.

3.3.2 Pre processing blocking scenario

When a station has multiple predecessors, it is possible for blocking to be caused between any of these predecessor stations under the first NIS scenario. An example is shown in Figure 1(b). The station precedence network is (a).

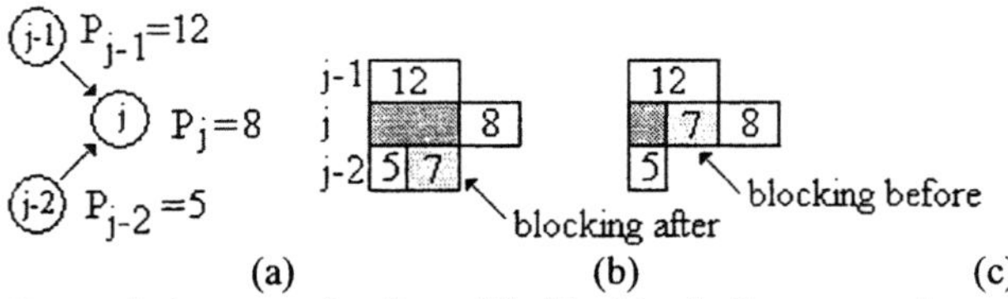

Figure 1. An example of possible blocking before processing.

The first scenario upholds the occurrence shown in (b), which is that for a given stage, all stations processing that stage and that have a common successor station have the same departure time. In such a circumstance however it is possible to do better. The way to do this is to allow blocking to occur before processing as shown in (c). Hence when the first predecessor finishes (that is the operation performed on station *j*-2), the item is sent to the successor station *j* to wait for all the other predecessor components (that is on station *j*-1). Therefore what happens is that the station becomes temporarily blocked before the processing can begin. Therefore in this new occurrence blocking may occur before and after a station task has been preformed.

The new mathematical model that incorporates this is as follows.

$$\textit{Minimise} \quad C\max = \sum_{k=1}^{L}\left(I_{M,k} + B_{M,k}^{before} + \sum_{i=1}^{N} P_{i,M} X_{i,k} \right) \tag{17}$$

Subject to

$$\sum_{k'=1}^{k}\left(I_{j,k'}+B_{j,k'}^{before}+\sum_{i=1}^{N}P_{i,j}X_{i,k'}+B_{j,k'}^{after}\right)+W_{j,k}$$

$$=\sum_{k'=1}^{k-1}\left(I_{j',k'}+B_{j',k'}^{before}+\sum_{i=1}^{N}P_{i,j'}X_{i,k'}+B_{j',k'}^{after}\right)+I_{j',k}+B_{j',k}^{before}$$

$$k=2,..L;\ \forall j,j':succ_j=j' \quad (18)$$

$$I_{j,1}+B_{j,1}^{before}+\sum_{i=1}^{N}P_{i,j}X_{i,1}+W_{j,1}=I_{succ_j,1}+B_{succ_j,1}^{before}$$

$$for\ j=1,..,M\text{-}1 \quad (19)$$

$$B_{j,k}^{before}=0 \quad \forall j:\#pred_j\leq 1;\ k=1,..,L \quad (20)$$

$$B_{j,k}^{before}=\max_{\forall j':j'=pred_j}\{W_{j',k}\}\ \forall j:\#pred_j>1 \quad (21)$$

$$B_{j,1}^{after}=0 \quad j=1,..,\mathrm{M} \quad (22)$$

$$B_{j,k}^{after}=0 \quad \forall j:\#succ_j=0;\ k=1,..,L \quad (23)$$

$$W_{j,k}=0\ \forall j:\#pred_{succ_j}=1 \quad (24)$$

$$W_{j,k}=0\ \forall j:\#succ_j=0;\ k=1,..,L \quad (25)$$

This second alternative scenario is a simple concept that is also a logical progression from the first scenario. It is in reality, however, a much more complicated case to be modelled (as can be seen from the model above) as it also poses a number of unique difficulties. These difficulties in particular centre around the development of a balance equation that correctly upholds the rules of this proposed flowshop. In formulating this model, extra waiting time and blocking time before variables have had to be introduced.

In this model, constraint (15) remains unchanged from the first scenario. The objective function (17) is much the same also except that an extra blocking before term is added that gives the time that the last station is used as an intermediate storage (IS) area for the outputs of predecessor operations. Equation (22) is a new constraint that was not in the previous model. In particular in the previous model, blocking was often caused between rival predecessor operations that had a common successor. In this case, however, no first stage item is blocked on a station after it has been processed because the successor station can be used as a temporary IS area. Equation (21) calculates the blocking before processing on a station as the maximum waiting time spent by any predecessor operations. If there is only one or less predecessor then blocking before cannot occur on the successor station. This is also given by equation (20). In this model there are two particular balance equations. The first is (19), which is the balance equation that allows the pre first stage idle time to be calculated correctly. This expression is similar to expression (14) of the previous model, however blocking before variables have been introduced as well as a waiting time aspect.

The second balance equation is (18) which performs the same function as (13) of the previous model. While this equation has a few extra blocking before and waiting time variables in the expression for which (13) did not, the main difference that should be noted is the addition of summation terms. A standard balance equation is not suitable because of the uncertainty of which blocking before quantity should be used in the equation. An example of this is displayed in Figure 2(c). Note that blocking before quantities are shaded in light grey and the idle time in dark grey. By adding the summation term in equation (18) however, everything is balanced from time zero, thus ensuring the rules of this particular flowshop are upheld. Balance equation (18) contains a large number of terms and may be simplified in certain circumstances. In particular equation (18) may be simplified to the following when $\#pred_{succ_j} \leq 1$.

$$I_{j,k} + B^{before}_{j,k} + \sum_{i=1}^{N} P_{i,j} X_{i,k} + B^{after}_{j,k} + W_{j,k}$$

$$= B^{before}_{j',k-1} + \sum_{i=1}^{N} P_{i,j'} X_{i,k-1} + B^{after}_{j',k-1} + W_{j,k-1} + I_{j',k}$$

$$k=2,..L;\ \forall j, j' : succ_j = j' \text{ and } \#pred_{j'} \leq 1 \qquad (26)$$

Hence, an added restriction for (18) is that $\#pred_{succ_j} > 1$.

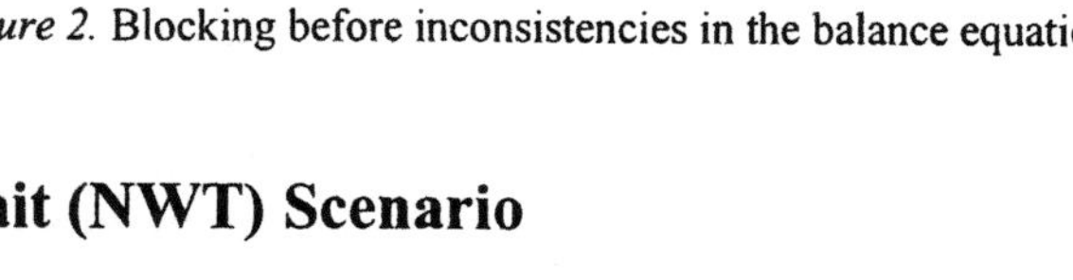

(a) (b) (c)

Figure 2. Blocking before inconsistencies in the balance equation.

3.4 No Wait (NWT) Scenario

The only extra constraints required for the no-wait scenario are as follows:

$$I_{j,k} + \sum_{i=1}^{N} P_{i,j} X_{i,k} = \sum_{i=1}^{N} P_{i,j'} X_{i,k-1} + I_{j',k}$$

$$\forall (j, j') : succ_j = j';\ k=2,..,L; \qquad (27)$$

$$I_{j,1} + \sum_{i=1}^{N} P_{i,j} X_{i,1} = I_{j',1} \quad \forall j, j' : succ_j = j' \tag{28}$$

The balance equation (27), is the same as for the serial case, except for the modification of the $j+1$ subscript to j', the successor station of j. Constraint (28) is an extension to the original model and defines the exact idle time before the first stage operations.

4. OPERATIONS SCHEDULING

4.1 Recurrence Equations

For a given sequence we would like to be able to determine when each product should be processed on each station, and when it should be moved from one station to another. While the mathematical models thus far developed could be used, it is common practice to use a set of recurrence equation. Recurrence equation are in general fairly simple to define and are also generally more computationally efficient.

4.1.1 Intermediate Storage (IS) Scenario

A backward sequencing approach to scheduling is more efficient for the IS scenario because the station precedence graph is divergent in reverse. This means in particular that each stage has only one predecessor and hence the maximum departure time of all the predecessors is not required as it is for the forward approach. The backward scheduling equation is as follows:

$$D_{M,L} = P_{s(L),M} \tag{29}$$

$$D_{j,L} = D_{j',L} + P_{s(k),j} \quad \forall j,\ j' = succ_j \tag{30}$$

$$D_{M,k} = D_{M,k+1} + P_{s(k),M} \quad for\ k=L-1,..,1 \tag{31}$$

$$D_{j,k} = \max\left(D_{j',k}, D_{j,k+1}\right) + P_{s(k),j} \quad j' = succ_j,\ k=L-1,..,1 \tag{32}$$

4.1.2 No-intermediate Storage (NIS) Scenario

For the first NIS scenario, it is more efficient to backward schedule with the following recurrence equations.

$$D_{M,L} = P_{s(L),M} \tag{33}$$

and $D_{j,L} = D_{j',L} + P_{s(k),j}$

$$\forall j : j \neq M \text{ and } j' = succ_j \quad (34)$$

$$D_{M,k} = \max\left\{ D_{M,k+1} + P_{s(k),M}, \max_{j' \in pred_M} \{D_{j',k+1}\} \right\} \quad k{=}L\text{-}1,..,1 \quad (35)$$

$$D_{j,k} = \max\left\{ D_{j',k} + P_{s(k),j}, \max_{\hat{j} \in pred_j} \{D_{\hat{j},k+1}\} \right\}$$

$$\forall j:\ j \neq M, \#pred_j \neq 0\ ,\ j' = succ_j, k{=}L\text{-}1,..,1 \quad (36)$$

$$D_{j,k} = D_{j',k} + P_{s(k),j} \qquad \forall j : \#pred_j = 0,\ j' = succ_j \quad (37)$$

It should be noted that for backwards scheduling, $D_{j,k}$ is the departure time from the final completion time, as opposed to the departure time after the initial beginning time. Backward scheduling is superior for this scenario because of the fact that $D_{j,k} = D_{j',k} \quad \forall j, j' : succ_j = succ_{j'}$, and the process is non-serial and convergent. By performing the calculations backward, all predecessor operations can be calculated without interference, since the process in reverse is divergent. Note that the makespan for the backward scheduling case is as follows, $C\max = \max_{\forall j:\#pred_j=0} \{D_{j,1}\}$.

For the second NIS scenario, which has pre processing blocking, forward scheduling is more efficient than backward scheduling and can be performed by applying the following equations:

$$D_{j,1} = P_{s(1),j} \qquad \forall j : \#pred_j = 0 \quad (38)$$

$$D_{j,1} = \max_{\forall j' \in pred_j} \{D_{j',1}\} + P_{s(1),j} \quad \forall j : \#pred_j \neq 0 \quad (39)$$

$$D_{j,k} = \max\{D_{j,k-1} + P_{s(k),j}, D_{j',k-1}\} \ \forall j : \#pred_j = 0,$$

$$j < M \ \ j' = succ_j, \textit{ and } k{=}2,..,L \quad (40)$$

$$D_{j,k} = \max\left\{ \max_{\forall \hat{j} \in pred_j} \{D_{\hat{j},k}\} + P_{s(k),j}, D_{j',k-1} \right\} \quad \forall j : \#pred_j \neq 0,$$

$$j < M\ ,\ j' = succ_j, \textit{ and } k{=}2,..,L \quad (41)$$

$$D_{M,k} = \max_{\forall j' \in pred_M} \{D_{j',k}\} + P_{s(k),M} \ \textit{for } k{=}2,..,L \quad (42)$$

The backward scheduling approach for the second scenario is more complicated than the first scenario because the departure times of all predecessors do not have to equal. These equations are very similar to those for the serial case. The main difference is that the maximum departure time of the predecessors is required.

4.1.3 No Wait (NWT) Scenario

The recurrence equations approach for the NWT scenario are as folows.

$$D_{j,0} = 0\,, D_{0,k} = 0 \tag{43}$$

$$D_{j,k} = \max\left(D_{j,k-1} + P_{s(k),j}, D_{succ_j,k-1}\right)$$

$$\forall j : \#pred_j = 0;\ k=1,..,L \tag{44}$$

$$D_{j,k} = \max\left(\max_{\forall \hat{j}:\hat{j}=pred_j}\left\{D_{\hat{j},k}\right\} + P_{s(k),j}, D_{succ_j,k-1}\right)$$

$$\forall j : \#pred_j > 0;\ k=1,.,L \tag{45}$$

$$D_{j,k} = D_{succ_j,k} - P_{s(k),succ_j} \quad j=M-1,..,1\ ;\ k=1,..,L \tag{46}$$

This is a forwards/backwards algorithm which means that for each stage that is scheduled, all operations are scheduled from the first station to the last and then in reverse order from the last to the first. Equation (15) and (16) constitutes the forward component which it should be noted are actually the same equations required as for the no-intermediate storage (NIS) scenario. The backward component on the other hand, which is equation (17), simply draws all operations together in a no-wait fashion starting from the last station. Since the precedence graph is divergent in reverse each station operation has only one predecessor and hence the backward component can be implemented efficiently. Note that by performing consecutive forwards and backward passes, operations are only shifted once to ensure no waiting time with predecessor and successor operations.

4.1.4 Directed Graph Approach

By finding the longest path through a particular directed graph for the serial flowshop the makespan could be calculated and operations could also be scheduled correctly as an alternative to recurrence equations. Such an approach is also possible for each of the non-serial flowshop scenarios. This theory in particular while allowing the makespan to be obtained for a given sequence and all operations to be scheduled correctly, also allows an alternative and maybe simpler style of mathematical formulation to be developed. This alternative model also allows extra resource constraints to be incorporated as extensions, whereas it is not possible to do so with the first set of formulations. This theory is however not presented here.

5. NUMERICAL INVESTIGATION

The numerical investigation in this section is provided so that a comparison of the two proposed NIS scenarios can be made. It is expected from the theory in Section 3, that Scenario 2 will be more efficient than Scenario 1 for non-serial flowshops. This is because it does not restrict all predecessor operations to have the same departure time, thus resulting in less blocking. By more efficient we mean with respect to having the minimum makespan when scheduling all operations (that is by recurrence equation). Given that we expect Scenario 2 to give smaller makespans, we would like to know by just how much smaller the resulting difference would be. In particular we wish to know which scenario gives the minimum makespan for a variety of different problems of various sizes and for different sequences. So that a reasonable comparison could be made, two sets of test problems were generated and evaluated. The first set was artificially generated for the purpose of investigating the effect of different numbers of stations and different station precedence network (SPN) non-seriality levels. The second source of test instances are real world and are evaluated for their practical relevance.

5.1 Test Problems 1

The first set of data consists of 480 randomly generated problems. These problems consisted of non-serial NIS flowshops with 5, 10, 15, 20, 25, 30, 35, 40, 45, 50, 55, and 60 stations, and sequences with 25, 50, 75, and 100 jobs. For each of these 48 problem sizes, (12x4), processing times were generated randomly so that the processing times of each operation were between 1 and 100 minutes in value. For each of these 48 sized problems, a set of 10 extra trials at this size was also evaluated. A different sequence and station precedence network was created for each.

Each SPN consisting of M stations was randomly generated by adding edges (arcs) between stations based upon the following expression, and the assumption that each node has a single successor and the graph is numbered in ascending order:

$$arc(j, random(j+1, M)) \quad for\, j=1,..,M\text{-}1 \tag{47}$$

Since the level of non-seriality of each SPN is also expected to contribute to the comparison, we need to define some suitable measure of non-seriality. The non-seriality indicator (NSI) is defined as follows, $nsi = \#levels / M$, which is based upon the number of levels in the SPN. For example, if the graph is serial, then the number of levels is M and the NSI is one. In general the more non-serial the graph is, the closer the NSI value is to zero. In particular the minimum number of levels is 2 which means that the graph is as non-serial as possible. The number of levels in the SPN is determined by the following recurrence equation.

$$\#levels = levels_M = 1 + \max_{\forall j' : j' \in pred_M} \{levels_{j'}\} \tag{48}$$

$$levels_j = 1 + \max_{\forall j' : j' \in pred_j} \{levels_{j'}\} \qquad j=1,..,M \tag{49}$$

Lastly, the results are expressed in terms of the relative error (RE) of the Scenario 2 makespan with respect to the Scenario 1 makespan. That is the RE is defined as follows,

$$\text{RE}=[C\max(Scenario2) - C\max(Scenario\ 1)] / C\max(Scenario\ 1)$$

5.2 Test Problems 2

Unlike the first 480 test problems that were generated randomly, the second set of 800 test problems is based upon a particular real-life industry situation. The particular station precedence network for this industry situation is displayed in Figure 8 below. There are 59 stations in this process and the non-seriality level as defined by (48) and (49) is 0.4406, that is it has 26 levels.

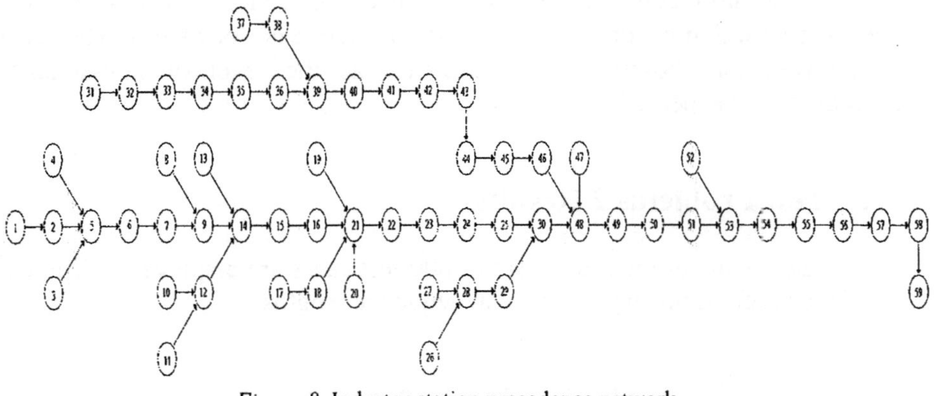

Figure 8. Industry station precedence network

Together with this SPN are a set of processing times for each of the 59 stations and for N=10 products. Since the number of stations is fixed, the 800 test problems consisted of sequences of varying lengths. Each of the 800 generated sequences was based upon 80 randomly generated demand vectors. The 80 demand vectors were generated so that the sum of the individual demand for each of the N products varied from 15, 20, 30, 50, 75, 100, 200, and 300 products. Ten different demand vectors were also generated for each given sequence length.

5.3 Test Problem 1 Results

The results of the first set of 480 test cases are displayed in Table 1 below. From Table 1 we can see that at worst the second scenario gives at least a 1 percent improvement in makespan over Scenario 1 for each of the test cases. The maximum improvement however was as much as 9 percent for some test cases. We can also see that the length of the sequence gives little extra improvement for this set of test problems. This would then mean that a longer sequence does not guarantee a higher and more substantial reduction in makespan than a smaller sequence.

To see what effect the level of non-seriality had, we refer to the graph in Figure 9. This graph shows the NSI value for each of the 480 test problems plotted against the percentage improvement (ie RE multiplied by -100). From this graph we see that for a particular level of non-seriality there is quite a difference in the range of possible improvement, that is any value from 0-10 percent. However, for a particular sequence length it appears that the level or range of improvement is much the same regardless of the level of non-seriality. It should also be noted that on average the level of improvement is reasonably constant regardless of the level of non-seriality. The range of improvement for each sequence length however, does decrease and become more consistent as the sequence length increases. For example the most variation occurs for L=25, and with the least for L=100. The level of improvement however, does decrease as the level of non-seriality decreases as is expected.

5.4 Test Problems 2 Results

The results of the second set of 800 test cases are displayed in Table 2 below for each of the eight particular sequence lengths.

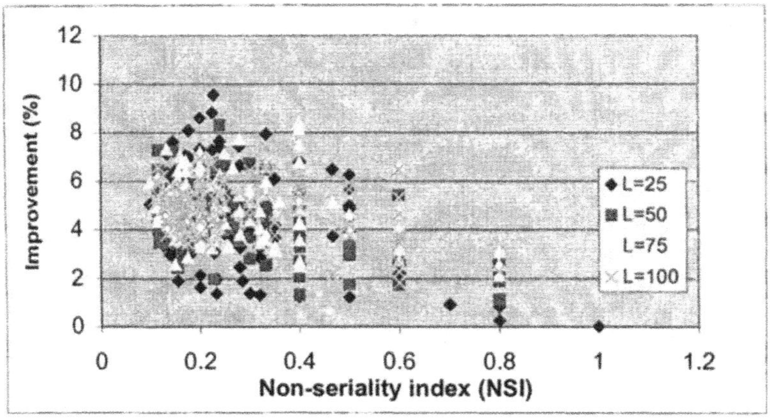

Figure 9. Non-seriality versus percentage improvement.

Table 1. Relative error values of Scenario 2 makespans with respect to Scenario 1 and non-seriality indicator values for each randomly generated test case.

L,M	RE			NSI			L,M	RE			NSI		
	MIN	AVG	MAX	MIN	AVG	MAX		MIN	AVG	MAX	MIN	AVG	MAX
25,5	0	-0.02	-0.03	1	0.68	0.4	**75,5**	-0.02	-0.04	-0.08	0.8	0.64	0.4
25,10	-0.01	-0.04	-0.07	0.7	0.5	0.4	**75,10**	-0.03	-0.06	-0.08	0.6	0.44	0.4
25,15	-0.01	-0.05	-0.08	0.47	0.38	0.33	**75,15**	-0.02	-0.05	-0.07	0.47	0.37	0.27
25,20	-0.01	-0.04	-0.06	0.4	0.3	0.2	**75,20**	-0.03	-0.05	-0.07	0.4	0.3	0.2
25,25	-0.01	-0.05	-0.08	0.32	0.28	0.24	**75,25**	-0.03	-0.05	-0.08	0.36	0.29	0.2
25,30	-0.01	-0.04	-0.07	0.27	0.22	0.17	**75,30**	-0.03	-0.05	-0.07	0.33	0.23	0.2
25,35	-0.02	-0.05	-0.1	0.29	0.21	0.14	**75,35**	-0.03	-0.05	-0.07	0.26	0.2	0.17
25,40	-0.02	-0.05	-0.09	0.25	0.2	0.15	**75,40**	-0.03	-0.04	-0.05	0.23	0.19	0.15
25,45	-0.02	-0.05	-0.08	0.18	0.16	0.13	**75,45**	-0.04	-0.05	-0.07	0.2	0.16	0.13
25,50	-0.03	-0.04	-0.05	0.2	0.16	0.12	**75,50**	-0.04	-0.05	-0.07	0.16	0.14	0.1
25,55	-0.03	-0.05	-0.08	0.2	0.15	0.13	**75,55**	-0.05	-0.05	-0.07	0.18	0.15	0.11
25,60	-0.03	-0.05	-0.06	0.17	0.14	0.1	**75,60**	-0.04	-0.05	-0.06	0.17	0.14	0.12
50,5	-0.01	-0.03	-0.05	0.8	0.72	0.6	**100,50**	-0.02	-0.05	-0.09	0.8	0.62	0.4
50,10	-0.02	-0.04	-0.07	0.5	0.42	0.3	**100,10**	-0.02	-0.04	-0.06	0.6	0.51	0.4
50,15	-0.01	-0.04	-0.06	0.4	0.36	0.27	**100,15**	-0.02	-0.06	-0.07	0.47	0.38	0.33
50,20	-0.03	-0.05	-0.07	0.4	0.28	0.2	**100,20**	-0.03	-0.05	-0.06	0.35	0.3	0.25
50,25	-0.03	-0.05	-0.08	0.4	0.28	0.2	**100,25**	-0.04	-0.05	-0.06	0.28	0.23	0.2
50,30	-0.03	-0.05	-0.06	0.33	0.23	0.17	**100,30**	-0.04	-0.04	-0.06	0.3	0.22	0.17
50,35	-0.02	-0.04	-0.06	0.26	0.2	0.14	**100,35**	-0.04	-0.05	-0.07	0.23	0.2	0.17
50,40	-0.03	-0.05	-0.06	0.25	0.19	0.13	**100,40**	-0.03	-0.05	-0.06	0.2	0.17	0.13
50,45	-0.02	-0.04	-0.05	0.22	0.16	0.13	**100,45**	-0.03	-0.05	-0.07	0.22	0.17	0.13
50,50	-0.03	-0.05	-0.05	0.18	0.15	0.12	**100,50**	-0.04	-0.05	-0.05	0.22	0.17	0.14
50,55	-0.03	-0.05	-0.07	0.2	0.16	0.13	**100,55**	-0.03	-0.05	-0.06	0.18	0.15	0.11
50,60	-0.04	-0.06	-0.07	0.18	0.13	0.12	**100,60**	-0.04	-0.05	-0.07	0.17	0.13	0.12

Table 2. Relative error values of Scenario 2 makespans with respect to Scenario 1.

L(M=59)	RE (min,avg,max)	L(M=59)	RE (min,avg,max)
15	0, -0.00475, -0.024	75	0, -0.0018, -0.0081
20	0, -0.0049, -0.019	100	0, -0.0013, -0.0057
30	0, -0.0034, -0.0176	200	0, -0.0008, -0.0033
50	0, -0.0032, -0.011	300	0, -0.0002, -0.0014

From Table 2 it can be seen that the level of improvement is decreasing with respect to the increasing sequence length. This is a different result to that obtained in Section 5.3 for which no real difference was detected for different sequence lengths. The reason for this is as yet unknown. We can also note that there is a fair reduction in the average and maximum improvement level obtained here as opposed to that obtained previously in Section 5.3. For example, the maximum here was only 2.4% as opposed to 9% before. This is partly to do with the fact that there are only 10 products in the second set of test problems and the particular non-seriality level of this particular industry SPN. The particular values of the processing times would however most probably be the main cause for this difference as in this industry they are between 0 and 10 hours (that is not 0-100 minutes as in the first set of problems). The processing times for a particular station are generally also comparable among all products (or jobs) whereas they would not have been for the first set of randomly generated problems. Hence, a smaller reduction is expected under such circumstances.

6. CONCLUSION

In this research the non-serial intermediate storage, no-intermediate storage and no-wait permutation flowshop with convergent operations was investigated. For each of these scenarios a suitable mathematical formulation was developed and recurrence equations for forward and backward sequencing where more efficient have also been developed.

For the NIS flowshop in particular, two specific scenarios were developed. The first enforces that the departure time of all predecessor operations is the same. Hence blocking occurs not only between adjacent stages but also between stations that have the same successor station. The second scenario has no such restriction, and hence the possibility of blocking also occurs on a station before processing may begin for a particular job. That is, the successor station is used as a temporary intermediate storage area for the previous station's outputs.

From our numerical investigation, Scenario 2 has been found to consistently give schedules of reduced makespan for over 1280 test problems with varying sequences, sequence lengths, processing times and precedence networks to that of Scenario 1. The expected reduction however appears to be problem dependent. That is, the reduction is dependent on the particular processing times. However, reductions of up to 10 percent are possible. It should also be noted that as the level of non-seriality decreases the level of improvement that Scenario 2 achieves over Scenario 1 also decreases. On average however the level of improvement appears to be reasonably constant regardless of the level of non-seriality. It may also be true that the level of improvement that Scenario 2 has over Scenario 1 decreases as the sequence length increases. This factor however needs to be verified from a further study of industrial problems.

7. ACKNOWLEDGEMENTS

This work was partly supported by the Australian Research Council under a SPIRT grant and by Volvo Truck Australia. This support is greatly acknowledged.

References

Agrawal A., Harkalakis G., Minis I. and Nagi R. Just-in-time production of large assemblies. IIE Transactions 1996; 28: 653-667

Anwar M. F. and Nagi R. Integrated lot-sizing and scheduling for just-in-time production of complex assemblies with finite set-ups. International Journal of Production Research 1997; 35(5): 1447-1470

Bard J. F., Dar-El E. and Shtub A. An analytic framework for sequencing mixed model assembly lines. International Journal of Production Research 1992; 30(1): 35-48

Bolat A., Savsar M. and Al-Fawzan M. A. Algorithms for real-time scheduling of jobs on mixed model assembly lines. Computers and Operations Research 1994; 21(5): 487-498

Bolat A. Stochastic procedures for scheduling minimum job sets on mixed model assembly lines. Journal of the Operational Research Society 1997; 48(5): 490-501

Burdett R. L. and Kozan E. Sequencing and scheduling within a mixed flowshop environment. Under review Mathematical and Computer Modelling 2000

Cambell H.G., Dudek R. A. and Smith M. L. A heuristic algorithm for the n job, m machine sequencing problem. Management Science 1970; 16(16)

Dauzere-Peres S., Roux W. and Lasserre J. B. Multi-resource shop scheduling with resource flexibility. European Journal of Operations Research 1998; 107: 289-305

Ding F. Y. and Cheng L. A simple sequencing algorithm for mixed model assembly lines in just-in-time production systems. Journal of Operations Management 1993; 11: 43-50

Duplaga E. A., Hahn C. K. and Hur D. Mixed-model assembly line sequencing at Hyundai Motor Company. Production and Inventory Management Journal 1996; 37: 20-26

Groeflin H, Luss H., Rosenwein M. B. and Wahls E. T. Final assembly sequencing for just-in-time manufacturing. International Journal of Production Research 1989; 27(2): 199-213

Kim J. U. and Kim Y. D. Simulated annealing and genetic algorithms for scheduling products with multi-level product structure. Computers and Operations Research 1996; 23(9): 857-868.

Koulamas C. A new constructive heuristic for the flowshop scheduling problem. European Journal of Operational Research 1998; 105: 66-71

Lalsare P and Sen S. Evaluating backward scheduling and sequencing rules for an assembly shop environment. Production and Inventory Management Journal 1995; 36(4): 71-77

Murata T., Ishibuchi H. and Tanaka H. Multi-objective genetic algorithm and its applications to flowshop scheduling. Computers and industrial Engineering 1996; 30(4): 957-968

Nawaz M., Enscore E. E. and Ham I. A heuristic algorithm for the m-machine n-job flow shop sequencing problem. Omega 1983; 11(1): 91-95

Ogbu F. A. and Smith D. K. The application of the simulated annealing algorithm to the solution of the N/M/Cmax flowshop problem. Computers and Operations Research 1990; 17(3): 243-253

Osman I. H. and Potts C. N. Simulated Annealing for permutation flow-shop scheduling. Omega 1989: 17; 551-557

Ozdamar L. and Ulusoy G. A note on an iterative forward/backward scheduling technique with reference to a procedure by Li and Wills. European Journal of Operations Research 1996; 89(2): 400-407

Patterson J. H., Talbot B. F., Slowinski R. and Weglarz J. Computational experience with a backtracking algorithm for solving a general class of precedence and resource constrained scheduling problems. European Journal of Operations Research 1990; 49: 68-79.

Reeves C. A genetic algorithm for flowshop sequencing. Computers and Operations Research 1995; 22: 5-13

Roach A. and Nagi R. A hybrid GA-SA algorithm for just-in-time scheduling of multi-level assemblies. Computers and Industrial Engineering 1996; 30(4): 1047-1060

Sumichrast R. T. and Russell R. S. Evaluating mixed model assembly line sequencing heuristics for just-in-time production systems. Journal of Operations Management 1990; 3: 371-390

Taillard E. Some efficient heuristic methods for the flowshop sequencing problem. European Journal of Operations Research 1990; 47: 65-74

Taillard E. Benchmarks for basic scheduling problems. European Journal of Operational Research 1993; 64: 278-285

Widmer M. and Hertz A. A new heuristic method for the flow shop sequencing problem. European Journal of Operations Research 1989; 41: 186-193

IMPROVED MULTIPLE ANT COLONIES SYSTEM FOR TRAVELING SALESMAN PROBLEMS

Hidenori Kawamura, Masahito Yamamoto, and Azuma Ohuchi

Laboratory of Harmonious Systems Eng., Division of Complex Systems Eng., Institute of System and Information Eng., Graduate School of Eng., Hokkaido University,
Nishi 8, Kita 13, Kita-ku, Sapporo, Hokkaido 060-8628, Japan
{ kawamura, masahito, ohuchi}@complex.eng.hokudai.ac.jp

Abstract Recently, many kinds of approximate optimization methods have been proposed. The ant system (AS), which is originally proposed by Dorigo et al., is one such algorithm. To improve the basic performance of the AS algorithm, we developed the AS into a multiple ant colonies system (MACS) by introducing multiple colonies and colony-level interactions. MACS showed better performance compared with ACO [Kawamura (2000)]. In this study, we implemented no special heuristic technique as is often used in approximate optimization methods; therefore, it is necessary to investigate the performance of MACS with some heuristics for further development of MACS. In this paper, we implement 2-opt heuristic to the MACS for more powerful performance for solving TSPs.

Keywords: Ant Colony Optimization, Ant System, Multi-agent Systems, Combinatorial Optimization Problems, Traveling Salesman Problems

1. INTRODUCTION

Recently, many kinds of approximate optimization methods for solving difficult combinatorial optimization problems have been proposed. Some of these methods have been inspired from analogies of nature, such as a genetic algorithm inspired from evolution of real life, Hopfield-type neural network modeled on simplified characteristics of real neurons, simulated annealing based on the analogy of an annealing process, and tabu search inspired from human memory. Ant colony optimization (ACO), which is an optimization method based on the foraging behavior of real ants, is one such method, and Dorigo et al. proposed an ant system (AS) as the first ACO algorithm for solving Traveling Salesman Problems (TSPs) in the papers Colorni (1991) and Colorni (1992).

Given a set of cities and distances between two cities, a TSP is a combinatorial optimization problem of finding the minimal length of a closed tour that passes through each city just once.

In an AS algorithm, many ant agents are initially placed on given TSP cities, and each ant agent continuously moves between cities to generate a tours as a solution of the TSP. In addition, the use of pheromone trail plays an important role in finding the shortest tour in the TSP. Pheromone is well known as a simple communication method widely used by in real life. At each step in the AS, ant agents decide to move from one city to a nearby city according to the pheromone trail and the distance between cities. After all of the ants have completed their tours, the pheromone trail is updated and part of it is evaporated.

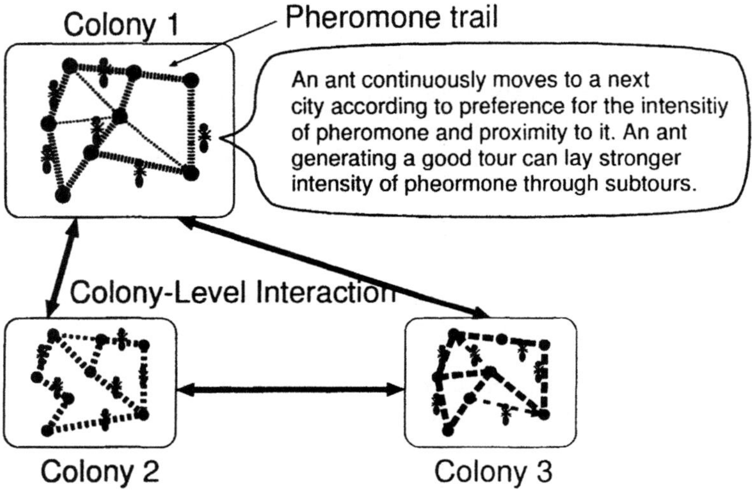

Figure 1. An image of the multiple ant colonies system for TSPs with three colonies.

An AS algorithm is very simple and has the potential for realizing a large parallel optimization. In fact, an AS has been applied to several combinatorial optimization problems, and the effectiveness of an AS in various applications has been investigated in Colorni (1994), Costa (1997), Dorigo (1997) and Gambardella (1995). However, as we discussed in a previous paper Kawamura (2000), one disadvantage of an AS is that an AS algorithm has a tendency to become trapped in specific local optima and is not good at escaping from local optima because an AS algorithm is based on only a positive feedback mechanism by the use of pheromone.

Therefore, we developed an AS algorithm to a multiple ant colonies system (MACS) that has several ant colonies (see Figure 1). In the MACS algorithm, each colony almost corresponds to the AS colony, but it is different from an AS algorithm in that it includes the colony-level interaction between colonies of the MACS. By colony-level interactions in the MACS, both positive feedback and negative feedback can be controlled, and the performance of the MACS is therefore better than that of an AS algorithm [Kawamura (2000)].

In our previous work, we focused only on the extension of basic search performance and did not consider possible refinement of the MACS, such as the implementation of heuristic techniques, which are often used in approximate methods. Therefore, it is necessary to try to refine the algorithm by implementing some heuristic techniques for development of the MACS. In this paper, we focus on TSPs and 2-opt heuristic technique. The reason why we chose TSPs and 2-opt is that a TSP and 2-opt are widely known as a typical combinatorial optimization problem and heuristic technique, and a TSP is the first problem to which an AS was applied. The implementation of 2-opt to the MACS algorithm is described in section 2. In section 3, the results of using a combination of MACS and 2-opt heuristics in several computer experiments are shown. conclusions are given in section 4.

2. MULTIPLE ANT COLONIES SYSTEM

A TSP is first introduced very briefly [Reinelt (1994)]. Given a set of n cities and distances d_{ij} between two cities, a TSP is a problem of finding the minimal length of a closed tour that passes through each city just once. TSPs become asymmetric in the case of $d_{ij} = d_{ji}$ and asymmetric in the case of $d_{ij} \neq d_{ji}$. MACS can be applied to both symmetric and asymmetric cases, but only symmetric cases are considered in this paper.

Next, an outline of the improved MACS with 2-opt heuristic is introduced. Let M be the number of colonies of ant agents, and let m be the number of ant agents belonging to each colony. The k-th ant agent belonging to the h-th colony is denoted as the (h, k) ant. At each time t, the number of $M \cdot m$ ants move between cities to search TSP tours. Each ant agent is a simple agent with the following characteristics:

1 It continuously moves from the current city i to the next city j in the time interval between t and $t + 1$. The next city j is chosen according to the probability that is a function of the distance d_{ij} and the intensity of the pheromone present on the connecting edge in each colony.

2 It visits only cities that it has not previously visited. This property, implemented to force an ant agent to make legal tours, holds until its tour is completed. The ant's memory is then reset to begin its tour again.

3 After completing its tour, (i.e., after visiting all cities), the tour is reconstructed by the 2-opt heuristic.

4 It lays pheromone on the edges of the reconstructed tour. The intensity of the pheromone is decided according to the total distance [Reinelt (1994)].

5 It behaves continuously until stop time t_{max}.

Let $\tau_{ij}^h(t)$ be the intensity of the pheromone on the edge (i, j) in the h-th colony at time t. The intensity of the pheromone is initialized to a small positive value at the initial time. After each ant has completed its tour by n time intervals, the intensity of the pheromone $\tau_{ij}^h(t)$ becomes

$$\tau_{ij}^h(t+n) = \rho \cdot \tau_{ij}^h(t) + \sum_{k=1}^{m} \Delta\tau_{ij}^{hk}, \tag{1}$$

where ρ is a coefficient such that $(1-\rho)$ represents the evaporation rate of the pheromone between time t and $t+n$. The value of ρ must be set to a positive value less than 1 to avoid unlimited accumulation of the pheromone. The value of $\Delta\tau_{ij}^{hk}$ represents the intensity per unit of length of the edge (i, j), and this is given as

$$\Delta\tau_{ij}^{hk} = \begin{cases} Q/L^{hk} & \text{if ant } (h,k) \text{ uses edge } (i,j) \text{ between } t \text{ and } t+n \\ 0 & \text{otherwise} \end{cases} \tag{2}$$

Q is constant and scarcely affects the behavior of the algorithm. L^{hk} is the total tour length of ant (h, k); i.e., the ant generating a shorter tour can lay a larger intensity of the pheromone on its tour. The pheromone laid on probably better sub tours would take large value by the result of such rule.

The transition probability from city i to city j for ant (h, k) is defined as

$$p_{ij}^{hk}(t) = \begin{cases} \pi_{ij}^h(t) / \sum\limits_{l \notin \boldsymbol{tabu}^{hk}(t)} \pi_{il}^h(t) & \text{if } j \notin \boldsymbol{tabu}^{hk}(t) \\ 0 & \text{otherwise} \end{cases} \tag{3}$$

$$\pi_{ij}^{h}(t) = \left\{ \prod_{l=1}^{M} \left[\tau_{ij}^{l}(t) + C(h) \right]^{\alpha(h,l)} \right\} \cdot [1/d_{ij}]^{\beta(h)} \qquad (4)$$

Here, $\boldsymbol{tabu}^{hk}(t)$ indicates the tabu list of ant (h, k). This list consists of cities that have already been visited until time t, and the ant is forbidden to choose such cities repeatedly. This is reset to ϕ when the ant visits all cities and completes its tour, so the ant agents can usually keep legal tours.

$\pi_{ij}^{h}(t)$ means the degree of preference for an edge connected to city j. If this value is large, ant (h, k) tends to choose city j as the next one to visit. The parameter $\alpha(h, l)$, which must be set in advance, determines the type of pheromone effect and the degree of influence from colony l. If $\alpha(h, l)$ is set to a positive value, i.e., a positive pheromone effect, ant (h, k) tends to prefer sub tours indicated by the pheromone in the l-th colony. Conversely, if $\alpha(h, l)$ is negative, i.e., a negative pheromone effect, ant (h, k) tends to avoid choosing edges that have been frequently used in the l-th colony. If the value of $\alpha(h, l)$ is zero, the pheromone in the l-th colony has no effect on the behavior of the h-th colony. The absolute value of $\alpha(h, l)$ indicates the degree of the pheromone effect. The effect becomes stronger as the value increases and becomes weaker as the value decreases. In particular, $\alpha(h, h)$, the effect from itself, must be positive in order to cause a positive feedback in the colony. $C(h)$ represents the degree of insensitivity to the pheromone, and this parameter is introduced so that the ant agents will avoid overly sensitive effects toward the pheromone near zero. The parameter $\beta(h)$ represents how important the proximity is for the ants in the h-th colony.

Next, the 2-opt heuristics is explained. A 2-opt move consists of eliminating two edges and reconnecting the two resulting paths in a different way to obtain a new tour. Note that there is only one way to reconnect the paths (if we do not use the eliminated edges again). For improving a tour this way, the two eliminated edges do not necessarily have to cross. We implemented the 2-opt improvement heuristic in the following way.

Let T be the tour generated by an ant agent after visiting all cities. Until failure is obtained, it is performed to examine all 2-opt moves involving the edge between every city and its successor in the tour. If it is possible to improve the tour length this way and then choose the best such 2-opt move and update T. If no better move can be found, then declare failure. By the 2-opt heuristic, it is expected to improve the performance of MACS for solving TSPs. The whole algorithm is shown in Figure 2.

1. *Initialize*
   Set $t := 0$
   For $h := 1$ to M do
     For every edge (i, j)
       Set $\tau_{ij}^h(t) :=$ *initialvalue*
2. *Set all ants at the starting city*
   Set $s := 0$
   Empty all tabu lists.
   For $h := 1$ to M do
     For $k := 1$ to m do
       Place the starting city of the (h, k) ant, and insert it in $\mathbf{tabu}^{hk}(t)$.
3. *Generate the tour based on pheromone feedback*
   For $h := 1$ to M do
     For $k := 1$ to m do
       For $s := 1$ to $n - 1$ do
         Choose the city j to move to, with probability $p_{ij}^{hk}(t + s)$
          given by equation (3).
         Move the (h, k) ant to the city j.
         Insert the city j in $\mathbf{tabu}^{hk}(t + s)$.
4. *2-opt improvement heuristic*
   For $h := 1$ to M do
     For $k := 1$ to m do
       Repeat the following until failure is obtained.
         For every node $i := 1$ to n do
           Examine all 2-opt moves involving the edge between i and its successor
             in the tour of the ant (h, k).
           If it is possible to decrease the tour length this way, update the tour.
           If no better move could be found, then declare failure.
6. *Compute changes of the pheromones in every colony*
   For $h := 1$ to M do
     For every edge (i, j)
       Update the intensity of pheromone on edge (i, j) by equation (1).
7. *Check stop time*
   If $(t < t_{max})$
     then Goto step 2.
   else Print shortest tour and stop.

Figure 2. Algorithm of multiple ant colonies system.

Table 1. Experimental parameter settings of MACS. These parameter settings were derived from the results of a previous study [8] and from the results of preliminary experiments.

	Negative effect MACS				
	$\alpha(i,1)$	$\alpha(i,2)$	$\alpha(i,3)$	$\beta(i)$	$C(i)$
Colony1 $(i=1)$	1	−0.01	0	2	0.5
Colony2 $(i=2)$	0	1	−0.01	2	0.5
Colony3 $(i=3)$	−0.01	0	1	2	0.5

	Positive effect MACS				
	$\alpha(i,1)$	$\alpha(i,2)$	$\alpha(i,3)$	$\beta(i)$	$C(i)$
Colony1 $(i=1)$	1.5	0.2	0	2	1
Colony2 $(i=2)$	0	1.5	0.2	2	1
Colony3 $(i=3)$	0.2	0	.51	2	1

3. COMPUTER EXPERIMENTS

To examine the performance of the improved algorithm, we prepared better parameter settings for three colonies ($M = 3$) of the improved MACS and original MACS by preliminary experiments. Table 1 shows two kinds of parameter settings, that is, a positive pheromone effect and a negative pheromone effect. The positive pheromone effect causes ant colonies to utilize pheromone information of other colonies, whereas the negative pheromone effect causes ants to avoid search region indicated by other colonies' pheromone information. The difference between these effects possibly causes the difference in performances. In addition, we prepared the settings of $t_{max} = 100, n = 10, \rho = 0.4$ for the improved MACS and $t_{max} = 1000, n = 100, \rho = 0.9$ for the original MACS respectively. These settings seem natural because the improved MACS requires more computational cost for 2-opt than does the original MACS, and the values of t_{max} and n for the original MACS are set to larger values than those for the improved MACS. In fact, the computational times of both were clocked as almost the same values.

Table 2 shows the results of various settings for some TSPs known as benchmarks in TSBLIB [Reinelt (1994)]. The rows in the table correspond to the names of problems, of which the dimensions (number of cities) are all 100. The columns indicate the methods, namely, *random, negative-effect MACS with 2-opt, positive-effect MACS with 2-opt, negative-effect MACS without 2-opt* and *positive-effect MACS without 2-opt.* The random method was prepared for comparison to MACS with 2-opt. It generated TSP solutions by application of 2-opt heuristic to

Table 2. Results of experiments in which each algorithm was applied to 100 dimension TSPs. All problems are quoted from TSPLIB. In each cell, the upper value represents the distance of the best obtained solution. The middle value is standard deviation. The lower is the quality from a known global optimum solution. These results were averaged by 20 trials. The parts of "no 2-opt" are the results of the number of 100 times solutions compared with 2-opt for almost the same computational time. In these results, positive-effect MACS could perform best in all methods.

	2-opt			no 2-opt	
Problem	Random	Negative effect MACS	Positive effect MACS	Negative effect MACS	Positive effect MACS
KroA100	21287.85 9.64 100.03%	21285.2 5.98 100.02%	**21282** 0 **100%**	23028.7 367.90 108.21%	22947.95 492.43 107.83%
KroB100	22245.30 44.12 100.47%	22243.8 57.12 100.46%	**22179.85** 39.62 **100.18%**	24005.60 285.99 108.42%	23922.75 372.45 108.05%
KroC100	**20761.75** 18.83 **100.06%**	20776.05 19.63 100.13%	20775.50 21.08 100.13%	22002.65 194.36 106.04%	22026.40 166.94 106.16%
KroD100	21453.75 55.11 100.75%	21464.65 56.98 100.80%	**21392.15** 42.0 **100.46%**	22946.65 308.04 107.76%	22962.45 312.07 107.84%
KroE100	22193.50 43.51 100.57%	22172.65 59.97 100.47%	**22146** 32.02 **100.35%**	23730.15 293.63 107.53%	23575.20 390.68 106.83%

100×10 randomly generated initial solutions. In regard to methods with 2-opt, positive-effect MACS almost could perform better in three methods. This indicates that 2-opt heuristic and positive pheromone effect is a good combination, and ant colonies effectively interact with each other to exchange information on search regions. The random method does not seem worse, but it needed about 8.99-times longer computational time for discovering the best solutions compared with positive-effect MACS with 2-opt. This was because the randomly generated initial solutions were often apart from good solutions, and required more computational cost to be improved by 2-opt heuristic. However, the improved MACS generated acceptable solutions, and easily improved these solutions by 2-opt.

A Comparison of the results shows that the original MACS without 2-opt heuristic is inferior to the improved MACS. As pointed out

Table 3 . Results of positive-effect MACS with 2-opt for various TSPs. These results were also averaged by 20 trials. The time was measured by a Sun UltraSPARC-IIi 440Mhz without a specific load.

Name	Dim.	Best Avg.	Std. Dev.	Quality	Time	Std. Dev.
eil51	51	426.85	0.745	100.20%	23 sec	12
st70	70	676.4	0.60	100.21%	44 sec	32
pr107	107	44433.35	66.38	100.29%	314 sec	167
ch130	130	6173.15	15.60	101.03%	311 sec	189
ch150	150	6574.4	9.62	100.71%	375 sec	207
tsp225	225	3999.7	16.15	102.06%	2068 sec	711
pr264	264	49162.95	20.12	100.06%	829 sec	307
lin318	318	42922.8	112.33	102.13%	5793 sec	3045

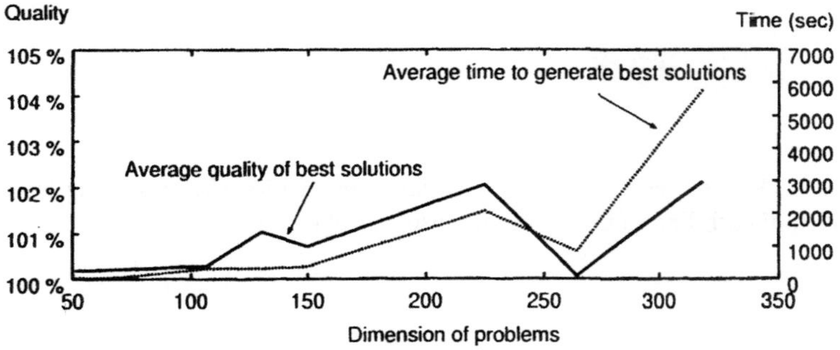

Figure 3. The averaged time and the quality of discovered best solutions according to the number of problem dimensions. The X axis indicates the number of dimensions of TSP, the left Y axis indicates the quality from a known global optimum solution, and the right Y axis corresponds to the time for finding the best obtained solutions.

in Kawamura (2000), the original MACS often generates TSP tours including crossing sub tours, and the quality of solutions is inferior to that of the improved MACS. Thus, the improved MACS is more effective for solving TSPs.

Table 3 shows the results of positive-effect MACS with 2-opt for other various dimension problems. In the table, "BestAvg" is the average distance of the best obtained solutions, "StdDev" is the standard deviation, and "Quality" is the quality from the known optimal solution. "Time" is the average time to discover best solutions, which was clocked using a Sun Ultra 10 workstation with an UltraSPARC-IIi 440MHz Processor. The graph in Figure 3 shows the quality plotted against average time (Y axis) according to the dimension of the problems (X axis).

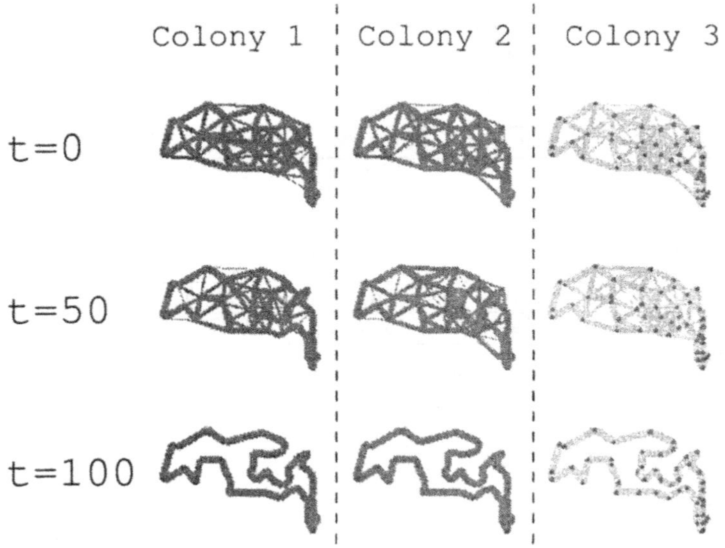

Figure 4. The screen shot of the pheromone intensities after updating in each time. The width of each line indicates the intensity of pheromone.

The results show that performance of the improved MACS slightly decreases according to the dimension of the problem because the number of generated solutions is kept constant, i.e., $t_{max} = 100$. To obtain better performance, t_{max} should be set to a larger value. In particular, the time and quality are very good in the problem pr264. The reason for this result is thought to be related to the locations of cities. In fact, pr264 has biased city location such as there are many clusters of cities. Namely, the structure of the problem tends to influence the performance of MACS, and we need to investigate the relation between the improved MACS and the structure of TSPs.

Figure 4 shows the transition of intensities of pheromone in 3 colonies in the case applied to the problem att48. The width of the line corresponds to the intensity of pheromone on each edge of att48. It is shown that each colony uses many same sub tours and some different sub tours. It is thought that important sub tours were emphasized by the effect of pheromone and these colonies could search around a critical region in the solution space. According to time progress, the search region were specified by the feedback effect of pheromone, and these colonies could find the global optimum solution.

Table 4 . Results for solving other TSPs derived from TSPLIB. These results were averaged in 20 trials.

Name	Dimension	Best Avg.	Quality	Std. Dev.
att48	48	33523.6	100.00%	7.155
berlin52	52	7542	100.00%	0
eil76	76	538.85	100.16%	0.745
rat99	99	1212.3	100.11%	1.593
pr107	107	44433.35	100.29%	66.38
bier127	127	118321	100.03%	59.362
pr144	144	58574.25	100.06%	10.533

Additionally, we carried out some experiments for solving other TSPs cited from TSPLIB. Table 4 shows the qualities of obtained solutions. It is shown that the improved MACS could generate good solutions in various TSPs.

4. CONCLUSIONS

In this paper, we implemented the 2-opt heuristic to the MACS. The performance of the improved MACS is better than that of the original MACS, and it was confirmed that this algorithm generates good solutions of TSPs with reasonable time cost. It was shown that a positive pheromone effect is better than a negative pheromone effect for the improved MACS. This algorithm can be easily developed to parallel computing by allocating a processor to each colony. As further work, we will apply this algorithm to very large TSPs.

References

Agosta, W. C. (1992). "Chemical Communication - The Language of Pheromone -," W. H. Freeman and Company, New York.

Colorni, A., M. Dorigo and V. Maniezzo (1991). "Distributed Optimization by Ant Colonies," Proc. ECAL91-European Conf. on Artificial Life, Paris, France, F. Varela dn P. Bourgine (Eds.), Elsevier Publishing, 134-142.

Colorni, A., M. Dorigo and V. Maniezzo (1992). "An Investigation of Some Properties of an Ant Algorithm," Proc. the Parallel Problem Solving from Nature Conf., Brussels, Belgium, R.Manner and B. Manderick (Eds.), Elsevier Publishing, 509-520.

Colorni, A., M. Dorigo, V. Maniezzo and M. Trubian (1994). "Ant System for Job-shop Scheduling," Belgian Journal of Operations Research, Statistics and Computer Science, 34, 39-53.

Costa, D. and D. Snyers (1997). "Ants can Colour Graphs," Journal of the Operational Research Society, 48, 295-305.

Dorigo, M. and L. M. Gambardella (1997). "Ant Colonies for the Travelling Salesman Problem," BioSystems 43, Elsevier, 73-81.

Gambardella L. M. and M. Dorigo (1995) "Ant-Q: A Reinforcement Learning Approach to the Traveling Salesman Problem," Proc. ML-95, Twelfth International Conf. on Machine Learning, 252-260.

Kawamura, H., M. Yamamoto, K. Suzuki and A. Ohcuhi (2000). "Multiple Ant Colonies Algorithm Based on Colony Level Interactions," Publication in the IEICE Transactions, Fundamentals, Vol. E83-A, No. 2, 372-379.

Reinelt, G. (1994). "The Traveling Salesman - Computational Solutions for TSP Applications," Lecture Notes in Computer Science 840, G. Goos and J. Hartmanis (Eds.), Springer-Verlag.

THE VIDEO PLACEMENT AND ROUTING PROBLEM

I. Ouveysi, L. Sesana and A. Wirth

Department of Mechanical and Manufacturing Engineering,
The University of Melbourne, Victoria, 3010 Australia.
wirth@mame.mu.oz.au

Abstract The placement and transmission of videos subject to storage and transmission constraints in a video on demand network is investigated. A heuristic for the general problem of cost minimization is developed and its effectiveness demonstrated through extensive numerical testing. Whilst the general problem is shown to be non totally unimodular, an O(n^2) time solution for a special case is found. The model is then adapted to a hierarchical architecture.

Keywords: Routing, storage, transmission, multi-commodity problem.

1. INTRODUCTION

Video on Demand (VoD) is a service that provides a set of videos to users through a network. Users select any video programs they like, and then after a short setup time, receive the video programs through the network. As for videotape, they have greater flexibility in scheduling the viewing time and have fine-grained control: enabling them to pause, resume, fast rewind and fast-forward the video.

Storage location and routing strategy are important issues in cost minimization of such a service. Ouveysi et al. (1997) and Tanaka and Berlage (1996) have proposed integer programming models to determine where to place the video programs so as to minimize the total cost of storage and transmission, subject to storage and transmission capacity constraints. These authors also proposed heuristic approaches to the solution of this problem, as the problem appears to be NP-hard. This paper firstly shows that this problem is at least not totallyunimodular. Polynomial time algorithms are then provided for some special cases and finally integer programming solutions are considered.

2. PROBLEM FORMULATION

2.1 The Multi-commodity Problem (P0)

In general, the notation in Ouveysi et al. (1997) is followed. A fully meshed network represented by the undirected graph $G=<V,E>$ is considered, where $V=\{1,2,...,N\}$ is the set of nodes and E is the link set of the $N(N-1)/2$ links of the network . The nodes denote the locations of the potential locations for video storage and also the locations of the demands. We also assume for the moment that video transmission is along one-hop paths only, that is, by single links. In Ouveysi et al. (1997) this is slightly generalized to one and two hop paths. However, Tanaka and Berlage (1996) also restrict their routes to one-hop paths. We denote the link connecting nodes i and j by e_{ij} and define the following variables and parameters:

$C_s(l)$ The storage capacity of the server located at site $l \in V$;

$C_t(i,j)$ The transmission capacity of $e_{ij} \in E$;

P The set of programs to be stored in the servers, $P=\{1,...,p\}$;

m_k The capacity requirement for storing program $k \in$ P at any node;

μ_k The bandwidth requirement for transmission of program $k \in P$ in the network;

$s_k(l)$ The cost of storing program $k \in P$ at site $l \in V$;

$t_k(i,j)$ Transmission cost of program $k \in P$ from node i to node j using link e_{ij};

x_{ij}^k The binary variable which is equal to 1 if program $k \in$ P is transmitted using link e_{ij} and 0 otherwise;

y_l^k The binary variable which is equal to1 if program $k \in P$ is stored at site l and 0 otherwise;

Given the demand forecast of available programs, the objective of this research is to determine at which nodes to store what programs such that the overall cost of storage and transmission of the programs in the network is minimized. This optimization problem can be formulated as the following integer linear program:

Problem (P0)

$$\min \sum_{k \in P} \sum_{l \in V} s_k(l) y_l^k + \sum_{k \in P} \sum_{i \in V} \sum_{\substack{j \in V \\ j \neq i}} t_k(i,j) x_{ij}^k$$

subject to:

(i) $$\sum_{k \in P} m_k y_l^k \leq C_s(l) \qquad \forall l \in V$$

(ii) $$\sum_{k \in P} \mu_k x_{ij}^k \leq C_t(i,j) \qquad \forall e_{ij} \in E$$

(iii) $$\sum_{\substack{i \in V \\ i \neq j}} x_{ij}^k + y_j^k = 1 \qquad \forall j \in V,\ \forall k \in P$$

(iv) $$y_i^k \geq x_{ij}^k \qquad \forall i,j \in V, i \neq j, \forall k \in P$$

(v) $$x_{ij}^k, y_l^k \in \{0,1\} \quad \forall i,j,l \in V, \forall k \in P$$

Constraints *(i)* and *(ii)* are storage and transmission capacity constraints, respectively. Constraints *(iii)* require that each video must either be stored at a node or be transmitted from another node to that node. Constraints *(iv)* require that a video may only be transmitted from a node if it is stored at that node.

The above model is essentially the same as that formulated in Ouveysi et al. (1997) except for the fact that we do not for now allow two-hop paths. The model should be compared with that of Tanaka and Berlage (1996). They include fixed establishment costs for nodes and links but ignore link bandwidth and node capacity constraints *(i)* and link bandwidth constraints *(ii)*.

2.2 The Single-commodity Problem (P1)

The above formulation of Problem (P0) as an integer linear program suggests that P0 may be NP-hard. Ouveysi et al. (1997) proposed a heuristic which relies on the solution of a sequence of single-commodity problems. Details of the modified version of this heuristic are provided later. For the moment we consider the following integer program which is the engine for the heuristic.

Problem (P1)

$$\min \sum_{l \in V} s(l) y_l + \sum_{i \in V} \sum_{\substack{j \in V \\ j \neq i}} t(i,j) x_{ij}$$

subject to:

(i)	$\sum_{\substack{i \in V \\ i \neq j}} x_{ij} + y_j = 1$	$\forall j \in V$
(ii)	$y_i \geq x_{ij}$	$\forall i,j \in V,\ i \neq j$
(iii)	$y_i = 0$	$\forall i \in V \setminus Q$
(iv)	$x_{ij} = 0$	$\forall e_{ij} \in E \setminus S$
(v)	$x_{ij}, y_l \in \{0,1\}$	$\forall i,j,l \in V$

Where:

- $Q \subseteq V$ is the set of still available nodes in the network.
- $S \subseteq E$ Is the set of links e_{ij} for which there is an available transmission capacity.

If the matrix of the linear programming relaxation of (P1) were totally unimodular, that is, if each vertex of the relaxed problem were integer valued, then clearly (P1) would be solvable in polynomial time. Unfortunately, this is not the case, as we show below. We then discuss (P2), a special case of (P1), which is polynomial time.

***Proposition 1**: For $n \geq 3$, $Q = V$ and $S = E$ the feasible region of the linear programming relaxation of (P1) has non-integer vertices, that is, the problem matrix is not totally unimodular.*

Proof: Suppose n=3, Q=V and S=E. We can represent a point in the feasible region of the linear programming relaxation of (P1) as

$$\begin{bmatrix} y_1 & x_{12} & x_{13} \\ x_{21} & y_2 & x_{23} \\ x_{31} & x_{32} & y_3 \end{bmatrix}$$

with the following constraints:

$y_1 + x_{21} + x_{31} = 1$, $y_2 + x_{12} + x_{32} = 1$, $y_3 + x_{13} + x_{23} = 1$

$y_1 \geq x_{12}$, $y_1 \geq x_{13}$, $y_2 \geq x_{21}$, $y_2 \geq x_{23}$, $y_3 \geq x_{31}$ and $y_3 \geq x_{32}$

We shall prove that $\underline{x}=\begin{bmatrix}\frac{1}{2} & 0 & \frac{1}{2}\\ \frac{1}{2} & \frac{1}{2} & 0\\ 0 & \frac{1}{2} & \frac{1}{2}\end{bmatrix}$, which is clearly non-integral, is a vertex of the feasible region. Write $\underline{x}=\left(\frac{1}{2},0,\frac{1}{2},\frac{1}{2},\frac{1}{2},0,0,\frac{1}{2},\frac{1}{2}\right)$. We recall that $\underline{x}$ is a vertex if

$$\underset{-}{x}=\lambda\ \underset{-}{u}+(1-\lambda)v, 0\leq\lambda\leq 1$$

for some feasible $\underline{u}$ and $\underline{v}$ then $\underline{u}=\underline{v}=\underline{x}$.

It will suffice to show that if $\underline{x}=\frac{1}{2}\underline{u}+\frac{1}{2}\underline{v}$ for feasible $\underline{u}$ and $\underline{v}$ then $\underline{u}=\underline{v}=\underline{x}$. Clearly $u_2=v_2=0$, $u_6=v_6=0$ and $u_7=v_7=0$.

Suppose $u_1=\frac{1}{2}+\varepsilon$ for some $\varepsilon>0$ then $v_1=\frac{1}{2}-\varepsilon$ also $u_4=\frac{1}{2}-\varepsilon$ and $v_4=\frac{1}{2}+\varepsilon$. Now $u_4\leq u_5, v_4\leq v_5$ and $u_5+v_5=1$, so $u_5=\frac{1}{2}-\varepsilon$ and $v_5=\frac{1}{2}+\varepsilon$. Also $u_5+u_8=1, u_8\leq u_9$ and $u_9+v_9=1$. Thus $v_9\leq\frac{1}{2}-\varepsilon$ and since $v_3+v_9=1$. We have $v_3\geq\frac{1}{2}+\varepsilon$ but $v_3\leq v_1=\frac{1}{2}-\varepsilon$ a contradiction. The following summarizes the situation:

$$\underline{u}=\begin{bmatrix}\frac{1}{2}+\varepsilon & 0 & \frac{1}{2}-\varepsilon\\ \downarrow & & \uparrow\\ \frac{1}{2}-\varepsilon\rightarrow & \frac{1}{2}-\varepsilon & 0\\ & \downarrow & \uparrow\\ 0 & \frac{1}{2}+\varepsilon\rightarrow & \frac{1}{2}+\varepsilon\end{bmatrix},\ \underline{v}=\begin{bmatrix}\frac{1}{2}-\varepsilon & 0 & \frac{1}{2}+\varepsilon\\ \downarrow & & \uparrow\\ \frac{1}{2}+\varepsilon\rightarrow & \frac{1}{2}+\varepsilon & 0\\ & \downarrow & \uparrow\\ 0 & \frac{1}{2}-\varepsilon\rightarrow & \frac{1}{2}-\varepsilon\end{bmatrix}$$

Similar arguments, starting with a non-zero component of $\underline{u}$ other than u_1 also lead to contradictions. Thus the result is true for n= 3.

For $n > 3$ consider the point
$$\begin{bmatrix} \frac{1}{2} & 0 & \frac{1}{2} & 0 & 0 & \dots & 0 \\ \frac{1}{2} & \frac{1}{2} & 0 & & & & \\ 0 & \frac{1}{2} & \frac{1}{2} & & & & \\ 0 & 0 & 0 & 1 & 0 & 0 & 0 \\ 0 & 0 & 0 & 0 & 1 & 0 & 0 \\ \dots & \dots & \dots & 0 & 0 & \dots & 0 \\ 0 & \dots & \dots & 0 & 0 & 0 & 1 \end{bmatrix}$$

The same argument can be used to show that the above point is a vertex. This proves the result for all $n \geq 3$. (Note that, trivially, the problem for n = 2 is totally unimodular.)

2.3 A polynomial Time Case of (P1)

We now assume that storage and transmission costs are constant, that is $s(l) = s \ \forall l \in V$ and $t(i,j) = t \ \forall e_{ij} \in E$. Let $\frac{s}{t} = k$. Then, after deleting a multiplicative constant from the objective function (P1) reduces to:

Problem (P2)

$$\min z = k\sum_{l\in V} y_l + \sum_{i\in V}\sum_{\substack{j\in V\\ j\neq i}} x_{ij}$$

subject to:

(i)	$\sum_{\substack{i\in V\\ i\neq j}} x_{ij} + y_j = 1$	$\forall j \in V$
(ii)	$y_i \geq x_{ij}$	$\forall i,j \in V$ *and* $i \neq j$
(iii)	$y_i = 0$	$\forall i \in V \setminus Q$
(iv)	$x_{ij} = 0$	$\forall e_{ij} \in E \setminus S$
(vi)	$x_{ij}, y_l \in \{0,1\}$	$\forall l \in Q, \forall e_{ij} \in S$

Proposition 2: *Problem (P2) is* $O(n^2)$ provided $k \leq 1$.

Proof: The following procedure which is $O(n^2)$ clearly solves (P2), provided $k \leq 1$.

Step1: $\forall i \in Q$ set y_i:=0, $\forall e_{ij} \in S$ set x_{ij}:=0, for example:

$$\begin{bmatrix} 0 & x & 0 & 0 \\ 0 & x & 0 & x \\ x & 0 & 0 & 0 \\ 0 & 0 & x & x \end{bmatrix}$$ where x indicates an element not in $Q \cup S$

Step2:

- $\forall i \in Q$ set y_i:=1 for the example above: $$\begin{bmatrix} 1 & x & 0 & 0 \\ 0 & x & 0 & x \\ x & 0 & 1 & 0 \\ 0 & 0 & x & x \end{bmatrix}$$

- $\forall i \in (V \setminus Q)$ do:

  Set $\hat{j}$:=i

  if in the $\hat{j}^{\text{th}}$ column there are rows with a "0" element and a "1" element on the principal diagonal of that row

  then choose the first raw ($\hat{i}$) and set $x_{\hat{i}\hat{j}}$:=1

  so $$\begin{bmatrix} 1 & x & 0 & 1 \\ 0 & x & 0 & x \\ x & 1 & 1 & 0 \\ 0 & 0 & x & x \end{bmatrix}$$

is an optimal solution for the example. If this process does not produce a feasible solution, then clearly the problem is infeasible.

2.4 A Hierarchical Model (P3)

In this section we indicate how the above models, based essentially on a fully meshed network may be modified to deal with a hierarchical architecture. The network is organized in a multi-layered structure as proposed in Nussbaumer et al. (1995), Brubeck and Rowe (1996), Bisdikian and Patel (1996), Berzsenyi et al. (1998), Jia et al. (1999), Milenkovic (1998), and more precisely in 3 levels as in Li et al. (1996): central, metropolitan and local, as shown in Figure 1.

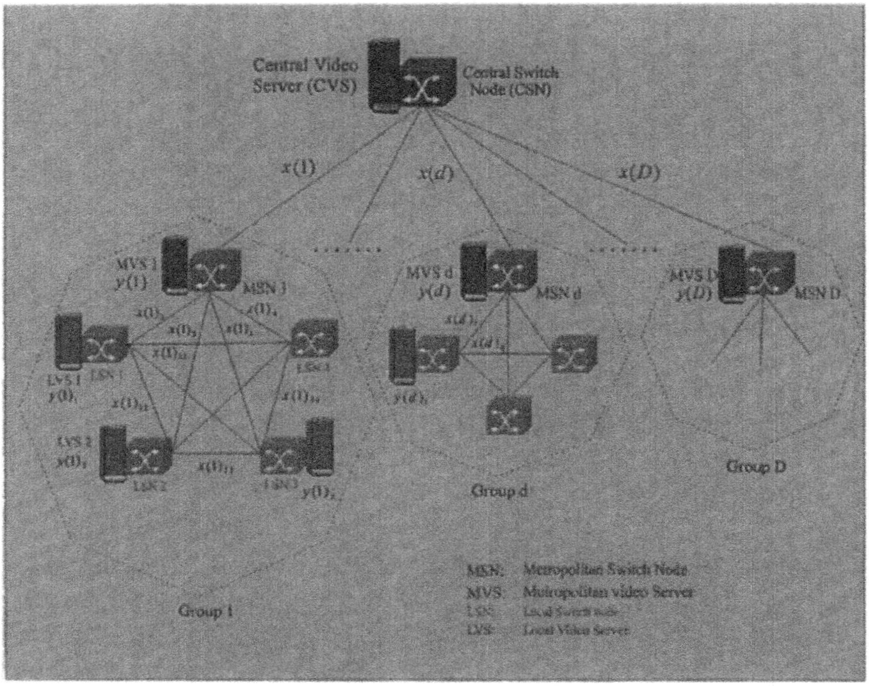

Figure 1. A generic hierarchical structure.

It is assumed that there is one Central-Video-Server (*CVS*), which can be considered as a remote archive. It contains all the programs. Through its central switch node it can transmit a program to one of the metropolitan switch nodes.

We have D metropolitan switch nodes. Each metropolitan switch node has an associated Metropolitan-Video-Server (*MVS*). We denote the metropolitan video servers by $y(d), d = 1,\dots,D$. Through a metropolitan switch node we can transmit a program, which is coming from the associated *MVS* or from the central switch node, to one or more of the local switch nodes. Each metropolitan switch node with its local switch nodes defines a group. Each group is a fully meshed network. If a program is stored in one of the local video servers, it will be stored in its *MVS* too.

In every d^{th} *group (d=1,...,D)* we have N_d local switch (access) nodes. Every user accesses the network through a local switch node and every user has a default switch node. A local switch node does not necessarily have an associated Local-Video-Server *(LVS)*. When a user requires a program from their default node, it could come from its associated LVS (if exists) or from another LVS or from the MVS (via its metropolitan switch node) or from the CVS (via the central switch node and the metropolitan switch node). In each dth *group (d=1,...,D)* there are L_d local video servers, denoted by $y(d)_i$. Every program must be accessible from each local switch node. A local switch node provides the initial connection establishment to the user's STB (set-top-box), a menu of choices to the subscriber, interconnection establishment between the STB and a selected video server and management of such a connection, and disconnection of a session for any reason and potentially billing and maintenance.

We introduce the following variables and parameters:

$$CVS^k = \begin{cases} 1 & \text{if program } k \text{ stored in } CVS \\ 0 & \text{otherwise} \end{cases}$$

$$y(d)^k = \begin{cases} 1 & \text{if program } k \text{ stored in } y(d) \\ 0 & \text{otherwise} \end{cases}$$

$$y(d)_i^k = \begin{cases} 1 & \text{if program } k \text{ stored in } y(d)_i \\ 0 & \text{otherwise} \end{cases}$$

L_d: The total number of local video servers in the group number *d*.

N_d: The total number of local switch nodes in the group number *d* (in general $L_d \leq N_d$).

$x(d)$: The link connecting the central switch node to the dth metropolitan switch node.

$$x(d)^k = \begin{cases} 1 & \text{link } x(d) \text{ used to transmit program } k \\ 0 & \text{otherwise} \end{cases}$$

$x(d)_i$: The link connecting the dth metropolitan switch node to the ith local switch node.

$$x(d)_i^k = \begin{cases} 1 & \text{if link } x(d)_i \text{ used to transmit progam } k \\ 0 & \text{otherwise} \end{cases}$$

$x(d)_{ij}$: The link connecting the ith local switch node to the jth local switch node (where both nodes are in the same group number *d*).

$$x(d)_{ij}^k = \begin{cases} 1 & \text{if link } x(d)_{ij} \text{ is used to transmit program } k \\ 0 & \text{otherwise} \end{cases}$$

m_k: The capacity requirement, for storing program $k \in P$ at any local, metropolitan, central video server.

μ_k: The bandwidth requirement for transmission of program $k \in P$ in the network.

$C_s(CVS)$: The storage capacity of the Central Video Server.

$C_s(d)$: The storage capacity of the dth *MVS*.
$C_s(d)_i$: The storage capacity of the ith *LVS* in the dth group.
$C_t(d)$: The transmission capacity of the link.
$C_t(d)_i$: The transmission capacity of the link $[x(d)_i]$.
$C_t(d)_{ij}$: The transmission capacity of the link $[x(d)_{ij}]$.

Λ^k_C: The cost of storing program $k \in P$ at the *CVS*.

Λ^k_M: The cost of storing program $k \in P$ at any of the *MVSs*.

Λ^k_L: The cost of storing program $k \in P$ at any of the *LVSs*.

Π^k_C: The transmission cost of program $k \in P$ using the central switch node.

Π^k_M: The transmission cost of program $k \in P$ using a metropolitan switch node.

Π^k_L: The transmission cost of program $k \in P$ using a local switch node.
z(k): Number of requests for program $k \in P$ in the network
z(k,d): Number of requests for program $k \in P$ in the dth group.
z(k,d,i): Number of requests for program $k \in P$ at the ith local node in the dth group.

$\delta_C, \delta_M, \delta_L$: Parameters which take into account the delay caused by transmitting a program from the central, metropolitan and local level respectively.
In general $\delta_C > \delta_M > \delta_L$.

$\alpha_C, \alpha_M, \alpha_L$: Maximum tolerable delays (quality of service) at the central, metropolitan and local levels, respectively.

The objective function is

$$\min \sum_{k \in P} \left\{ \Lambda^k_C CVS^k + \Lambda^k_M \sum_{d=1}^{D} y(d) + \Lambda^k_L \sum_{d=1}^{D} \sum_{i=1}^{L_d} y(d)^k_i \right.$$

$$+ \Pi^k_C \sum_{d=1}^{D} x(d)^k + \Pi^k_M \sum_{d=1}^{D} \sum_{i=1}^{N_d} x(d)^k_i + \Pi^k_L \sum_{d=1}^{D} \sum_{i=1}^{N_d} \sum_{\substack{j=1 \\ j \neq i}}^{N_d} x(d)^k_{ij}$$

Transmission cost:

The transmission of a program takes place only between switch nodes. In Wu et al. (1997) the communication cost of a video connection is defined as *size* x *distance*, where *size* is the size of the video that is transferred from the server to the user, and *distance* is the number of switches on the video path from the server to the user.

Let τ_k be the duration of program k and let Π_C, Π_M, Π_L be the costs per unit of storage, at the central, metropolitan and local levels, respectively. Then,

$$\Pi_C^k = \Pi_C \mu_k \tau_k z(k,d)\delta_C, \qquad \Pi_M^k = \Pi_M \mu_k \tau_k z(k,d,i)\delta_M,$$
$$\Pi_L^k = \Pi_L \mu_k \tau_k z(k,d,j)\delta_L.$$

Storage cost:

Let Λ_C, Λ_M, Λ_L represent the unit storage cost per at the central, metropolitan and local levels, respectively. Therefore the storage costs are: $\Lambda_C^k = \Lambda_C m_k$, $\Lambda_M^k = \Lambda_M m_k$ and $\Lambda_L^k = \Lambda_L m_k$.

Notice that as shown in Wu et al. (1997), Barnett and Anido (1996), and Liu et al. (1996) the cost of a video server depends mainly on the *server throughput*, which is defined as the number of *simultaneous video streams* which the video server can support for the guaranteed quality of Service. The number of simultaneous streams multiplied by the bandwidth requirement of each stream will give an approximate value for the server throughput.

The constraints are:

i) $\sum_{k \in P} m_k CVS^k \le C_s(CVS)$

ii) $\sum_{k \in P} m_k y(d)^k \le C_s(d) \quad \forall d = 1,\dots,D$

iii) $\sum_{k \in P} m_k y(d)_i^k \le C_s(d)_i \quad \forall d = 1,\dots,D; \ \forall i = 1,\dots,L_d$

iv) $CVS^k \ge x(d)^k \qquad \forall k \in P; \ \forall d = 1,\dots,D$

v) $y(d)^k + x(d)^k \ge x(d)_i^k \quad \forall k \in P; \ \forall d = 1,\dots,D; \ \forall i = 1,\dots,N_d$

In general, the number of local video servers in d th group (L_d) is smaller than the number of local switch nodes in the same group (N_d). A local switch node does not necessarily have an associated video server and a program cannot be stored in a site that does not have a video server. It is assumed without loss of generality that the numbering of local switch nodes begins in a group with the ones that have an associated video server. These remarks lead to the following two constraints:

vi) $y(d)_i^k + x(d)_i^k \geq x(d)_{ij}^k \;\; \forall k \in P;\; \forall d = 1,\ldots,D;$

$\forall i = 1,\ldots,N_d;\;\; \forall j = 1,\ldots,N_d \;\; \text{and} \;\; j \neq i$

vii) $y(d)_i^k = 0 \qquad \forall i = L_d + 1,\ldots,N_d;\; \forall k \in P;\; \forall d = 1,\ldots,D$

viii) $\sum_{\substack{i=1 \\ i \neq j}}^{N_d} x(d)_{ij}^k + y(d)_j^k + x(d)_j^k = 1 \quad \forall k \in P;\; \forall d = 1,\ldots,D;$

$\forall j = 1,\ldots,N_d$

ix) $\sum_{k \in P} \mu_k z(k,d) x(d)^k \leq C_t(d) \qquad \forall d = 1,\ldots,D$

x) $\sum_{k \in P} \mu_k z(k,d,i) x(d)_i^k \leq C_t(d)_i \qquad \forall d = 1,\ldots,D;$

$\forall i = 1,\ldots,N_d$

xi) $\sum_{k \in P} \mu_k z(k,d,j) x(d)_{ij}^k \leq C_t(d)_{ij} \qquad \forall d = 1,\ldots,D;$

$\forall i = 1,\ldots,N_d; \forall j = 1,\ldots,N_d \;\; \text{and} \;\; j \neq i$

xii) $y(d)^k \geq y(d)_i^k \qquad \forall k \in P;\; \forall d = 1,\ldots,D; \forall i = 1,\ldots,L_d$

xiii) $CVS^k \geq y(d)^k \qquad \forall k \in P;\; \forall d = 1,\ldots,D$

xiv) $\sum_{k \in P} \delta_C z(k,d) x(d)^k \leq \alpha_C \qquad \forall d = 1,\ldots,D$

xv) $\sum_{k \in P} \delta_M z(k,d,i) x(d)_i^k \leq \alpha_M \quad \forall d = 1,\ldots,D;\; \forall i = 1,\ldots,N_d$

xvi) $\sum_{k \in P} \delta_L z(k,d,i) x(d)_{ij}^k \leq \alpha_L \quad \forall d = 1,\ldots,D;\; \forall i = 1,\ldots,N_d;$

$\forall j = 1,\ldots,N_d \;\; \textit{and} \;\; j \neq i$

xvii) $CVS,\; y(d),\; y(d)_i,\; x(d),\; x(d)_i,\; x(d)_{ij} \in \{0,1\}$

$\forall d = 1,\ldots,D;\; \forall i=1,\ldots,N_d;\; \forall j=1,\ldots,N_d \;\; \textit{and} \;\; j \neq i$

3. A HEURISTIC FOR P0

Tanaka and Berlage (1996) solve their problem by first dealing with each possible network topology separately. Since they do not consider node capacity and link bandwidth constraints they can decompose each of these problems in turn to a sequence of P1 problems, one for each video. They apply a greedy heuristic for each of these problems. They then sum the costs to provide heuristic solutions for each topology and finally use a genetic algorithm to find the lowest cost topology and thus provide an approximate solution for P0.

A modified version of the heuristic first given in Ouveysi et al. (1997) is now presented. The heuristic discussed below improves on the earlier one in Ouveysi et al. (1997) in that it does not assume that individual nodes and links have sufficient capacity, but rather temporarily deletes any such nodes and links, which have insufficient capacity before solving P1. The temporarily deleted node or link is the added back at the next iteration unless it does not have enough capacity to store or transmit any of the programs in the queue in which case it is permanently deleted.

The Heuristic Outline

In summary, the steps of the heuristic are:

1. Sort the programs in descending popularity order.
2. Permanently delete any nodes and links which have insufficient capacity for any program yet to be allocated.

Temporarily delete any nodes and links which have insufficient capacity for the most popular and as yet unallocated program.

Solve (P1) for the most popular and as yet unallocated program.

3. Decrease the capacities of the nodes and links used in the optimal solution of (P1) above thereby allocating capacity for this program. Delete this program from the list of unallocated programs.
4. Add back any temporarily deleted nodes and links.
5. Return to step 2 if there are still any programs with unallocated capacity.

The following are the details of the heuristic.

Let $Q \subseteq V$ ($S \subseteq E$) be the set of available nodes (links) in the network. Let Q_t (Q_p) be the set of temporarily (permanently) deleted nodes. Similarly let S_t (S_p) be the set of temporarily (permanently) deleted links.

Step 1: Given $G=<V,E>$; $C_s(l)\ \forall l \in V$; $C_t(i,j)\ \forall e_{ij} \in E$;

$P=\{1,2,\ldots,p\}$; $\mu_k, m_k\ \forall k \in P$; $s_k(l)\ \forall k \in P, \forall l \in V$;

$t_k(i,j),\ \forall k \in P, \forall e_{ij} \in E$,

sort the programs in P in descending order in terms of program popularity (demand) and let $\widetilde{P} = (\upsilon_1, \upsilon_2, \ldots, \upsilon_p)$ represents this ordered set.

Set $Q=V$, $S=E$, $h=1$, $k=0$ and $Q_t = Q_p = S_t = S_p = \varnothing$

Step 2: Set $k = \upsilon_h$

$\forall$ node $l \in Q$ do: if $C_s(l) < \min_{k=h}^{p} m(v_h)$

then permanently delete node l from Q and add it to Q_p.

$\forall e_{ij} \in S$ do: if $C_t(i,j) < \min_{k=h}^{p} \mu(v_h)$

then permanently delete link e_{ij} from S and add it to S_p.

$\forall$ node $l \in Q$ do: if $C_s(l) < m(v_h)$

then temporarily delete node l from Q and add it to Q_t.

$\forall e_{ij} \in S$ do: if $C_t(i,j) < \mu_k$

then temporarily delete link e_{ij} from S and add it to S_t.

Solve (P1) for program k. Let (x^*, y^*) represent the optimal solution to (P1).

Step 3: $\forall e_{ij} \in S$ and $\forall l \in Q$ do:

if $y_l^* = 1$ then $C_s(l) \leftarrow C_s(l) - m_k$ and let $y_l^k = 1$

if $x_{ij}^* = 1$ then $C_t(i,j) \leftarrow C_t(i,j) - \mu_k$

and let $x_{ij}^k = 1$

Step 4: Add back the temporarily deleted nodes and links to sets Q and S. That is,

$\forall$ node $l \in Q_t$ delete node l from Q_t and add it to Q,

$\forall e_{ij} \in S_t$ delete link e_{ij} from S_t and add it to S.

Step 5: Set $h \leftarrow h+1$, and if $h=p+1$ go to Step 6, otherwise go to Step 2.

Step 6: Stop.

It is clear that, provided that at Step 2 we have a feasible solution to each of the P1 problems then, we have constructed a solution to P0. We discuss the effectiveness and efficiency of the heuristic in the next section.

We note that this heuristic may be easily adapted to the hierarchical model P3. Details are given in Sesana (2000).

4. SOME NUMERICAL RESULTS

In this section some numerical results obtained for the problems discussed above are reported. The GAMS interface with CPLEX is used as the solver. The symbol * below indicates that execution was terminated after two hours.

4.1 Exact Solution for P0

It is assumed that storage cost is constant with $s = 100$, transmission cost *t(i,j)* is normally distributed with $\mu = 100$ and $\sigma = 5$, the number of nodes $n = 30$ and k, the number of programs to be allocated varies from 5 to 80. Note that the integer program consists of n^2k variables and $n^2(k+1)$ equations. A two-hour elapsed time limit was imposed. As can be seen from the table below, it was not possible to solve the $k = 80$ problem in that time.

Table 1. Exact solution for P0.

k	Model dim. (MB)	No. of equations	No. of variables	Generation Time (sec)	Execution Time (sec)	Elapsed Time (sec)	No. of iteration	Optimal solution z
5	3	5400	4500	0.82	0.93	31.63	509	13963
10	4.5	9900	9000	1.45	1.65	106.18	1085	27882
20	7.7	18900	18000	2.96	3.18	280.12	2063	55767
40	14.5	36900	36000	6.65	7.2	1714.06	4314	111519
80	27.1	72900	72000	61.36	62.5	7200(*)	8293	-

4.2 Exact Solution for P1

It is assumed as in the section above that storage cost is constant with $s = 100$ and transmission cost *t(i,j)* is normally distributed with $\mu = 100$ and $\sigma = 5$. The number of nodes, *n* now varies from 3 to 200 and *k*, the number of programs to be allocated is of course, 1. In the table below, we show the results obtained for problem (P1) varying the number of nodes from 3 to 200. A one-hour elapsed time limit was imposed for these problems, and thus it was not possible to solve the $n = 200$ case.

Table 2. Exact solution for P1

n	Model dim. (MB)	No. of equations	No. of variables	Generation Time (sec)	Execution Time (sec)	Elapsed Time (sec)	No. of iteration	Optimal solution Z
3	1.4	9	9	0.01	0.01	0.35	2	291
10	1.4	100	100	0.05	0.05	0.38	28	951
20	1.4	400	400	0.05	0.05	2.31	75	1879
30	1.4	900	900	0.16	0.16	6.26	122	2800
40	1.4	1600	1600	0.17	0.22	9.01	170	3705
50	1.9	2500	2500	0.33	0.33	33.67	211	4604
60	2.4	3600	3600	0.5	0.55	109.09	309	5523
70	2.4	4900	4900	0.71	0.77	145.44	341	6412
80	3	6400	6400	0.94	1.05	327.8	442	7303
90	3.5	8100	8100	1.21	1.32	303.57	500	8202
100	4	10000	10000	2.97	3.35	751.54	609	9095
150	8.2	22500	22500	5.76	6.31	2599.29	1186	13566
200	20.1	40000	40000	8.16	9.72	3600(*)	2341	-

Nevertheless, it is clear from the above that splitting P0 into k separate P1 problems is very time effective. For example, for $n = 30$ and $k = 40$, it requires about 28 minutes to solve P0 but only about 4 minutes to solve 40 P1 problems.

4.3 Heuristic Performance

The efficiency of the heuristic outlined above, which effectively replaces P0 by k separate P1 problems, was analyzed in Ouveysi et al. (1997). We reproduce their results, for the problem, which incorporates two-hop paths as well as direct links, below.

The heuristic was evaluated by applying it to a 5-node fully meshed network, which is shown in Figure 2, with $k = 30$. The video server capacity at nodes 1 and 5 was 5 units and at nodes 2, 3 and 4 it was 20 units. The links have the capacity of 40 units.

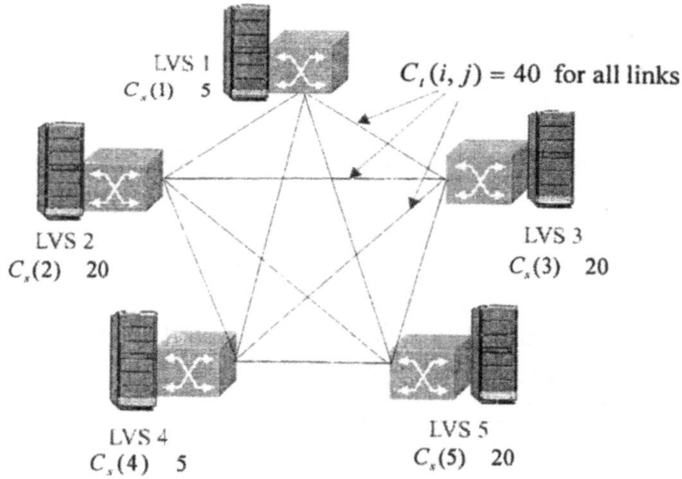

Figure 2. A 5-node VoD network

The transmission cost t_k is modeled as:

$$t_k(i,j) = \varphi_t \mu_k T_k \upsilon_j^k d_{ij}$$

where T_k is the total transmission time of program k, υ_j^k is the number of viewers of program k at node j, d_{ij} is the distance between node i and node j, and φ_t is the transmission cost per unit bandwidth per unit time and per viewer. Since the product of μ_k and T_k is equivalent to the storage size of program k, m_k, the transmission cost can be written as $t_k(i,j) = \varphi_t m_k \upsilon_k^j d_{ij}$ and the storage cost $s_k(l)$ is modeled as

$s_k(l) = \varphi_s m_k$ where φ_s is the storage cost per unit size. Here we introduce the parameter n, which is defined as φ_s / φ_t and $s_k(l)$ becomes $s_k(l) = n\varphi_t m_k$. We assume that the total bandwidth demand at node j, $m_k \upsilon_j^k$, is normally distributed with mean equal to 1.0 and variance equal to 0.2. Also, it is assumed $\varphi_s = 160$ and $s_k(l)$ is uniformly distributed between 0.8 and 1.2. The product of $\upsilon_j^k d_{ij}$ is approximately 1.0.

Table 3 below contains the total cost generated by the heuristic for the network with different φ_t. In each case, they also attempted to find the optimal placement of video programs by solving problem (P0) using CPLEX.optimizer. In those cases when CPLEX was unable to produce an optimal solution after running for 5 hours, it is stopped and the best solution obtained thus far was taken as the near-optimal solution. Such cases are denoted by a ⋆ in the optimal column.

Table 3. The optimal and heuristic solution result for a 5-node fully meshed network.

φ_t	Optimal	Heuristics	% Differences
30	9115.6(⋆)	9207.2	1.0
40	10380.1	10644.0	2.5
50	11751.0	12076.0	2.8
60	13090.3	13429.3	2.6
70	14359.5(⋆)	14670.3	2.2
80	15573.0(⋆)	15837.1	1.7
90	16588.7(⋆)	16923.9	2.0
100	17574.4	17921.2	2.0
110	18440.3	18771.0	1.8
120	19278.2	19851.1	3.0
130	20084.3	20980.5	4.5
140	20882.8	21927.4	5.0
150	21645.3	22765.2	5.2
160	23779.1(⋆)	23814.9	0.2

It is noted again that the above results refer to models for P0 and P1, which use two-hop paths in addition to direct links. Also Sesana (2000) has numerically solved the hierarchical version of P1 with up to 30 metropolitan video servers and 30 nodes per metropolitan video server.

5. CONCLUSION AND SOME OPEN QUESTIONS

In this paper the problem of the placement and transmission of videos in a fully meshed network has been formulated, with the objective of cost minimization. The problem was shown to be non totally unimodular and an efficient and effective heuristic was developed for the general case. Furthermore, a special case was solved exactly in polynomial time. The general problem was then modified to deal with a hierarchical architecture. The paper also reported on extensive numerical testing of the heuristic.

However, many open problems remain. The models in this paper assume one-hop paths only. Our analysis should be extended to two-hop and, indeed, to multi-hop paths. The efficacy of other heuristics for these problems should also be investigated. Furthermore, the computational complexity of problem P2 for $s > t$ and more generally the complexity of P1 and any special cases need to be considered. For example, are there other special cases which are solvable in polynomial time? The model can be further generalized by incorporating setup costs for nodes and links as in Tanaka and Berlage (1996).

References

Ouveysi I., Chan S. and Wong K.C. A Video placement algorithm for video-on-demand networks. Proceedings APCC'97, Sydney, Australia 1997: 1448-1452

Li V.O.K., Liao W., Qiu X. and Wong E.W.M. Performance model of interactive video-on-demand systems. IEEE Journal on Selected Areas in Communications 1996; 14(6): 1099-1108

Nussbaumer J.P., Patel B.V., Schiaffa F. and Sterbenz J.P.G. Networking requirements for interactive video on demand. IEEE Journal on Selected Areas in Communications 1995; 13(5): 779-787

Brubeck D.W. and Rowe L.A. Hierarchical storage management in a distributed VoD system. IEEE Multimedia 1996: 37-47

Bisdikian C.C. and Patel B.V. Cost based program allocation for distributed multimedia-on-demand systems. IEEE Multimedia 1996: 62-72

Wu T.H., Korpeglu I. and Cheng B.C. Distributed interactive video system design and analysis. IEEE Communications Magazine 1997; 35(3): 100-108

Barnett S.A. and Anido G.J. A cost comparison of distributed and centralized approaches to video-on-demand. IEEE Journal on Selected Areas in Communications 1996; 14(6): 1173-1182

Berzsenyi M., Vajk I. and Zhang H. Design and implementation of a video-on-demand system. Computer Networks and ISDN Systems 1998; 30: 1467-1473

Liu J.C.L., Hsieh J., Du D.H.C. and Lin M.J. Performance of a storage system for supporting different video types and qualities. IEEE Journal on Selected Areas in Communications 1996; 14(7): 1314-1331

Tanakata Y. and Berlage O. Application of genetic algorithms to VOD network topology optimization. IEICE Transactions on Communications 1996; E79-B(8): 1046-1053

Jia X., Zhang Y., Pissinou N. and Makki K. A distributed multicast routing protocol for real-time multicast applications. Computer Networks 1999; 31(1-2): 101-109

Milenkovic M. Delivering interactive services via a digital TV infrastructure. IEEE Multimedia 1998: 34-43

Sesana L. The video placement and routing problem. Unpublished doctoral dissertation, University of Bergamo, Italy 2000

A MARKOV CHAIN TABU SEARCH APPROACH TO THE EVOLUTIONARY ALLOCATION PROBLEM

Hussein A. Abbass[1], Michael Towsey[2], Erhan Kozan[3], and Julius Van der Werf[4].

[1]School of Computer Science, University of New South Wales, ADFA campus, Canberra ACT, 2600, Australia. h.abbass@adfa.edu.au

[2]School of Computing Science, Queensland University of Technology, GPO 2434, Brisbane Qld4001, Australia. m.towsey@qut.edu.au

[3]School of Mathematical Sciences, Queensland University of Technology, GPO 2434, Brisbane Qld 4001, Australia. e.kozan@qut.edu.au

[4]Division of Animal Science, School of Natural Resources and Rural Science, University of New England, Armidale, NSW 2350, Australia. jvanderw@metz.une.edu.au

Abstract Mate-selection is the problem of deciding which animal should be culled and which should be mated in a breeding program. In addition, if the animal is to be mated, which animal from the other sex should it be mated with? With a linear objective function, integer linear programming was successfully used to solve the problem. Also, when the relation between the animals' traits is additive, a strategy based on the classical index theory for selection with random allocation, results in the optimal solution. Today, non-linear and non-additive objectives are introduced and the problem is becoming increasingly complex. In this paper, we formulate the problem as a multi-stage quadratic transportation model. We then introduce a new version of Tabu Search that we call *Markov Chain Tabu Search* (MCTS) for solving the mate-selection problem. We then compare MCTS against five conventional heuristic techniques; these are, random search, hill climbing, simulated annealing, sequential genetic algorithms, and simultaneous genetic algorithms. MCTS is found statistically significantly better than the other heuristics for handling this large scale mate-selection problem.

Keywords: Tabu search, mate--selection, dairy industry.

1. INTRODUCTION

Designing a *breeding program* is an important issue in animal breeding research. The mating decision in a breeding program has two steps, *selection* and *allocation*, and is usually referred to as *mate-selection*. Selection is the decision which animals will be selected for mating and which for culling. Allocation is the choice of the suitable mating pairs among those animals selected for mating. A general definition of the mate-selection problem, called the *evolutionary allocation problem* (EAP), is introduced in Abbass *et al.* (1999) as: "Given a set of males, *M*, females, *F*, a number of years, *Y*, and criteria *C*, assuming the individuals are considered mature after *G* years from their date of birth; find the optimal set of mating-pairs between *M* and *F* for each year in *Y* in order to maximise *C*, so that the progeny can be of either sex and they are available for allocation in subsequent years only when their age is greater than or equal to *G*".

The problem is very complex, both in reality and in its mathematical representation. In this paper, we attempt to solve the mate selection problem in dairy cattle over five stages. We will simplify the problem somewhat by three assumptions. These are that the farmer uses only *artificial insemination sires* (AI-sires), the time horizon is two years, and the progeny are assumed to be females. In general, these assumptions are not severe as most farmers nowadays depend on the artificial insemination program. It is worth mentioning here that the time horizon is taken to be two years to reduce the complexity of the model, by eliminating those transition and state variables which keep track on the animal age. The purpose of the multi-stage model is to penalise the current mate-selection schedule with possible future impact. That is, the model is solved for a number of years (say 5) but the farmer is given the first year only. The reason for that is, giving the farmer a mate-selection schedule for the next ten or even two years is associated with high risk such as the gender of progeny, death of animals, and failure of conception.

This paper provides a new version of *Tabu Search*, called *Markov Chain Tabu Search* (MCTS), for solving the mate-selection problem. We then compare MCTS with five heuristics. In Section 2, the mathematical formulation of the mate-selection problem is presented followed by the experimental design and objectives in Section 3. Preparing the problem to be handled by MCTS is presented in Section 4 followed by the Markov Chain Tabu Search algorithm in Section 5. Section 6 presents a summary of the best result found in each experiment and an overall comparison between the proposed solution strategies.

2. MATHEMATICAL FORMULATION

Parameters List:

T	Total number of stages.
S	Total number of sires in all stages.
D	Total number of dams in all stages.
NT	Number of traits.
$S_{max}(t)$	The maximum number of mating per sire at time t.
$D_{cull}(t)$	Number of dams to be culled at stage t.
$N_{mate}(t)$	The number of mating required at stage t.
$c(t)$	A dummy index used to indicate a to-be-culled animal.
λ_{sinbr}	The cost of 1 unit inbreeding.
λ_{linbr}	The cost of 1 unit coancestry.
b_k	The price of one unit of trait k.
a_k	The economic value of trait k.
1	A column vector of ones.
A	The additive genetic numerator relationships matrix over all generations.
G_{ki}	The additive genetic merit of trait k for animal i.
P_{ki}	The phenotypic value of trait k for animal i.

Decision Variables List:

$SP_i(t)$	Bull i at stage t is productive.
$DP_j(t)$	Cow j at stage t is productive.
$X_i(t)$	A binary variable set to 1 if bull i is alive in stage t; otherwise 0.
$Y_j(t)$	A binary variable set to 1 if cow j is alive in stage t; otherwise 0.
$M_{ij}(t)$	A binary variable set to 1 if the ith sire is mated with the jth dam in time t; otherwise 0.
$DD_{ij}(t)$	A binary variable set to 1 if dam i is the mother of a new born cow j at stage t; otherwise 0.
$SD_{ij}(t)$	A binary variable set to 1 if sire i is the father of a new born cow j at stage t; otherwise 0.

2.1 Objective Functions

In the design for the breeding objectives, we have considered a compromise between short and long term objectives. Two short term objectives are considered: the profit of not culling an animal in the current period taken to be its expected production profit next year, and the short term inbreeding depression represented with the average inbreeding (in reproducing animals) multiplied by the cost of 1 unit inbreeding depression. Two long-term objectives are defined; genetic gain and long term inbreeding (coancestry) (*ie.* the level of relationship between the animals within each year). The four sub-objectives for short and long term planning can be formulated as follows:

1. *O1*: the first short term sub-objective representing the production index:

$$\text{Maximise} \quad O1 = \sum_{t=1}^{T}\sum_{j=1}^{D}\left(\sum_{k=1}^{NT} b_k \times P_{kj}(t)\right) \times Y_j(t) \qquad (1)$$

In this sub-objective a production index is calculated by multiplying the price of trait k, b_k, by the phenotypic value of this trait for animal j, $P_{kj}(t)$, at time t. The index's value is significant only if the animal is alive and reproductive; that is, $Y_j(t)=1$. We may note that the value of production is meaningful only for the female and consequently, the males are not included in this sub-objective.

2. *O2*: the second short term sub-objective representing short term inbreeding:

$$\text{Minimise} \quad O2 = \sum_{t=1}^{T} \lambda_{\sin b} \times \left(\left[X^t(t), Y^t(t)\right]^t \times (Diag(A) - 1)\right) \qquad (2)$$

In this sub-objective, *Diag(A)* returns the diagonal elements of the numerator relationship matrix as a column vector and by subtracting from it a vector of ones, **1**, the result is the inbreeding coefficient for each animal which, if multiplied by the vector $[X^t(t), Y^t(t)]^t$, the result will be the total inbreeding in generation t. Multiplying this value by the cost of one-unit inbreeding, λ_{sinb}, results in the total short-term inbreeding cost/penalty. The matrix A can be built using the algorithm of Henderson (1975) and the matrices *SD* and *DD*. We may add here that the computational time for evaluating the matrix A is large and dominate the computational time for all the algorithms.

3. *O3*: the first long term sub-objective representing the breeding values:

$$\text{Maximise} \quad O3 = \sum_{t=1}^{T}\sum_{j=1}^{D} Y_j(t) \times \sum_{k=1}^{NT}\left(a_k \times G_{kj}(t)\right) \qquad (3)$$

In this sub-objective, a selection index is calculated by multiplying the economic value of trait k, a_k, by the expected genotypic value of this trait for animal j, $G_{kj}(t)$, at time $_t$. The index value is significant only for alive females, $Y_j(t)=1$.

4. *O4*: the second long term sub-objective representing long term inbreeding:

$$\text{Minimise}\, O2 = \sum_{t=1}^{T} \lambda_{linb} \times \left[\sum_{j=1, j\neq c(t)}^{D} M_{ij}(t), \sum_{i=1, i\neq c(t)}^{S} M_{ij}(t)\right] \times A \times \left[\sum_{j=1, j\neq c(t)}^{D} M_{ij}(t), \sum_{i=1, i\neq c(t)}^{S} M_{ij}(t)\right]^t \qquad (4)$$

In this sub-objective, the coancestry relationship is calculated by multiplying a row vector of the proportional contribution of each animal, $\left[\sum_{j=1, j\neq c(t)}^{D} M_{ij}(t), \sum_{i=1, i\neq c(t)}^{S} M_{ij}(t)\right]^t$, times the numerator relationship matrix, times the transpose of the vector representing the contribution of each animal. If this value is multiplied by the cost/penalty of coancestry divided by twice the total number of mating, λ_{linb}, the resultant value is the cost of long term inbreeding.

The problem can also be formulated as a single objective stochastic multi-stage decision model by introducing a set of preferential weights, α, β, γ, and δ. The negative weights are assigned to transform a minimization sub-objective to a maximisation one. In this case, the objective function will be,

$$\text{Maximise} \qquad \alpha O1 + \beta O2 + \gamma O3 + \delta O4 \tag{5}$$

2.2 Constraints

1. The number of mating for each sire (dam) is not allowed to exceed the maximum number of mating allowed for a sire (dam) in stage t.

$$\sum_{i=1, i \neq c(t)}^{S} M_{ij}(t) \leq 1, j = 1 \cdots D, t = 1 \cdots T \tag{6}$$

$$\sum_{j=1}^{D} M_{ij}(t) \leq S_{\max}(t), i = 1 \cdots S, t = 1 \cdots T \tag{7}$$

2. The total number of allocations should equal the total number of mating required.

$$\sum_{i=1, i \neq c(t)}^{S} \sum_{j=1}^{D} M_{ij}(t) = N_{mat}(t), t = 1 \cdots T \tag{8}$$

3. The total number of cows allocated for culling should equal the total number of culling required.

$$\sum_{j=1}^{D} M_{c(t)j}(t) = D_{cull}(t), t = 1 \cdots T \tag{9}$$

4. A dam that is allocated for mating cannot be allocated for culling in the same stage.

$$M_{c(t)j}(t) \times \sum_{i=1, i \neq c(t)}^{S} M_{ij}(t) = 0, j = 1 \cdots D, t = 1 \cdots T \tag{10}$$

5. Each sire/dam mated at the beginning of stage t has a number of progeny in the end of the stage equal to the number of its mating.

$$\sum_{i=1, i \neq c(t)}^{S} M_{ij}(t) - \sum_{k=1}^{D} DD_{jk}(t) = 0, j = 1 \cdots D, t = 1 \cdots T \tag{11}$$

$$\sum_{j=1}^{D} M_{ij}(t) - \sum_{k=1}^{D} SD_{ik}(t) = 0, i = 1 \cdots S, t = 1 \cdots T \tag{12}$$

6. If a mating M_{ij} occurred, the offspring gender in the stage for both the sire and the dam is the same. (Group A2)

$$M_{ij}(t)-\sum_{k=1}^{D}DD_{jk}(t)\times SD_{ik}(t)=0, i=1\cdots S, j=1\cdots D, t=1\cdots T \quad (13)$$

7. A reproductive female child in stage t cannot be a new calf of any dam or sire in stage t.

$$DP_j(t)+\sum_{l=1}^{D}DD_{lj}(t)\leq 1, j=1\cdots D, t=1\cdots T \quad (14)$$

$$DP_j(t)-\sum_{l=1}^{S}SD_{lj}(t)\leq 1, j=1\cdots D, t=1\cdots T \quad (15)$$

8. Any reproductive cow in the next stage is either a born cow in the current stage, or it is a reproductive cow in the current stage and has not been culled. This also forces a cow to be either reproductive or not.

$$DP_j(t+1)=\sum_{k=1}^{D}\sum_{l=1}^{S}DD_{kj}(t)SD_{lj}(t)+Y_j(t), j=1\cdots D, t=1\cdots(T-1) \quad (16)$$

9. A flag constrains $Y_j(t)$ to refer to dams that are not culled in stage t. If dam j had a real mating in stage t then we must force $Y_j(t)$ to be **1**.

$$(1-Y_j(t))\times\sum_{i=1, i\neq c(t)}^{S}M_{ij}(t)=0, j=1\cdots D, t=1\cdots(T-1) \quad (17)$$

$$-Y_j(t)+\sum_{i=1, i\neq c(t)}^{S}M_{ij}(t)\geq 0, j=1\cdots D, t=1\cdots(T-1) \quad (18)$$

10. The cows are available for allocation in stage t if and only if they are reproductive in stage t. In the first constraint, if cow j is mated or culled, it must be productive and $DP_j(t)$ forced to **1**. In the second constraint, if cow j is not mated or culled then it is unproductive and $SD_j(t)$ is forced to *0*.

$$(1-DP_j(t))\times\sum_{i=1}^{S}M_{ij}(t)=0, j=1\cdots D, t=1\cdots(T-1) \quad (19)$$

$$-DP_j(t)+\sum_{i=1}^{S}M_{ij}(t)\geq 0, j=1\cdots D, t=1\cdots(T-1) \quad (20)$$

11. The progeny in stage t must be guaranteed not to exist as individuals in previous stages (*ie.* 0 ... *(t-1)*).

$$\sum_{k=1}^{D}DD_{kj}(t)\times\sum_{l=1}^{(t-1)}Y_j(l)=0, j=1\cdots D, t=1\cdots(T-1) \quad (21)$$

$$\sum_{k=1}^{D}SD_{kj}(t)\times\sum_{l=1}^{(t-1)}Y_j(l)=0, j=1\cdots D, t=1\cdots(T-1) \quad (22)$$

2.3 Model Complexity

If the model is expanded as a single-stage non-linear binary model, the number of variables will be $T \times D^2 + 2 \times D \times T + 2 \times D \times S \times T$ with $12 \times D \times T + 2 \times S \times T + D \times S \times T + 2 \times T - D$ constraints. The search space will be $2^{T \times D^{**}2 + 2 \times D \times T + 2 \times D \times D \times T}$. For example, if we have 200 dams and 10 sires, there will be 222,000 variables in all stages with 11,910 constraints, and an approximate search space size of $10^{66,828}$.

3. EXPERIMENTAL DESIGN AND OBJECTIVES

The experiment's objective is to evaluate different heuristics for mate-selection over 5 stages. It is common practice in animal genetics to simulate a herd with characteristics derived from real life situations. Consequently, we need to simulate a herd that endorses a relatively real life scenario and incorporates some of the potential aspects of the problem. The experimental setup is divided into two parts. First, we need to simulate AI-sires and an initial starting herd. Second, we need to optimise the process of selection and allocation between the cows in this herd and the AI-sires. The experimental design for these two parts is discussed in the following two sub-sections.

3.1 The Simulation Model for the Initial Herd

Two herds were simulated. The first was used to generate AI-sires for each year for the second herd to use. The following model was used to simulate an Australian Holstein Friesian breed in both herds to generate the initial set of animals. The records were generated by sampling from the following equation.

$$Y_{jk} = \mu + Ye + S_i + A_j + PE_j + E_{jk} \tag{23}$$

where μ is the herd mean, *Ye* is the year effect, S_i is the season fixed effects where $i=1...4$, A_j is the additive genetic effect for animal j, PE_j is the permanent environmental effect for animal j, and E_{jk} is the residual effect for the kth record of animal j. The heritability (h^2) and repeatability (r) were 0.25 and 0.40 respectively. The herd mean (μ) was 5500. The four season effects were -100, 0, 100, 200. The year effect was sampled using a uniform distribution between 0-200. The phenotypic (σ_P), additive (σ_A), permanent environmental (σ_{PE}), and residual (σ_E) standard deviations were calculated using the following equations:

$$\sigma_p = 0.1 \times \mu \tag{24}$$

$$\sigma_A = \sqrt{h^2} \times \sigma_p \tag{25}$$

$$\sigma_{PE} = \sqrt{(r - h^2)} \times \sigma_p \tag{26}$$

$$\sigma_E = \sqrt{(1 - r)} \times \sigma_p \tag{27}$$

The matrix σ^2(BV,EBV) is a symmetric matrix with the main diagonal's elements (1,1) = (2,2) = σ^2_a and the elements (2,1), (1,2) = $r^2 \times \sigma^2_a$. The EBVs for the sires and dams were sampled by taking a Cholesky decomposition (Mrode, 1996) of this matrix and multiplying the resultant triangular matrix with a vector of two random variables - each of them sampled independently from a Gaussian distribution ($N(0,1)$). Each dam was allocated a season effect at random, where the pseudo-random number was sampled from a uniform distribution $U(0,1)$. The year effect was sampled from $U(0,200)$ for each year. The permanent environmental and residual effects were sampled from $N(0, \sigma^2_{PE})$ and $N(0, \sigma^2_E)$ respectively. Each dam selected for mating was assigned a bull at random. The EBVs for the progeny, assumed to be heifer, were calculated using the following equation,

$$\text{Progeny EBV} = 0.5 \times (\text{Sire EBV} + \text{Dam EBV}) + \phi \tag{28}$$

where the Mendelian sampling, ϕ, is $N(0, \sigma^2(\phi))$. $\sigma^2(\phi) = 0.5 * \sigma^2_A * (1 - 0.5 (F_s+F_d))$, and where F_s and F_d are the parents' inbreeding coefficients.

A nucleus herd was simulated to generate AI-sires required for breeding commercial herds. To mimic a practical genetic relationship structure, simulation of both a nucleus herd and a commercial herd was required. In year 0, 500 dams and 250 sires were generated independently using the previous model. For each year, the best 10 sires were saved in the AI-sires' list and mated with 100 dams at random. Additionally, 100 sires, not including any AI-sire, and 100 dams were selected at random for culling. The mating ratio between sires and dams was 1:10 and the simulation was carried out for 21 years. Each mating resulted in two progeny, one male and one female, for replacement. The AI-sires' list in each year was saved in a list to be available for mating for the cows in the commercial herd. In year 0, 200 dams were simulated using the previous model for a commercial herd. The population was mated at random for 20 years. In each year, half of the dams was selected at random for culling and the other half was mated at random with the AI-sires generated previously. One may notice that the number of years for the nucleus herd is 1 year more than the initial herd. The AI-sires generated in the additional year were saved for the optimisation model. Each dam selected for mating was assigned an AI-sire at random and the progeny was assumed to be a heifer.

3.2 Experimental Design for the Optimisation Model

The experiments to be presented here are very computationally expensive because of the relationship matrix's generation. A complete run of length 30,000 objective-evaluations took around 30 hours on a 400Mhz PC. Two super-computers and 1 Sun-Sparc server were utilized to distribute the experiments. In solving the multi-stage model, it is not practical to wait until the algorithm converges to a solution since the farmer requires to reach a solution within a reasonable time. Consequently, the number of objective function evaluations is used as a criterion to terminate the search. The main objective of this experiment therefore is to select the heuristic that results in the best solution within a maximum of 30,000 objective evaluations. Additionally, the search is terminated if the algorithm does not improve the best solution found so far within 3,000 objective evaluations (Michalewicz, 2000). This number is obtained by randomly searching for solutions. It is found that the average number of steps needed by random search to reach the best solution found is less than 3,000 objective evaluations. Also, the farmer needs to terminate the run when he/she needs to. Therefore, it is decided that an algorithm which does not improve the solution within 3,000 objective evaluations is to be terminated. The variance of the best solutions found over 10 different runs with different seeds, is used as a measure for the stability of these heuristics.

A nucleus and a commercial herd were simulated using the model discussed in the previous sub-section. The total number of animals required for calculating pedigree information for the herd was 45 sires and 1250 dams. The commercial herd contained 200 dams still alive (*ie.* available for mating) and the nucleus herd contained 10 AI-sires. The maximum number of mating allowed per sire was 20, the number of required culling 100, and the total number of required mating 100. Each experiment is repeated ten times with ten different seeds. The same ten seeds are used for all experiments to guarantee a fair comparison among the algorithms since all algorithms will start from the same point in the search space.

4. PROBLEM PREPARATION

4.1 Representation of the Multi-Stage Solution

In the proposed strategies, a solution is represented using a two-dimensional array where each row represents a single stage. Each cell in each row is associated with a dam, which is available for allocation, as shown in Figure 1. It is obvious in this representation that the pedigree information is missing. To illustrate how the pedigree information exists in the proposed representation, in stage t, the progeny of stage $t-1$ take the culled animals' places in stage t in the same order of which the mating in stage $t-1$ and culling in stage $t-1$ occurred. It is worth mentioning that, which progeny occupy which cell does not affect the solution, since each index in the chromosome has a significant meaning only when it is associated with an animal. Having this representation, the logical constraints connecting the stage with each others are satisfied. Additionally, the pedigree information for all possible solutions can be generated, by the heuristic defined above, when required. This representation reduces the search space to

$$\binom{T \times D}{T \times D - C} \times S^{(T \times D - C)} \tag{29}$$

where T is the number of stages, D the number of dams, S the number of sires, and C the number of culling. When $T=5$, $D=200$, $S=10$, and $C=100$, the search space is approximately 10^{1000}!

Dam ID → / Sire ID ↓	0	1	2	...	...	...	...	...	...	...	D-1
	-1	3	0	4	3	-2	-2	-1	3	-1	5
	4	0	4	-1	0	-1	4	-2	-1	4	-2
	5	3	0	4	5	3	3	0	-1	-1	-1

Figure 1. The data structure for a solution in the multi-stages model.

It is important to emphasise here that this representation removes redundancy from the optimisation model without excluding any possible optimal solutions. Therefore, the whole search space of the original model is still preserved in this representation, but in a compact way while it is guaranteed that the optimal solution is reachable by this representation (that is, it is one of the 10^{1000} solutions).

4.2 Generation of Neighbour Solutions

A solution x is generated in the neighbourhood of another solution x' by randomly changing up to ζ cells in each stage of x', where ζ is the maximum *neighbourhood length.* The neighbourhood length is measured in terms of the number of cells with different values in both solutions, and is equivalent to the hamming distance between both solutions if they were mapped to the mating matrix M. Figure 2 presents the algorithm used for generating neighbour solutions. We use "*neighbourhood size*" to represent the number of solutions examined in a neighbourhood. Therefore, a neighbourhood size of 20 requires calling the algorithm, 20 times.

```
function neighbour(x', ζ)
x ← x'
neighbour_length = ζ × random()
for each row i in x and for k = 1 to neighbour_length
j= random() × x
x[i][j] = random() × number_of_sires
repair(x)
return x
end function
```

Figure 2. Generation of neighbour solutions.

4.3 Constraint Satisfaction Method

The *repair operator* is used in all experiments to satisfy the constraint system. It is important to note here that the additional constraints connecting the animals in different stages are satisfied by the solution's representation.

Procedure repair(x')
  For each stage g
    count the number of allocations, $alloc_i$, per each sire, i, and the total number of culling, *Cull*
    foreach sire i
      if $alloc_i$ > *maximum_number_of_sire_allocations* **then**
        while $alloc_i$ ≠ *maximum_number_of_sire_allocations* **do**
          randomly select one of the dams allocated to i
          assign the value *-2* to it, $alloc_i = alloc_i - 1$
    if *Cull* > *number_of_allowed_culling* **then**
      while *Cull* ≠ *number_of_allowed_culling* **do**
        randomly select a dam j
        if j has a value *-1* **then** assign *-2* to j, Cull = Cull - 1
    else if *Cull* < number_of_allowed_culling **then**
      while *Cull* ≠ *number_of_allowed_culling* **do**
        randomly select a dam j
        if j has a value *-2* **then** assign *-1* to j
        else if j has a value > *-1* **then**
          $alloc_i = alloc_i - 1$, i is the sire allocated to j
          assign *-1* to j, *Cull = Cull + 1*
    count the total number of matings, *Mat*
    if *Mat* > *number_of_allowed_matings* **then**
      while *Mat* ≠ *number_of_allowed_matings* **do**
        randomly select a dam j
        if j has a value > *-1* **then**
          $alloc_i = alloc_i - 1$, i is the sire allocated to j
          assign *-2* to j, *Mat = Mat – 1*
    if *Mat* < *number_of_allowed_matings* **then**
      while *Mat* ≠ *number_of_allowed_matings* **do**
        randomly select a dam j
        if j has a value *-2* **then**
          randomly select a sire i, $alloc_i$ < *maximum_number_of_sire_allocations*
          assign i to j, *Mat = Mat + 1*
          $alloc_i = alloc_i + 1$
end procedure

Figure 3. The repair operator for the multi-stage model

Figure 3 presents the operator algorithm. The algorithm starts with counting the number of allocations for each sire. A loop is then constructed over all sires. If a sire exceeds the maximum number of allowed mating per sire, the excess is eliminated by randomly selecting a dam from those mated to this sire and assigning to it the symbol -2, denoting a *no action* decision. This process is repeated until all sires meet the maximum allowed mating per sire constraint. Then the total number of dams selected for culling is calculated. If it exceeds the maximum number of culling, a dam that is assigned to culling is randomly selected and the symbol -2 is assigned to it. The step is repeated until the total number of dams assigned for culling equals the number of culling required. If the total number of culling is less than that determined by the user, a dam is selected at random. If the value assigned to this dam is -2, it is set to -1 and the total number of culling is increased by one. If the value assigned to this dam is greater than -1 - the dam is selected for mating and is being assigned to a bull - the number of mating for the bull allocated to this dam is reduced by one, the total number of mating is reduced by one, and the dam is assigned -1. The process is repeated until the total number of culling equals the number of culling required by the user. Then the total number of mating in the chromosome is calculated. If the total number of mating is greater than the number of mating required by the user, a dam which is being assigned to a bull is selected at random and assigned -2. The total number of mating is then reduced by 1 and the algorithm loops until the total number of mating equals the number of mating required by the user. If the total number of mating is less than the number of mating required by the user, a dam which is assigned -2 is selected at random and a bull that has not reached maximum number of mating per bull, is chosen at random and is assigned to this dam. The total number of mating is increased by one and the algorithm loops until the total number of mating is equal to the number of mating required by the user.

4.4 Handling the Multi-Objective Problem

Since this is a multi-objective problem, one of the methods for handling multi-objective problems should be used. Similar to the single-stage model (Abbass et al., 2000), the weighted-sum method will be used for combining the objectives. Each objective is divided by a constant representing an estimation of the standard deviation of the values of this objective. Ten runs were undertaken for 200 objective-evaluations each and the variance of each objective was calculated. The constants are 160331 for phenotypic values, 14936.3 for genotype, 29.6864 for inbreeding, and 815.292 for coancestry. The following differential weights were then used: $\alpha = 1$, $\beta = 1$, $\gamma = -1$, and $\delta = -1$ for phenotypic, genotypic, inbreeding, and coancestry respectively. All the results are presented after unscaling the objectives.

5. MARKOV CHAIN TABU-SEARCH

5.1 Methods

Glover (Glover, 1989; Glover, 1990) introduced *Tabu search* (TS) as a method for escaping from a local optimum. The goal is to obtain a list of forbidden (Tabu) solutions/directions in the neighbourhood of a solution to avoid cycling between solutions while allowing a direction, which may degrade the solution although it may help in escaping from the local optimum. In essence, the Tabu list aims to constrain or limit the search scope in the neighbourhood while still having a chance to select one of these directions. The Tabu list can be the set of solutions visited in the last *n* iterations. However, this is a memory consuming process and it is a limited type of memory. Another possibility is to define the neighbourhood in terms of a set of moves. Therefore, instead of storing the solution, the reverse of the move is stored. Clearly, this approach prohibits, not only returning to where we came from, but also many other possible solutions. Notwithstanding, since the Tabu list is a short term memory, at some point in the search, the reverse of the move will be eliminated from the Tabu list, therefore, allowing to explore this part of the search space which was Tabu.

A very important parameter here, is the choice of the Tabu list size which is referred in the literature as the method's *adaptive memory*. On the one hand, a large list size would be inefficient in terms of memory capacity and the time required to scan the list. On the other hand, choosing the list size to be small would result in a cycling problem; that is, revisiting the same state again (Glover, 1989). In general, the Tabu list's size is a very critical issue for the following reasons:- First, the performance of Tabu search is sensitive to the size of the Tabu list in many cases. Second, there is no general algorithm to determine the optimal Tabu list size apart from experimental results. And last, choosing a large Tabu list is inefficient in terms of speed and memory.

Paulli (1993) found that the choice of the Tabu list's size has a large influence on the solution's quality and it varies considerably among problems. Furthermore, he claimed that SA is much better than TS for the quadratic assignment problem. Recalling from Abbass et al. (2000) that the single-stage problem is a quadratic transportation model, and the multi-stage model discussed in the current chapter is a dynamic transportation model, this gives an expected bad performance for Tabu search. The previous points motivated the development of a new version of TS which we called *Markov chain Tabu search* (MCTS). Here, the Tabu list will be referred to as the *probabilistic Tabu list* (PTL). There is almost no relation between MCTS as presented here and a version of Tabu search called *probabilistic Tabu search* (PTS) (Glover and Laguna, 1997). In MCTS, the PTL's size is always fixed and the structure depends on the representation. For example, in our problem a solution is represented using a two-dimensional array where one dimension, t, represents the number of stages and the second dimension, j, represents the dams. The PTL is represented using a three-dimensional array where in addition to the two-dimensional solution array, a third dimension, i, is added of size equal to the number of sires. Each cell in the three-dimensional array represents the probability that sire i is to be allocated to dam j in stage t. If we refer to this probability as π_{tij}, then the conventional probability conditions ($0 \leq \pi_{tij} \leq 1$ and $\sum_{i=1}^{S+1} \pi_{tij} = 1$) where S is the number of sires, should be satisfied all the time. The matrix is initialised with all values equal to $1/(S+1)$. The reward/penalty is initialised with 1% of the initial value used in the PTL; that is, $1/(100 \times (S+1))$. The algorithm is depicted in Figure 4.

Every time the MCTS algorithm selects a new solution from the neighbourhood, a reward and/or a penalty is assigned to each cell in the PTL, which is present in the solution. If the new solution is better than the previous best solution found, the assignments corresponding to the previous best solution found are penalized and those corresponding to the new solution are rewarded. That is, if sire 5 is allocated to dam 3 in stage 1 in the previous best solution found, the cell which corresponds to sire 5, stage 1, and dam 3 in the PTL is decreased by a constant. If the previous best solution found is still better than the new solution, the best solution found is rewarded and the new solution is penalised. Therefore, the algorithm uses a double reinforcement approach. In this way, instead of storing the set of directions, which are Tabu, they are indirectly penalised in the probabilistic Tabu list. The algorithm is based on Markov chain since each new PTL depends only on the previous one and represents a transition matrix between states. The update rule can be represented as a transition matrix as well, with all cells equal to 0 except those needing to be rewarded or penalised, which they are set to the penalty/reward constant. Note that continue to reward the best solution found as long as it does not change, helps to speed up the algorithm, although it may cause stagnation which will be overcome as will be explained in the discussion of the function ***choose***. Speeding up the algorithm is a necessity for two reasons; First, Markov chain based search techniques are very slow by nature and it can take a very long time to reach equilibrium. Second, the greediness of the algorithm is overcome by the repair operator which introduces disruption to the neighbourhood and therefore a probability to escape local optima. However, there remains a question here, how does the MCTS algorithm choose x^k using π^{k-1}; that is, the function **choose**(x^{opt},π^{k-1}), Figure 5. In the figure, the parameter α represents the threshold of rejecting the allocation of sire i to dam j. Now, let us assume that the initial probability for all sires was 0.1, $\alpha = 0.05$, the discount step = 0.01, the solution did not improve after 5 objective-evaluations from the start, and sire i was allocated to dam j in all last 5 stages. Therefore, π^5_{tij} =

0.05 since we kept penalising the probability. The algorithm illustrated in Figure 5 will force dam *j* to be allocated to a different sire. But, can sire *i* be allocated later on to dam *j*? The answer is yes, since the repair operator does not see the probability matrix π; therefore, even though that it is decided by the probability matrix that this sire is Tabu, the repair operator introduces some distortion to the influence of the probability matrix.

---

procedure MCTS
    Input: ζ
    initialise discount step to be $1/(100 \times (S+1))$
    generate π^0 with $\pi^0_{tij} = 1/(S+1)$
    generate solution array x^0 at random}
    repair(x^0)
    $f^0 \leftarrow$ *evaluate* x^0
    $x^{opt} \leftarrow x^0$ and $f^{opt} \leftarrow f^0$
    $k=1$
    while $M <$ maximum number of objective-evaluations
        $x^k \leftarrow$ **choose**(x^{opt}, π^{k-1})
        repair(x^k)
        $f^k \leftarrow$ *evaluate* x^k
        $M = M + 1$
        for l := 1 to Neighbourhood size
            $x^k_{new} \leftarrow$ **neighbour**(x^k, ζ)
            repair(x^k_{new})
            $f^k_{new} \leftarrow$ **evaluate** x^k_{new}, $M = M + 1$
            if $f^k_{new} > f^{opt}$, **then** $f^{opt} \leftarrow f^k_{new}$, $x^{opt} \leftarrow x^k_{new}$
            if $f^k_{new} > f^k$, **then** $f^k \leftarrow f^k_{new}$, $x^k \leftarrow x^k_{new}$
        if $f^k > f^{opt}$, **then**
            $f^{opt} \leftarrow f^k$, $x^{opt} \leftarrow x^k$
            for each stage t **and for each** cow j
                    add discount step to π^{k-1}_{tij} where $i = x^k_{tj}$
                    subtract discount step from π^{k-1}_{tij} where $i = x^{opt}_{tj}$
        else for each stage t **and for each** cow j
                    add discount step to π^{k-1}_{tij} where $i = x^{opt}_{tj}$
                    subtract discount step from π^{k-1}_{tij} where $i = x^k_{tj}$
        $\pi^k \leftarrow \pi^{k-1}$
    $k=k+1$
end procedure

Figure 4. The Markov chain Tabu search algorithm.

```
function choose(x^opt, π^(k-1))
        x' ← x^opt
        for each stage t
                randomly select J < number of dams
                for j = 1 to J
                        randomly select sire i
                        if π_tij > α then x'_tj = i
        return x'
end function
```

Figure 5. The generation of neighbourhood solutions in the MCTS algorithm.

To guarantee that the algorithm will not stagnate at some point (*ie.* a sire is always allocated to the same dam), we need to choose the value of α carefully. Assume that we need to run the algorithm for up to 1,000 objective-evaluations; the initial probability is 0.1, the discount factor is 0.01, $\alpha = 0.05$, and we have 10 sires (including a dummy sire for culling). In this case, after around 45 objective-evaluations without an improvement in the best solution found, there is a possibility that the algorithm will keep on this sire for this dam forever. We need to avoid stagnation if this occurs to all dams in the solution. Note that the solution does not change and the repair operator has no effect since the solution is still feasible. The obvious way to overcome this is to add the discount factor every some number of objective-evaluations to those cells in the probability matrix, which reached the minimum rejection level α. This would force the algorithm to change the current solution and escape the current local optimum.

It is important to mention here that the algorithm differs from any Tabu search since it starts all the time from the best solution found (the input to the function *choose*(x^{opt},π^{k-1})). This may imply that this algorithm is a greedy algorithm. However, this is not the case because the neighbourhood of the step generated by the function *choose* is searched regardless of whether this step generated a better solution than the best found so far or not. Also, every point in the search space is reachable from the best solution found (see the fourth line in the function *choose*). We can imagine the algorithm as a search technique, which starts with the best solution found and try to take steps in the space to find a better neighbourhood, while these steps themselves may downgrade the best solution found.

5.2 Results and Discussion

The three parameters that we may need to determine in the MCTS algorithm are: the rejection level, the neighbourhood size, and neighbourhood length. We keep the discount factor constant – since it is correlated to the rejection level - and change the rejection level instead. The initial probability is $1/11 \approx 0.091$ for each cell in the probability matrix, the discount factor is taken to be 0.001, and $\alpha \in \{0.02, 0.05, 0.08, 0.09\}$. Every 100 objective-evaluations, the discount factor is added to random number of elements in the matrix reached the rejection level and subtracted from random number of elements in the matrix exceeding the rejection level. Six neighbourhood sizes of 150, 100, 50, 10, 5, and 1 and neighbourhood lengths of 10%, 5%, and 1% were tested as shown in Table 1. A neighbourhood size of 1 means that the algorithm is one-step look ahead algorithm.

In Table 1, the best combination is found with the shortest neighbourhood size (1), the shortest neighbourhood length (1% of the chromosome size or equivalently 2 cells for each stage), and the highest rejection level (0.09). Note that we cannot increase the rejection level anymore since the rejection threshold is 0.1.

The previous results need further explanation. We may notice that there is a relation between the neighbourhood size and the number of points visited in the space. With a large neighbourhood size, the algorithm spends most of the time in exploring the neighbourhood (intensification of local search) while reducing the exploration of the overall search space (sacrificing with diversification of global search) because the number of objective evaluations is fixed. With a small neighbourhood size, the reverse is occurring. The probability matrix is different. Continuous update of the probability matrix emphasises the exploitation of the overall search space (intensification of global search). Therefore, the results found here suggest that the intensification of the search while globally exploring the search space through the probability matrix is more beneficial than the local intensification of the search through the neighbourhood. This result is consistent with the fact that the search space is huge; therefore, spending time in the neighbourhood may not be a good idea.

Table 1. The results for the Markov chain Tabu search method.

			Neighbourhood Size	
α	ζ		150	100
0.02	10%	Avg best solutions	2,357,290±28,400	2,323,650±21,000
		# of obj-evaluations	3,220±1,630	3,100±2,190
		Best solution found	2,397,500	2,352,200
0.05	10%	Avg best solutions	2,358,340±33,600	2,316,300±24,200
		# of obj-evaluations	3,450±1,800	2,390±1,140
		Best solution found	2,425,600	2,352,200
0.08	10%	Avg best solutions	2,370,950±36,800	2,345,720±68,300
		# of obj-evaluations	3,200±1,900	4,130±6,640
		Best solution found	2,445,900	2,535,400
0.09	10%	Avg best solutions	2,408,780±46,200	2,462,380±82,000
		# of obj-evaluations	4,040±2,490	9,910±5,630
		Best solution found	2,482,000	2,561,500
0.02	5%	Avg best solutions	2,453,970±30,500	2,399,320±33,600
		# of obj-evaluations	2,830±1,510	2,820±1,490
		Best solution found	2,489,600	2,444,000
0.05	5%	Avg best solutions	2,456,070±32,600	2,388,810±37,800
		# of obj-evaluations	2,980±1,900	2,170±1,830
		Best solution found	2,501,600	2,444,000
0.08	5%	Avg best solutions	2,461,330±44,100	2,418,240±80,900

		# of obj-evaluations	4,040±2,210	5,200±5,970
		Best solution found	2,520,800	2,562,200
0.09	5%	Avg best solutions	2,531,740±82,000	2,605,310±132,400
		# of obj-evaluations	10,090±9,260	12,600±7960
		Best solution found	2,653,100	2,735,000
0.02	1%	Avg best solutions	2,439,260±44,100	2,326,810±46,200
		# of obj-evaluations	4,740±2,310	4,030±2,900
		Best solution found	2,517,900	2,395,000
0.05	1%	Avg best solutions	2,436,110±43,100	2,337,320±43,100
		# of obj-evaluations	4,350±2,020	3,200±1,820
		Best solution found	2,503,100	2,395,000
0.08	1%	Avg best solutions	2,459,230±76,700	2,616,870±252,200
		# of obj-evaluations	5,150±4,340	13,220±10,580
		Best solution found	2,607,700	2,953,400
0.09	1%	Avg best solutions	2,800,790±288,000	3,033,050±33,600
		# of obj-evaluations	16,740±12,600	27,050±3,620
		Best solution found	3,085,000	3,087,400
			Neighbourhood Size	
			50	10
0.02	10%	Avg best solutions	2,357,290±28,400	2,837,570±45,200
		# of obj-evaluations	3,220±1,630	29,760±120
		Best solution found	2,397,500	2,887,200
0.05	10%	Avg best solutions	2,358,340±33,600	2,948,970±65,200
		# of obj-evaluations	3,450±1,800	28,540±3,940
		Best solution found	2,425,600	3,011,100
0.08	10%	Avg best solutions	2,370,950±36,800	3,156,010±46,200
		# of obj-evaluations	3,200±1,900	29,920±100
		Best solution found	2,445,900	3,263,300
0.09	10%	Avg best solutions	2,408,780±46,200	3,152,860±45,200
		# of obj-evaluations	4,040±2,490	28,780±1,510
		Best solution found	2,482,000	3,219,700
0.02	5%	Avg best solutions	2,453,970±30,500	2,869,100±63,100
		# of obj-evaluations	2,830±1,510	28,250±3,460
		Best solution found	2,489,600	2,947,600
0.05	5%	Avg best solutions	2,456,070±32,600	3,012,030±41,000
		# of obj-evaluations	2,980±1,900	29,640±460
		Best solution found	2,501,600	3,067,100
0.08	5%	Avg best solutions	2,461,330±44,100	3,267,410±33,600
		# of obj-evaluations	4,040±2,210	29,940±50
		Best solution found	2,520,800	3,324,000
0.09	5%	Avg best solutions	2,531,740±82,000	3,356,740±39,900
		# of obj-evaluations	10,090±9,260	29,880±170
		Best solution found	2,653,100	3,415,300
0.02	1%	Avg best solutions	2,439,260±44,100	2,998,370±42,000
		# of obj-evaluations	4,740±2,310	29,410±730
		Best solution found	2,517,900	3,064,500
0.05	1%	Avg best solutions	2,436,110±43,100	3,148,650±33,600
		# of obj-evaluations	4,350±2,020	29,530±540
		Best solution found	2,503,100	3,209,500
0.08	1%	Avg best solutions	2,459,230±76,700	3,489,160±27,300
		# of obj-evaluations	5,150±4,340	29,930±60
		Best solution found	2,607,700	3,537,500
0.09	1%	Avg best solutions	2,800,790±288,000	3,659,410±22,100
		# of obj-evaluations	16,740±12,600	29,950±30
		Best solution found	3,085,000	3,680,100
			Neighbourhood Size	
			5	1

0.02	10%	Avg best solutions	3,139,190±56,800	3,293,680±82,000
		# of obj-evaluations	29,760±140	20,850±2,310
		Best solution found	3,212,900	3,386,800
0.05	10%	Avg best solutions	3,260,050±58,900	3,190,690±70,400
		# of obj-evaluations	29,930±50	15,890±2,390
		Best solution found	3,346,700	3,338,500
0.08	10%	Avg best solutions	3,151,810±88,300	3,050,910±46,200
		# of obj-evaluations	24,620±3,190	7,590±2,030
		Best solution found	3,265,000	3,123,300
0.09	10%	Avg best solutions	3,091,900±36,800	2,962,630±74,600
		# of obj-evaluations	16,290±2,610	5,010±1,610
		Best solution found	3,155,100	3,063,700
0.02	5%	Avg best solutions	3,195,950±30,500	3,369,350±86,200
		# of obj-evaluations	29,710±250	26,030±3,860
		Best solution found	3,269,500	3,498,700
0.05	5%	Avg best solutions	3,118,170±88,300	3,345,180±72,500
		# of obj-evaluations	29,880±100	23,300±6,310
		Best solution found	3,346,100	3,442,000
0.08	5%	Avg best solutions	3,332,570±50,400	3,340,980±53600
		# of obj-evaluations	29500±720	22,590±4,210
		Best solution found	3,436,300	3,436,300
0.09	5%	Avg best solutions	3,360,940±38,900	3,274,770±87200
		# of obj-evaluations	26,440±4,290	17,140±8,740
		Best solution found	3,409,500	3,397,700
0.02	1%	Avg best solutions	3,263,210±39,900	3,642,600±147,100
		# of obj-evaluations	29,780±190	27,630±7,450
		Best solution found	3,320,800	3,715,300
0.05	1%	Avg best solutions	3,463,940±31,500	3,737,190±14,700
		# of obj-evaluations	29,860±120	29,990±10
		Best solution found	3,505,100	3,756,700
0.08	1%	Avg best solutions	3,703,550±23,100	3,800,240±10,500
		# of obj-evaluations	29,920±110	29,970±40
		Best solution found	3,744,100	3,815,800
0.09	1%	Avg best solutions	3,767,660±12,600	3,805,500±13,700
		# of obj-evaluations	29,970±30	29,910±110
		Best solution found	3,782,700	3,824,600

6. OVERALL DISCUSSION

We compared the performance of MCTS against five conventional heuristic techniques, viz *random search*, *hill climbing*, *simulated annealing*, *sequential genetic algorithms*, and *simultaneous genetic algorithms*. The details of the experiments undertaken for the five heuristics can be found in Abbass (2000). In Table 2 the best results obtained in each experiment is summarised together with the *t*-value obtained from comparing the experiment with the best strategy *Markov Chain Tabu Search* using the correlated paired t-test. It is found that the results in all experiments are statistically different from the MCTS strategy at confidence level 0.999. Therefore, it is a simple matter to conclude that MCTS was the best strategy for the mate-selection problem.

Table 2. A summary of the best combination found in each experiment and its corresponding t-value relative to the MCTS algorithm

Experiment	Average best solutions	# of objective-evaluations	Best solution found	t-value
MCTS	3,805,500±13,700	29,910±110	3,824,600	
GA	3,532,250±30,500	29,490±1,210	3,597,100	23.43
Hill climber	3,510,180±15,800	29,600±350	3,533,000	46.87
Sequential GA	3,429,260±30,500	N/A	3,483,700	47.97
SA	2,407,730±34,700	2,370±950	2,493,000	94.29
Random	2,013,620±23,100	2,940±1,710	2,058,400	199.10

The findings required some interpretation. MCTS can also be seen as a series of Markov chains, where the probabilistic Tabu list is updated in each objective-evaluation and the new one depends only on the previous one. The real difference between TS, and MCTS is that TS uses negative knowledge only with a positive search strategy (*ie.* prohibiting previously visited states and bias the search to previously un-visited and potentially good states), but MCTS uses two types of knowledge (positive and negative) with a positive search strategy (*ie.* rewarding good solutions, penalising bad solutions, and biasing the search towards potentially good areas). SA is somewhat different since it does not use any of this knowledge but instead uses a mixture of positive and negative search strategies with a certain probability. At the start of the search, SA emphasises a negative search strategy by accepting bad steps most of the time while the temperature is high. At the end of the search, the situation is reversed. However, there is no guarantee that this is always good. Perhaps a negative or positive knowledge is more important during the search in some situations. When the search space is huge, as in our case, any of these algorithms is expected to need so much time to reach equilibrium; a problem that less exists with Gas – because of its use of populations. However, MCTS performed the best due to the search intensification through a continuous reward for the best solution found and a continuous penalty for bad solutions. GAs easily outperformed SA since it survived the condition of an improvement every 3,000 objective evaluations. SA did not survive this condition because they were behaving like random search to explore the search space. The temperature of SA did not have time to cool. In GA, the selection pressure resulted from the modified tournament strategy proved to have a positive effect in our problem. Also, the 1-point crossover operator managed successfully to sustain good schemas while the repair operator adds some mutations from time to time when feasibility is broken by recombination.

7. CONCLUSION

In this paper, the mate-selection problem is formulated and a new version of Tabu search, viz Markov Chain Tabu Search (MCTS), was developed to solve the problem. MCTS was compared against five standard heuristic techniques, gave the best results overall, and achieved significantly better solutions than all other techniques. It was therefore chosen as our preferred strategy for solving the multi-stage mate-selection problem.

References

Abbass, H. Computational intelligence techniques for decision making: with applications to the dairy industry. PhD thesis, School of Computing, Queensland University of Technology, 2000.

Abbass, H., Towsey, M., der Werf, J. V., and Finn, G. Modelling evolution: the evolutionary allocation problem. In (Abbass, H. and Towsey, M.), The Application of Artificial Intelligence, Optimisation, and Bayesian Methods in Agriculture, QUT-Publication, 1999, pages 117-134.

Abbass, H., Towsey, M., Kozan, E., and der Werf, J. V. The performance of genetic algorithms on the one-generation mate-selection problem. Proceedings of the 2nd Joint International Workshop on Operations Research, Sapporo, Japan, 2000.

Glover, F. Tabu search: part 1. ORSA Journal on Computing, 1989, 1(3):190-206.

Glover, F. Tabu search: part 2. ORSA Journal on Computing, 1999, 2(1):4-32.

Glover, F. and Laguna, M. Tabu search. Kluwer Academic Publisher, Boston, 1997.

Henderson, C. Use of all relatives in intraherd prediction of breeding values and producing abilities. Dairy Science, 1975, 58:1910-1916.

Michalewicz, Z. and Fogel, D. How to solve it: modern heuristic. Springer verlag, 2000.

Mrode, R. Linear models for the prediction of animal breeding values. CAB International, 1996.

Paulli, J. A computational comparison of simulated annealing and Tabu search applied to the quadratic assignment problem. In (Vidal, R.), Applied simulated annealing. Springer-Verlag, 1993.

A GAME OF CORRUPTION IN A STYLIZED SOUTH-EAST ASIAN TIMBER INDUSTRY

A Genetic Algorithm Approach

Ryan McAllister[1] and Michael Bulmer[2]

[1]*Department of Economics*
University of Queensland
r.mcallister@mailbox.uq.edu.au

[2]*Department of Mathematics*
University of Queensland
mrb@maths.uq.edu.au

Abstract This paper analyzes the impact of corruption in a stylized timber industry representative of many South-East Asian nations, where corruption is common and market driven logging is the leading contributor to deforestation. Corruption is analyzed using a three-player open-loop difference game consisting of a commercial logging contractor, a central government, and a regional government intermediary responsible for enforcing the government's forest property rights. The intermediary, however, is assumed to either cooperate fully with the contract logger or the central government. The game is piecewise non-linear in both the control and state variables making analytical solutions hard to derive. A genetic algorithm approach is therefore used to solve the model numerically. The results indicate that corruption may negatively impact on an economy due to mis-management of forest resources.

Keywords: Genetic algorithms, discrete dynamic game, open loop

1. INTRODUCTION

While agricultural expansion and firewood collection contribute heavily to deforestation globally, market driven deforestation is the major contributor in South-East Asia (Dudley et al., 1995; Miller and Tangley, 1991). Some South-East Asian governments have sought to ban timber logging (for example, Thailand in 1998 and Laos in 1991) in recognition of the economic costs of deforestation. However, illegal logging has impeded the effectiveness of both logging bans and more modest attempts to reduce the pressure of logging on timber resources. Within South-East Asia, countries such as Cambodia, Indonesia, Thailand, The Philippines, Laos and Vietnam have documented illegal logging problems (Dudley et al. 1995; Miller and Tangley, 1991). Much illegal logging appears to be carried out in collaboration with corrupt government officials.

This paper analyzes the impact of corruption in a stylized three-tiered timber industry representative of many South-East Asian timber industries. The stylized industry consists of a commercial logging contractor, a central government, and a regional government intermediary. The commercial logger commits to logging a government issued quota, but may also choose to log illegally at levels above that stipulated by their contractual arrangements. The government chooses the level of funding on forest property rights enforcement. The role of the intermediary is to cooperate with the government and to enforce their property rights. However, the prevalence of corruption implies that the intermediary may not cooperate with the government completely. With chronic corruption, an intermediary may instead cooperate fully with the commercial logging contractor.

The impact of corruption is analyzed in this paper by a three-player difference game with one player (the intermediary) cooperating fully with either of the other players. The game is piecewise non-linear in both the state and control variables, making it hard to solve analytically. A range of numerical methods are available, including nondeterministic algorithms such as simulated annealing and tabu search. In this paper we will demonstrate the use of co-evolving *genetic algorithms* (GAs), an approach pioneered in Özyildirim, 1996.

2. MODEL

This paper pursues a market approach to explaining deforestation, where three players are treated as profit maximizers. *Player One* represents the commercial logger who makes profits by logging for the government. *Player Two* represents the intermediary responsible for enforcing the government's forest property rights. *Player Three* represents the central government who makes profits from forest rental services[1] and by contracting the commercial logger to harvest state forests and deliver them to a third party. The third party represents a buyer who buys at a fixed price.

The government owns the property rights on the trees. Once the trees are logged by the commercial contractor, they are delivered to the third party who then pays the government an amount p per unit of logged forest. At the same time the government pays the commercial logger ϑ per unit of logging for their services. The third party, however, is assumed to be willing to buy any number of logs for the set price p. There is therefore incentive for the commercial contractor to log an additional amount, h_t, where t is the time period, above the government quota and sell these extra logs directly to the third party. However, since the forests belong to the government, they may wish to enforce their property rights. The government chooses a level of expenditure on property rights enforcement k_t. The commercial logger then faces a chance of being detected and fined $\tilde{f}$ for the illegal logging. All players discount future profits to account for their time preference of money. In each discrete time period t, *Player One* makes a profit of

$$u_t^1(h_t) = h_t(p-c) + Q(x_t)\,(\vartheta - c) - h_t\Phi(h_t, k_t, x_t)\tilde{f} \tag{1}$$

while *Player Three* makes a profit of

$$u_t^3(k_t) = Q(x_t)(p-\vartheta) - k_t + B(x_t) + h_t\Phi(h_t, k_t, x_t)\tilde{f} \tag{2}$$

where x_t is the remaining forest stock in the region, c the fixed cost of harvesting a unit of forest, $\Phi(\cdot)$ the chance of being caught and fined for illegal logging, $Q(x_t)$ the unit quota level set by the government[2], and $B(x_t)$ the forest rental services. All unit measurements refer to an area.

2.1 Allowing for Corruption

With zero corruption the functions (1) and (2) may be used to solve for optimal levels of h_t and k_t. However, empirical evidence suggests that no corruption is an unlikely scenario in both South East Asian and industrialized countries alike. The exogenous parameter θ is introduced to account for the degree to which the intermediary cooperates with the commercial loggers (that is, the level of corruption). Allowing a corrupt intermediary then requires the endogenization of the imposed fine level. The term $\tilde{f}$ in (1) is used to denote the monetary penalty enforced by the intermediary. However, with corruption the intermediary may choose to enforce $f_t \in [0, \tilde{f}]$.

The actual objective functions pursued by each player will consist of the weighted payoffs indicative of the cooperative arrangements. *Player One* will only pay the proportion θ of monetary penalties. *Player Three*, on the other hand, will only receive the proportion θ of the monetary fines imposed. *Player Two* does not generate payoffs from any form of logging, and instead chooses f_t in pursuit of either government or logger profits (presumably in return for bribe payments) depending on which player they cooperate with. Incorporating the weights for various degrees of cooperation, the objective functions for the logger and government may be expressed as

$$\textit{One:} \max_{h} \sum_{t=0}^{\infty} \overbrace{\{h_t(p-c) + Q(x_t)\,(\vartheta - c) - \theta h_t f_t \Phi(h_t, k_t, x_t)\}}^{\pi_{1,t}} \Omega_t, \quad (3)$$

$$\textit{Three:} \max_{k} \sum_{t=0}^{\infty} \overbrace{\{Q(x_t)(p-\vartheta) - k_t + B(x_t) + \theta h_t f_t \Phi(h_t, k_t, x_t)\}}^{\pi_{3,t}} \Omega_t, \quad (4)$$

subject to $x_{t+1} - x_t = R(x_t) - h_t - Q(x_t), 0 \leq x_t \leq K, 0 \leq h_t \leq \tilde{h}, 0 \leq f_t \leq \tilde{f}$. Here Ω_t are the discount factors used in respective decision processes, K is the maximum forest capacity, $\tilde{h}$ is the maximum level of illegal harvests, and $R(x_t)$ is the forest regrowth. The objective function for the intermediary weights the profits per time period of the other players by the level of θ:

$$\textit{Two:} \max_{f} \sum_{t=0}^{\infty} \{(1-\theta)\,\pi_{1,t} + \theta \pi_{3,t}\}\, \Omega_t, \quad (5)$$

In particular, if θ is 1 then the intermediary is completely cooperative with the government and so has the same objective function as *Player One*. Similarly, if θ is 0 then the intermediary has the same objective as the logger. Only these two extreme cases are analyzed in this paper. Modelling partial corruption is a more complicated problem beyond the scope of this report.

As a result of the full corruption ($\theta = 0$), the game can be simplified (see Section 4). However, the model is presented in its general three-player form to aid the discussion of the nature of the GA and to introduce partially cooperative solutions ($0 < \theta < 1$) as an avenue for future work.

2.2 Chance of Being Caught and Fined for Illegal Logging

Assuming *Players One* and *Two* are risk neutral[3], the chance of being caught and fined $\Phi(\cdot)$, combined with the fine level f_t, represents an expected unit cost incurred by the commercial logger and an expected income made by the government. The function $\Phi(\cdot)$ is assumed[4] to have the form

$$\Phi(h_t, k_t, x_t) = \frac{h_t}{(x_t - h_t)} \frac{(k_t + \rho)\,\beta}{(x_t - h_t)}, \quad (6)$$

where β is a parameter representing the effectiveness of k_t worth of expenditure on property rights enforcement, and ρ some exogenous level of expenditure on enforcement. Given k_t, the commercial logger can expect greater expenses as h_t increases. Given h_t, the government can expect greater income from fines as k_t increases. However, as implied by 4, k_t also represents an expense to the government. Note that the function $\Phi(\cdot)$ implicitly assumes that property rights enforcement is a flow rather than a stock variable. This is consistent with Clarke et al., 1993.

2.3 Forest Regrowth, Quota Levels and Forest Rental Services

Regrowth is assumed to be a logistic function dependent on x_t. The maximum capacity of the forest in the region is defined above as K. Forest stock may therefore range between 0 and K. The regrowth function is

$$R(x_t) = \alpha x_t \left(1 - \frac{x_t}{K}\right), \tag{7}$$

where α refers to an intrinsic growth rate.

The quota levels $Q(x_t)$ are also assumed to be dependent on x_t. It is assumed the government would like to sustain a target level of forestation Υ. Therefore, if forest stock falls below Υ, quotas are set less than regrowth $R(x_t)$. Alternatively, if the forest stock is greater than Υ, then quotas are set greater than $R(x_t)$. The level of Υ may be more appropriately studied from a political economy perspective. For this reason, Υ is assumed exogenous to the model. The assumed function of $Q(x_t)$ is defined piecewise by

$$Q(x_t) = \begin{cases} \alpha x_t(1 - \frac{x_t}{K})\frac{x_t}{\Upsilon}, & x_t \leq \widetilde{x} \\ \frac{\alpha}{\Upsilon}\frac{4k^2}{27}, & x_t > \widetilde{x} \end{cases} \tag{8}$$

Here $\widetilde{x} = \frac{2k}{3}$ is the forest stock where $R(x_t)\frac{x_t}{\Upsilon}$ is maximal. (The maximum value is $\frac{\alpha}{\Upsilon}\frac{4k^2}{27}$.) We assume that quota levels will stay at this maximum for higher stock levels, as given in Equation 8.

Forest rental services $B(x_t)$ are defined $B(x_t) = x_t^\gamma$, where $0 \leq \gamma \leq 1$ such that $B'(x_t) > 0$ and $B''(x_t) < 0$. The function $B(x_t)$ is therefore consistent with Ehui et al., 1990.

3. GENETIC ALGORITHMS

The modelling problem presented in this paper belongs to a class of problems amenable to the maximum principle in the form of the Isaacs equation (Kamien and Schwartz, 1991). The first order and co-state equations are, however, third order polynomial, piecewise and non-linear in both the state and control variables, making analytical solutions hard to derive. Therefore, a genetic algorithm (GA) approach is explored as an avenue to analyzing the problem numerically.

3.1 Background

Genetic algorithms (GAs) are a non-deterministic approach to learning and optimization which draw on analogies with biological evolution (Holland, 1975). Unlike traditional mathematical methods which iterate a single solution, a GA evolves a population of solutions in the context of a fitness landscape, where fitness is given by the objective. Fitter solutions are made more likely to contribute to the next generation, giving the evolution its drive. At each stage genetic operators are used to simulate mating and mutation to produce a new population which inherits information from its parent population.

The solutions in a GA are captured by a series of "chromosomes". In traditional GAs these are typically binary strings (unlike biological chromosomes which are base 4) with the genetic operators combining parent strings to form children strings (Mitchell, 1996). However, as the strategies in the corruption game are real-valued, representing them as binary strings limits their accuracy and is somewhat artificial. In this paper strategies are represented as real-strategy sets, $\{h_1, h_2, .., h_T\}$ for *Player One,* $\{f_1, f_2, ..., f_T\}$ for *Player Two* and $\{k_1, k_2, ..., k_T\}$ for *Player Three.* That is, each chromosome corresponds to a strategy, with the T genes in the chromosome giving the decisions for each of T periods of the game. The GA must then be modified to use genetic operators which manipulate such real vectors. Such modifications are described in the following section.

3.2 Floating Point Operators

As stated earlier, the algorithm used in this paper was pioneered by Özyildirim, 1996. Several modifications to Özyildirim's approach have been introduced here to make the algorithm more suitable to the problem. As noted before, a floating point (FP) representation was used for each strategy rather than a sequence of binary digits.

FP strategies are "mated" using *arithmetical crossover* (Michalewicz, 1999, p112). This takes two parent strategies v_i and w_i, with i referring to the ith iteration, and introduces two new strategies v'_{i+1} and w'_{i+1}, into the population using

$$\begin{aligned} v'_{i+1} &= av_i + (1-a)\, w_i \\ w'_{i+1} &= (1-a)\, v_i + aw_i \end{aligned}$$

where a is random number in $[0,1]$. In fact, to help explore possible strategies, a is sometimes allowed to go beyond this range early in the evolution process.

This paper also employs *non-uniform mutation* (Michalewicz and Janikow, 1996). Controlling mutation in deriving empirical games using a GA is a significant problem. Each player learns which seemingly better strategies do not lead to optimization due to the other player's reactions. Mutation can therefore mimic a player forgetting that a given strategy, made available once again due to mutation, does not lead to an optimal outcome. On the other hand, mutation is still required to add diversity into the population. Non-uniform mutation provides balance. In this scheme a mutated gene can either move up (towards its specified upper limit) or down (towards 0), with the size of the movement given by:

$$1 - r^{\left(1-\frac{i}{G}\right)^b}$$

Here r is uniformly random on $[0,1]$, i is the number of generations performed, and G is the planned total number of generations. This effectively scales down mutation over the course of the evolution, with b controlling the rate of reduction. Michalewicz, 1999 reports that non-uniform mutation outperforms other mutation operators.

3.3 Using a GA to Approximate Steady-State Solutions

The player objectives require estimates of payoffs to perpetuity, which in turn require calculations of steady states. Alemdar and Özyildirim, 1998 have shown how a GA can be used to approximate infinite horizon open-loop steady-state Nash solutions by truncating an infinite time horizon game to T periods and discounting the final period payoffs via an annuity. Alemdar and Özyildirim's approach allows the GA maximum time periods T to converge towards the steady state, with consideration given to computational costs associated with increasing T. The discount factor in the last time period T is then modified to represent an appropriate infinite time annuity. Here the discrete time discount factor used is $\Omega_t = 1/\left(1+\delta\right)^t$. In the final period of the game, T, the payoffs are therefore discounted by the annuity

$$\Omega_T = \left(1+\delta\right)^{1-T}/\delta.$$

Modifying Equations 3 and 4 gives the finite objectives

$$\textit{Player One:} \qquad \max_{h_t} \sum_{t=0}^{T} \pi_{1,t}\Omega_t \tag{9}$$

$$\textit{Player Three:} \qquad \max_{f_t} \sum_{t=0}^{T} \pi_{3,t}\Omega_t \tag{10}$$

All variables are held constant from T onwards, allowing the GA to mimic maximization to perpetuity. Assuming a unique steady state solution in the model presented above, the values of x_t, k_t, and f_t can be assumed constant past T and we will take $x^* = x_T, k^* = k_T$, and $f^* = f_T$ to be approximations to the steady-state solutions. The value of h_t is set at a level consistent with the steady-state approximation of x_t, giving

$$h^* = R(x_T) - Q(x_T).$$

3.4 The Genetic Algorithm

The algorithm used in this paper consists of three GAs. Each GA is associated with a unique player. More generally, however, the approach uses a GA for each control variable, which in this paper turns out to mean each player. Each GA uses the following steps:

1 Initialize a random population of strategy sets, and update a shared memory with a random strategy set.

2 Calculate the fitness of each strategy set against the strategies of other players held in the shared memory.

3 Perform crossover and mutation to population.

4 Wait until all other GAs have completed their respective fitness calculations.

5 Update the shared memory with fittest strategy set.

6 If the maximum number of generations G is reached then stop; otherwise return to step 2.

Note that the number of generations used is unrelated to the number of discrete time periods used in the model, except in that using more time periods, to get a better idea of the steady state, will require more generations of evolution to achieve convergence of the algorithm.

The above algorithm is discussed by Özyildirim and Alemdar, 2000 under the theme of *divide and conquer*. This is based on the equivalence of centralized and decentralized (parallel) decision making when the same objective function is pursued. This means for a two-player game, each player with a single control, two GAs pursuing the same objective function will produce the same results as though solving a single player problem with two control variables. In a game setting, different players will presumably pursue the same objective functions only when they are in full cooperation. It is this property that is exploited to analyze corruption in this paper. When θ is zero (full corruption) the GA mimics a two player game, with *Player One* and *Two* combining to choose h_t and f_t in order to maximise $h_t(p-c) + Q(x_t)(\vartheta - c)$ and *Player Three* choosing k_t in maximizing $Q(x_t)(p-\vartheta)-k+B(x_t)$. When θ equals one (zero corruption) then the two player game consists of *Player One* choosing h_t in maximizing $h_t(p-c)+Q(x_t)(\vartheta-c)-f_t\Phi(h_t, k_t, x_t)$ and *Players Two* and *Three* combining to choose f_t and k_t in maximizing $Q(x_t)(p-\vartheta) - k_t + B(x_t) + f_t\Phi(h_t, k_t, x_t)$. In either case the problem amounts to a non-cooperative two-player game. GAs have been successfully applied to non-cooperative games (Mcallister and Bulmer, 2000b; Özyildirim, 1997) and cooperative games (Alemdar and Özyildirim, 1998).

4. RESULTS

The main aim of this paper is to analyze the *impact* of corruption in a stylized South-East Asian timber industry. This impact is measured by changes in total discounted payoffs, that is the discounted profits of the commercial logger (Equation 3) and the central government (Equation 4). The intermediary pursues either logger or government profits and therefore their objective function is not considered. The GA, discussed above, is used to solve the model assuming either full or zero corruption over varying levels of starting forest stock x_0. Other parameters in the model are assumed as detailed in Table 1. These are used to provide a plausible illustration of the theoretical model in practice. The emphasis here will be on exploring the qualitative behaviour of the optimal strategies rather than studying a particular timber industry.

With full corruption ($\theta = 0$) both $\Phi(\cdot)$ and f_t cancel from all objective functions. In addition the role of k_t is only as a cost to the government - its optimal value is then clearly zero[5]. Therefore, assuming a unique solution, the optimal solution should be for x_t to move to its steady-state as quickly as possible (that is, a *Most Rapid Approach Path*; see Kamien and Schwartz, 1991 for discussion). In this paper the full

Table 1. Assumed parameter values used in GA trials

Parameter	α	ϑ	c	γ	δ	ρ
Value	0.20	0.50	0.30	0.40	0.08	0.01

Parameter	β	p	K	$\tilde{f}$	$\tilde{h}$	Υ
Value	50	1	100	10	4	30

corruption case has been solved by applying the GA to the general model. It is reassuring that the theoretical characteristics are demonstrated by the model's convergence paths as shown in Figure 1. The control h_t is optimal at its maximum $\tilde{h}$ until x_t has approached its steady-state[6]. The control k_t remains at zero along the entire convergence path while f_t's value appears somewhat random owing to it having no impact in the model.

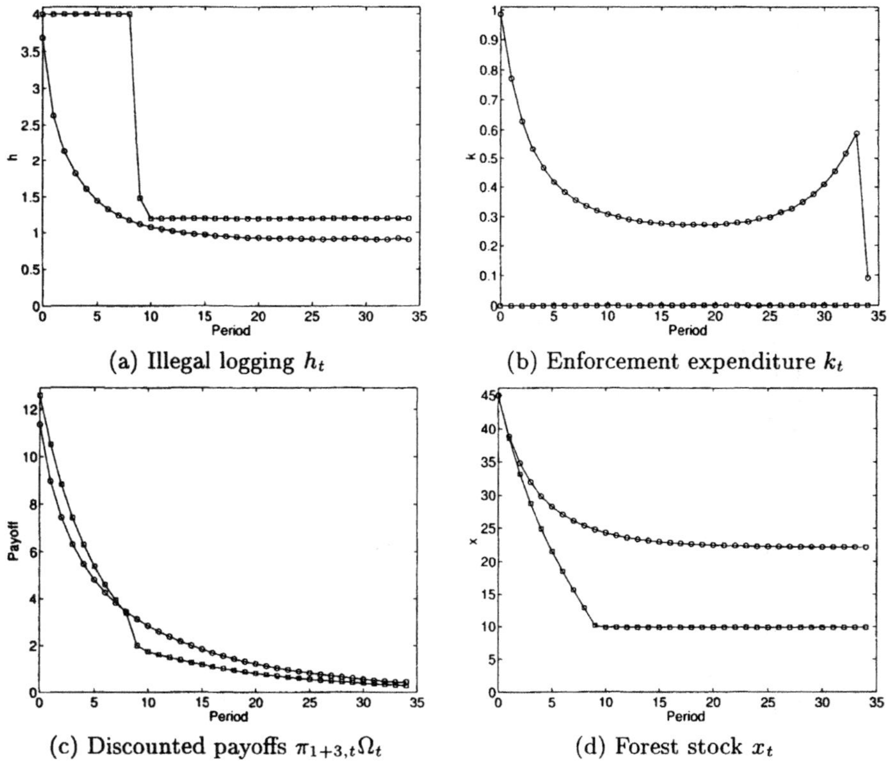

(a) Illegal logging h_t

(b) Enforcement expenditure k_t

(c) Discounted payoffs $\pi_{1+3,t}\Omega_t$

(d) Forest stock x_t

Figure 1. Convergence paths for zero corruption (○) and full corruption (□)

Table 2. Discounted payoffs

	Zero corruption		Full corruption		Impact of corruption (%)	
x_0	Government	Total	Government	Total	Government	Total
15	58.46	67.58	38.91	55.54	-33.4	-17.82
30	69.02	83.22	47.48	72.71	-31.2	-12.63
45	76.74	95.21	57.67	88.47	-24.9	-7.09
60	83.67	108.02	67.05	101.98	-19.9	-5.59

Figure 1 shows that the zero corruption ($\theta = 1$) case's convergence paths differ markedly from the case of full corruption. Firstly, while with full corruption f_t cancels from the model, with zero corruption f_t remains optimal at its maximum $\tilde{f}$ along the entire convergence path. Secondly, the convergence of h_t does not follow a Most Rapid Approach Path. This is signature to the added complexity of the zero corruption case. Thirdly, the path of k_t does not appear to converge smoothly onto its steady-state approximation. McAllister and Bulmer 2000a analyzed a similar model and found this *kink* characteristic to indicate that the game was given insufficient discrete time periods to converge onto a steady-state. However, with three controls and 35 periods, the search space is already large. In addition, any related inaccuracies will be concentrated towards the heavily discounted end of the game where the GA forces the approximation of steady-states. Therefore, no adjustments are made on the basis of such inaccuracies.

The payoffs along the entire convergence paths, with those in the final period T summed to perpetuity, are totalled to give the results displayed in Table 2. This shows that, regardless of the starting level of forest stock, the full corruption case is consistently associated with lower levels of total payoffs. The percentage loss due to corruption is also shown to be greater when smaller forest stocks are considered. Table 2 also shows government payoffs[7]. As may be expected, the model shows that the majority of payoff losses associated with corruption are felt by the government. Over the range of starting forest stocks considered, the government losses amount to between 33.4 and 19.9 percent.

4.1 Genetic Algorithm Accuracy

Part of the reason why a GA may succeed in solving complex problems, where an analytical approach fails, is that the GA works with the objective functions rather than any necessary conditions. The analytical difficulties presented by complex problems, however, relate to deriving solutions from the necessary conditions, not the derivation of the necessary conditions themselves. This means the necessary conditions can be used to derive a condition that should be equal to zero under optimality (McAllister and Bulmer, 2000c). How far this condition deviates from zero can then be used to test the accuracy of the numerical results and their theoretical consistency with derived open-loop Nash equilibria. Additionally, steady states can be derived analytically for the full corruption ($\theta = 0$) case, allowing a further check of the accuracy.

As will be discussed later, $G = 2.5$ million generations were used for each run of the GA. Figure 2 shows appropriate necessary conditions applied to the steady-state approximations taken every 12 500 generations. Conditions are applied to both the case of zero and full corruption with forest starting stock x_0 set at 45.

Figure 2(a) refers to the zero corruption case. The condition used here was derived by differentiating the relevant Lagrangian function with respect to k_t. This figure shows that as the GA iterates, the condition tends closer and closer to zero, with zero indicating optimality. The decreasing trend of the condition is confirmed by the exponential trend line in Figure 2(a). When the GA stopped after G generations, the condition was found to hold to within 0.0047 error.

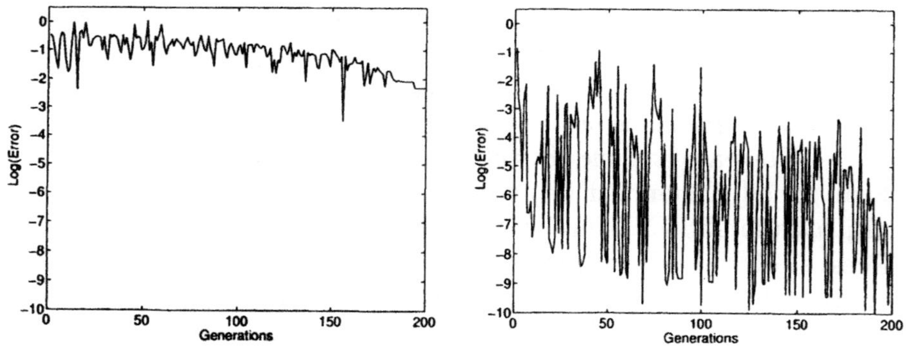

(a) Zero corruption (b) Full corruption

Figure 2. Accuracy of GA steady states during evolution (over 2.5 million generations)

Figure 2(b) refers to the full corruption case. The condition used here was derived by differentiating the relevant Lagrangian with respect to x_t giving a co-state equation. With $\theta = 0$ this co-state equation should equal zero in the event of optimality. This figure shows that as

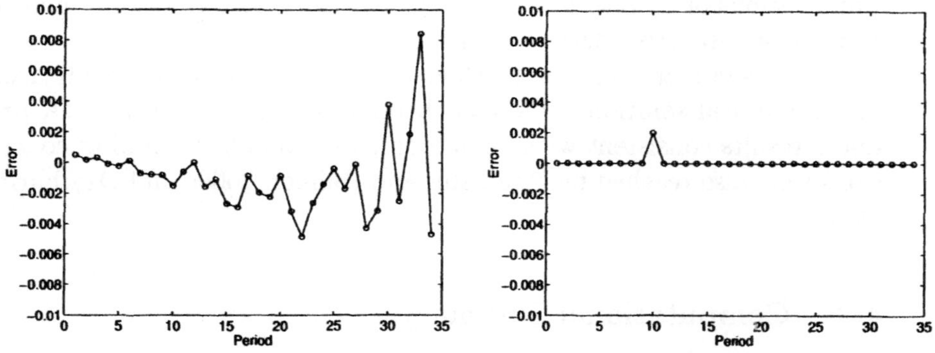

(a) Zero corruption

(b) Full corruption

Figure 3. GA accuracy over time periods

the GA iterates the optimality condition decreases, eventually falling to 2.2×10^{-10} error. In comparison to the zero corruption case, Figure 2(a) indicates that with full corruption more accurate approximations are produced. This is most likely due to the added complexity of the zero corruption case, as discussed above.

Figure 3 shows the same derived conditions used above applied to the final estimates (that is, after G generations) of the entire convergence paths for both the case of zero and full corruption. Figure 3(b) indicates highly accurate results for the full corruption case, a peak of 0.0020 error in time period 9 being the only noticeable deviation. Figure 3(a) shows a more noticeable and erratic error for the zero corruption case. At worst, the condition fails with 0.0145 error, the average error being 0.0045. Figure 3(a) also shows the largest errors for the zero corruption case occur towards the tail of its convergence path. This may have occurred for two reasons. Firstly, the GA evolves population strategies based on fitness. The later period approximations are more heavily discounted and therefore make a lessor contribution to a strategy's fitness. Secondly, in later periods the GA tries to force convergence onto a steady-state. Errors may be induced by forcing convergence within an insufficient time span.

An additional check of the GA's approximations can be obtained through the full corruption case. For the full corruption case, setting $\theta = 0$ simplifies the model substantially. In addition, as discussed above, the optimal values for k_t are clearly zero, effectively simplifying the model further. As a result, open-loop steady-states can be derived analytically as $h^* = 1.195785237$ and $x^* = 9.910820796$. The GA's numerical approximations of h^* and x^* deviate by -2.5$\times 10^{-7}$ and -3.4$\times 10^{-6}$ percent from their respective derived solutions.

In summary, the results show that the GA produces highly accurate numerical solutions. We can therefore conclude that the GA produces results consistent with open-loop Nash equilibria. Similar conclusions were also reached by Mcallister and Bulmer, 2000c and Özyildirim, 1997.

4.2 Computational Summary

The GA was run over G = 2.5 million generations, each GA evolving a population of 40 real-strategy sets. A mutation scaling factor b of 0.05 was used. However an *elitist* strategy was employed, whereby the fittest 10 of the 40 strategy sets were passed into subsequent generations unchanged. The GA was programmed in C and run as a batch on a multi-user SGI Origin 2000 computer with 64 MIPS R10000 CPU's and 16 Gigabytes of memory. While the algorithm treated each of the three control variables as an independent GA, parallelization of the code appeared to induce greater computational costs than benefits. Therefore the code was run as three *virtually* independent GAs, but from a single source code. The program allocated no more than 3 Megabytes of memory and on average took 4 hours 10 minutes to run, depending largely on other user activity.

5. IMPLICATIONS FOR SOUTH-EAST ASIA AND FURTHER WORK

While unsustainable agricultural practices, such as slash and burn, contribute to deforestation in developing South-East Asian countries, as discussed in the introduction, market driven logging is generally the major cause. Much of this activity is illegal and conducted in cooperation with corrupt bureaucrats (Dudley et al., 1995). From a government perspective, illegal logging may be regarded as rent dissipation (as it does not generate forest royalties (see, for example, the 1992 Laos Ministry of Agriculture and Forests framework for forestry management)). From an economy-wide perspective, the impact of illegal logging is less clear.

Providing the illegal logging is carried out by residents, which is not always the case in South-East Asia (Stuart-Fox, 1997), illegal logging may be considered as some form of transfer from a government monopoly to the illegal loggers and corrupt bureaucrats. Economics literature, however, argues that corruption may negatively impact on an economy due to its distortive effect on a government's presumably corrective role (Tanzi, 1997). The model results presented above agree with such a negative impact. In the stylized three-tiered timber industry not only does the government suffer rent losses when corruption is chronic, but total payoffs to the economy are also negatively affected.

The broad implication of the model is that South-East Asian nations may gain from pursuing anti-corruption policies within their forestry sectors. Many South East-Asian countries are in fact already pursuing such policies. However, corruption in South-East Asia is endemic, largely due to strong patronage and family ties. As such, the removal of corruption may, at best, be incremental. In light of this, further modelling avenues may be required to more fully capture the dynamics of the system. In its general form, the model was presented with the government intermediary weighting logger and government profits partially depending on their level of corruption. The numerical simulations, however, were only conducted for the cases of either full or zero corruption. The work was limited to these two extreme cases due to algorithmic issues. As an avenue for future work, modelling the impact of partially corrupt intermediaries would allow an analysis of an economy's transition from corrupt to less corrupt. This may prove useful in deriving specific policy implications.

6. CONCLUSION

This paper has analyzed the impact of corruption in a stylized industry structure representative of many South-East Asian nations. Corruption is analyzed using a three-player open-loop difference game consisting of a commercial logging contractor, a central government, and a regional government intermediary responsible for enforcing the government's forest property rights. The intermediary is assumed to either cooperate fully with the contract logger, signifying full corruption, or with the central government for the zero corruption case.

The game is piecewise non-linear in both the control and state variables. While necessary conditions can be derived from the model, obtaining analytical solutions proved difficult. A genetic algorithm approach was therefore used to solve the model numerically. The numerical solutions were then substituted into the necessary conditions to demonstrate their consistency with open-loop Nash equilibria.

The numerical results showed that corruption may negatively impact on both government and total payoffs due to mis-management of forest resources. The implication is that South-East Asian nations may gain from pursuing anti-corruption policies. The paper, however, suggests an analysis of partially corrupt intermediaries to help improve the understanding of the problem and derive specific policy implications.

7. ACKNOWLEDGEMENTS

We thank the anonymous referees for their helpful suggestions.

Notes

1. Forest rental services refer to non-timber forest benefits. In addition to timber, South-East Asian governments may also profit from other forest products such as tourism and hydropower. ? have modeled rental forest services in a dynamic setting.

2. The profit function (1) implies that the contract logger will always harvest the full quota $Q(x)$. In South-East Asia, most illegal logging by commercial loggers is likely to be concealed by mis-reporting (?). For example a contractor delivering a quota of 10 logs to a sawmill, may deliver the 10 logs plus an additional quantity of "illegal" logs. Thus, the model assumes that the commercial logger cannot undertake illegal logging without meeting its quota obligations.

3. The assumption of risk neutrality is a reasonable one for poorer Asian countries. This is changing for Thailand, but for Laos, Cambodia, and similar countries, the only risks involved, with respect to the government, are financial.

4. The probability of the loggers being active in an area is assumed to be $P(L) = h/(x-h)$. The denominator is $x-h$ rather than just x to better capture the relative scale of the illegal logging, since at the end of the discrete time period there will only be $x-h$ forest remaining. Similarly, the chance of enforcement activity occurring in the area is assumed $P(E) = (k+\rho)\beta/(x-h)$. Assuming $P(L)$ and $P(E)$ are independent, $P(L\text{and}E) = P(L)P(E)$.

5. In the presence of full corruption, the model suggests that the government should not spend anything on enforcing their property rights. The obvious alternative is to spend money to reduce corruption in some way. This possibility has not been included in the model here.

6. Despite no expenditure through k_t on enforcement, the exogenous parameter ρ does allow some chance of being caught (see Equation 6). This becomes significant when forest levels are nearly depleted or illegal harvesting rates are very high.

7. Net losses experienced by the commercial loggers cannot be analyzed by the model. This is because it is assumed that the loggers bribe the intermediariy in the case of corruption. The model does not attempt to quantify the bribe and therefore the discounted profits under corruption cannot be compared to the case of no corruption.

References

Alemdar, N. M. and Özyildirim, S. (1998). A genetic game of trade, growth and externalities. Journal of Economic Dynamics and Control, 22(6):811-832

Clarke, H. R., Reed, W. J., and Shrestha, R. M. (1993). Optimal enforcement of property rights on developing country forests subject to illegal logging. Resource and Energy Economics, 15(3):271-293.

Dudley, N., Jeanrenaud, J.-P., and Sullivan, F. (1995). Bad Harvest?: The Timber Trade and the Degradation of the World's Forests. Earth-scan, London.

Ehui, S., Hertel, T. and Preckel, P.V. (1990). Forest resource depletion, soil dynamics and agricultural productivity in the tropics. Journal of Environmental Economics and Management, 18(2):136-54.

Holland, J. H. (1975). Adaptation in Natural and Artificial Systems. University of Michigan Press, Ann Arbor.

Kamien, M. I. and Schwartz, N. L. (1991). Dynamic Optimization: The Calculus of Variations and Optimal Control in Economics and Management. Volume 31 of Advanced Textbooks in Economics. North-Holland, Amsterdam, 2nd edition.

Laos. Ministry of Agriculture and Forests (1992). New system of resource management - policy and institutional framework.

McAllister, R. and Bulmer, M. (2000a). Approximating Difference Game Open-Loop Steady States Using Reiterative Genetic Algorithms. Economics Discussion Papers. University of Queensland Press, Brisbane.

McAllister, R. and Bulmer, M. (2000b). Forest Property Rights Enforcement in Lao PDR: A Steady State Analysis Using Genetic Algorithms. Economics Discussion Papers. University of Queensland Press, Brisbane.

McAllister, R. and Bulmer, M. (2000c). Using genetic algorithms to solve open-loop difference games of illegal logging. In Kozan, E. and Beard, R., editors, 3rd Operations Research Conference, pages 82-97, Brisbane. ASOR, Qld Branch.

Michalewicz, Z. (1999). Genetic Algorithms + Data Structures = Evolution Programs. Springer, Berlin, 3rd edition.

Michalewicz, Z. and Janikow, C. Z. (1996). GENOCOP: A genetic algorithm for numerical optimization problems with linear constraints. Communications of the ACM, 39(12).

Miller, K. and Tangley, L. (1991). Trees of Life: Saving Tropical Forests and their Biological Wealth. Beacan Press, Boston.

Mitchell, M. (1996). An Introduction to Genetic Algorithms. MIT Press, Cambridge, Mass.

Özyildirim, S. (1996). Three-country trade relations: A discrete dynamic game approach. Computer and Mathematics with Applications, 32(5):43-56.

Özyildirim, S. (1997). Computing open-loop noncooperative solution in discrete games. Journal of Evolutionary Economics, 7(1):23-40.

Özyildirim, S. and Alemdar, N. M. (2000). Learning the optimum as a Nash equilibrium. Journal of Economic Dynamics and Control, 24(4):483-99.

Stuart-Fox, M. (1997). A History of Laos. Cambridge University Press, Cambridge.

Tanzi, V. (1997). The Economics of Organised Crime, chapter Corruption: Arms's Length Relationships and Markets, pages 161-182. Cambridge University Press, Cambridge.

A GENERAL FRAMEWORK FOR CONSTRUCTIVE META-HEURISTICS

Marcus Randall

School of Information Technology, Bond University, QLD 4229

Abstract Meta-heuristic search algorithms, by their very nature, are applicable across a range of optimisation problems. In practice however, meta-heuristics have been tailored to solve particular problems. Recent work by Randall and Abramson (2001b) has successfully shown that iterative meta-heuristics, such as simulated annealing and tabu search, can be successfully generalised to solve a range of problems without modification though the use of a uniform representation language. Constructive meta-heuristics, such as ant colony optimisation and generalised random adaptive search procedures, pose more substantial problems to achieve this same level of generalistaion. This paper investigates the issues involved and suggests some measures by which generalisation could be achieved.

Keywords: Constructive meta-heuristics, Ant colony optimisation, Greedy randomised search procedures.

1. INTRODUCTION

According to Abramson and Randall (1999:3), "For many years, the holy grail of operations research has been an efficient integer solver which can be applied to a wide range of problems with little modification. In this model, a user specifies a problem using some mathematical notation which is processed directly by the solver."

Such systems can take an arbitrarily specified optimisation problem and solve for an optimal/near optimal solution without the need for costly code modifications. In terms of techniques such as the simplex method and branch and bound, there are well-established commercial products. However, for integer optimisation problems (notably combinatorial optimisation problems (COPs)), these systems are inadequate for solving medium to large problems (Randall, 1999).

Recent work by Randall and Abramson (2001b) has shown that iterative meta-heuristic search techniques share a number of common features such that a generic solver system can be developed. It was found that an algebraic language based on linked lists closely modelled COPs as they are problems in which different arrangements of discrete objects must be explored. The solver system implemented by Randall and Abramson (2001b) consists of core meta-heuristic engines as well as common functions to handle constraints, evaluate cost functions and apply common neighbourhood transition operators. The implementation showed that the system could produce competitive solutions to difficult COPs in reasonable computational time. This system has the additional advantage that only a high level description (based on the well-known GAMS) of the problem needs to be provided. Normally, solving optimisation problems requires substantial development time in order to produce a competitive heuristic/meta-heuristic program. This system was shown to be able to represent a broad range of optimisation problems, from simple problems (such as the travelling salesman problem (TSP)) to real-world problems. While the representation scheme is general, it has not yet been proven to be able to model all problems.

Iterative techniques such as simulated annealing (SA) and tabu search (TS) share many similarities as they are based on neighbourhood search techniques. The question becomes whether constructive techniques can be generalised in a similar way. Unlike an iterative technique that continually refines a solution by using a defined set of local transition operators, a constructive technique gradually builds a solution by adding elemental components.

An integrated system offering a range of meta-heuristic search engines could be used as an alternative to products like ILOG's CPLEX and IBM's OSL, with the advantage that relatively large combinatorial problems could be solved efficiently. This paper will outline some strategies to help generalise constructive techniques such as ant colony optimisation (ACO) and greedy randomised adaptive search procedures (GRASP) with particular reference to COPs. As such, it may be considered as a first step toward the aforementioned goal. This paper is organised as follows. Section 2 describes the two most common constructive meta-heuristics while Section 3 discusses how generalisation can be achieved across these two techniques. Conclusions are drawn in Section 4.

2. CONSTRUCTIVE META-HEURISTICS

Iterative meta-heuristics such as SA, greedy/hill climbing search and TS are techniques that transform a feasible solution into another feasible solution using a neighbourhood transition operator. This process is repeated over a number of iterations. However, constructive meta-heuristics start from an empty solution and gradually build a complete solution. Due to the available standard local search neighbourhoods and the fact that feasibility can be preserved throughout the search process, iterative meta-heuristics are relatively easy to generalise. Constructive meta-heuristics, on the other hand, cannot make these claims.

The most widely used constructive meta-heuristics are ACO and GRASP. Both techniques are described in general terms so that generalisation can be discussed in subsequent sections.

2.1 ACO

ACO is an umbrella term encompassing a number of techniques including Ant System (AS) (Dorigo et al, 1996), *MAX-MIN* Ant System (Ramalhinho and Serra, 1998) and Ant Colony System (ACS) (Dorigo and Gambardella, 1997). Here ACS is examined as a representative ACO meta-heuristic.

ACS is a population based technique that is designed to simulate the ability of ant colonies to determine shortest paths to food. Although individual ants possess few capabilities, their operation as a colony is capable of complex behaviour.

Within each step of an ACO process, each ant adds an *element* to its solution until some *termination condition* is satisfied. In terms of the TSP (Lawler et al 1985), the QAP (Burkard, 1984) and the job sequencing problem (JSP) (Glover and Laguna, 1997), "element" and "condition" are interpreted according to Table 1.

Table 1. "Elements" and "condition" for the TSP, QAP, JSP.

Problem	*Element*	*Condition*
TSP	an edge	until a Hamiltonian tour is constructed
QAP	a facility	until all the facilities have been placed
JSP	a job	until all jobs have been assigned an order

Using this general terminology, the basic operations of ACS can be described thus. In discrete time steps, allow each ant to add one element to its solution until the termination condition is satisfied. After adding an element, the amount of *pheromone* on that element is modified. Pheromone is a scalar quantity that allows ants to communicate with one another about the utility of an element. The accumulated strength of pheromone on element i is denoted by $\tau(i)$.

At the commencement of each time step, Equation 1 is used to select the next element s to add to the solution. The elements that are still legal to add to the solution (i.e. those that won't break constraints) by ant k at step r are indexed by $J_k(r)$. $\eta(s)$ is the measure of how good element s would be for the solution, in other words an incremental cost measure (a general calculation strategy for which is given in Section 3.3). In terms of the TSP, this would correspond to the distance between two cities.

Equation 1(a) is a highly greedy selection technique favouring the best combination of pheromone level and cost. In order to ensure that other elements are incorporated into an ant's solution, Equation 2 selects elements probabilistically. This is essential to avoid premature convergence to a small group of elite solutions. The randomness of the selection process is governed by a uniformly distributed number q and a parameter q_0.

$$s = \begin{cases} \arg\max_{u \in J_k(r)} \{\tau(u)[\eta(u)]^{\beta}\} & \text{if } q \le q_0 \quad \text{(a)} \\ \text{Equation 2} & \text{otherwise} \end{cases} \tag{1}$$

$$p_k(r,s) = \begin{cases} \dfrac{\tau(s)[\eta(s)]^{\beta}}{\sum_{u \in J_k(r)} \tau(u)[\eta(u)]^{\beta}} & \text{if } s \in J_k(r) \\ 0 & \text{otherwise} \end{cases} \tag{2}$$

For minimisation (maximisation) problems, the parameter β is negative (positive) so that smaller (larger) costs of each s are favoured. The pheromone level of the selected element is updated according to the local updating rule 3.

$$\tau(s) \leftarrow (1 - \rho) \cdot \tau(s) + \rho \cdot \tau_0 \tag{3}$$

Where:

ρ is the local pheromone decay parameter, $0 < \rho < 1$.

τ_0 is the initial amount of pheromone deposited on each of the elements.

Upon conclusion of an iteration (that is all ants have constructed a feasible solution), elements that compose the best solution are rewarded with an increase in their pheromone level. Mathematically, Equation 4 relates the new pheromone level to the old.

$$\tau(s) \leftarrow (1 - \gamma) \cdot \tau(s) + \gamma \cdot \Delta\tau(s) \tag{4}$$

Where:

$\Delta\tau(s)$ is used to reinforce the pheromone level of the elements that compose the best solution found (see Equation 5). L is the cost of the best solution[1] and Q is a constant that is usually set to 100 (Dorigo et al, 1996).

γ is the global pheromone decay parameter, $0 < \gamma < 1$ and

$$\Delta\tau(s) = \begin{cases} \dfrac{Q}{L} & \text{if } s \in \text{solution by the best ant} \\ 0 & \text{otherwise} \end{cases} \tag{5}$$

Figure 1 shows the general algorithm for ACS.

```
Read Problem Model;
Get Parameters(num_ants, β, γ, Q, L, ρ, q0, τ0, termination_condition);
τi = τ0 (1 ≤ i ≤ |τ|);
While(termination_condition not met)
  % iteration level
  S = set of randomly ordered feasible elements;
  Xk1 = Sk (1 ≤ k ≤ num_ants);
  While(each solution is not complete)
    For(k = 1...num_ants)
      % step level
      Compute η(s) ∀s, s ∈ Jk(r);
      Xkj =Apply Equation 1;
      Compute C(Xk);
      τ =Apply Equation 3;
    End For;
  End While;
  best_ant =Determine the best solution cost from C(Xk)
  ∀k, 1 ≤ k ≤ num_ants;
  τ =Apply Equation 4;
  If((minimisation problem And C(Xbest_ant)< best_cost)
  Or (maximisation problem And C(Xbest_ant) > best_cost)
  Or (this is the first iteration))
    best_cost = C(Xbest_ant);
  End If;
End While;
Output best_cost;
End.
```

Where:

X_{ij} is j^{th} solution element for ant i.

$C(k)$ is the cost of the solution of ant k.

Figure 1. The pseudocode for ACS.

2.1.1 GRASP

GRASP is a search technique that consists of two distinct phases per iteration, namely a *construction* and *local search* phase (Feo and Resende, 1995; Glover and Laguna, 1997; Mavridou at al 1995). The first phase builds a new feasible solution from a list of elements, one element at a time. The list itself is produced by ordering the elements with respect to a greedy function. This list is similar to the $J_k(r)$ list in ACS except it is ordered by cost. A restricted candidate list (RCL) is then produced from the top y% of the candidate list, where y is a parameter of the process. Each element on the RCL has an equal probability of incorporation into the solution. In many instances, this solution will not correspond to a local optimum and can hence be improved. This is achieved by applying a local search procedure (such as hill climbing, SA and TS) to the solution as the second phase. Hence, GRASP may be considered a hybrid constructive/iterative technique. However, the attributes that distinguish GRASP from other meta-heuristics pertain to its construction phase. Figure 2 gives the general pseudocode for GRASP.

```
Read Problem Model;
Get Parameters(x, termination_condition);
While(termination_condition not met)
   % iteration level
   X = ∅;
   While(X is not complete)
      % step level
      S = Construct an element list ordered by element cost
      (decreasing order if maximisation, increasing order if
      minimsiation);
      RCL = The top x/100 * |S| elements of S;
      i = unif_rand(1, |RCL|);
      X = X ∪ RCL_i;
   End While;
   X =Execute local search(X);
   Compute C(X);
   If((minimisation problem And C(X)< best_cost) Or (maximisation
   problem And C(X) > best_cost) Or (this is the first iteration))
      best_cost = C(X);
   End If;
End While;
Output best_cost;
End.
```

Figure 2. The pseudocode for GRASP.

3. GENERALISATIONS

In order to build a constructive meta-heuristic system capable of solving arbitrary COPs, three aspects need to be considered. These are a) *Representational languages* - how an optimisation problem can be specified in a uniform way; b) *Constraint processing* - the way constraints are handled in constructive search and c) *Solution augmentation* - the different ways in which the utility/goodness of an element can be determined.

3.1 Representational Languages

In order to generalise any optimisation technique, there must be some way of specifying problems in a uniform way to be processed by a solver. General data representational languages have been studied and implemented for such applications as databases (with entity relationship in particular) and decision support systems. Within the context of optimisation problems (particularly COPs), Integer Linear Programming (ILP) and its extensions such as GAMS (Brooke et al, 1997) and AMPL (Fourer and Kernighan 1993) have been widely used.

Presented herein are two appropriate representational systems for constructive meta-heuristics. The first is based on a directed graph approach while the second uses linked lists.

3.1.1 Graph Based System

Gutjahr (2000) describes a general representation called *Graph Based Ant System* for COPs. This system uses a directed graph $C = (V, A, \Phi)$ where C is the construction graph, V is the set of nodes (the possible values or elements) and A is the set of arcs. The central idea is that an ant begins from a unique start node in the graph and walks to the end node. The path taken by the ant defines its solution. Φ is a function that defines the set of feasible walks through the construction graph. However, Φ is very problem specific and may be difficult to define for particular problems. However, V and A may be simply specified using either an adjacency matrix or list.

The generalised assignment problem (GAP) will be used to demonstrate the application of the construction graph technique. The aim of the GAP is to assign a number of jobs to a set of agents such that the cost of the assignment is minimised and the capacity of each agent is not violated. This is expressed in Equations 6-9.

$$\textit{Minimise} \sum_{i=1}^{N} \sum_{j=1}^{M} c_{ij} x_{ij} \tag{6}$$

Subject to:

$$\sum_{i=1}^{N} a_{ij} x_{ij} \le b_j \quad 1 \le j \le M \tag{7}$$

$$\sum_{j=1}^{M} x_{ij} = 1 \quad 1 \le i \le N \tag{8}$$

$$x_{ij} \in \{0,1\} \quad 1 \le i \le N \;\; 1 \le j \le M \tag{9}$$

Where:

x_{ij} is 1 if job i is assigned to agent j, 0 otherwise.

c_{ij} is the cost of assigning job i to agent j.

a_{ij} is the resource required by agent j to perform job i.
M is the number of agents.
N is the number of jobs.

There are many ways that the construction graph can be drawn from the one formulation. Figure 3 shows two different construction graphs for a small GAP in which $N = 3$ and $M = 2$. Each one will affect each meta-heuristic's performance differently. For instance, the construction graph of (a) forces the search technique to first assign the jobs to agent 1 and then move to agent 2. In this case, the jobs of agent 1 would have to be given biased probability weightings so that the search technique does not move too quickly from agent 1 without the chance to go back. This could lead to a situation in which agent 1 is not assigned enough jobs, potentially leading to poor quality solutions. The graph of (b) does not have this restriction. The former graph (a) allows fewer choices than (b) for every node (giving some efficiency gains), but is more restrictive. The number of arcs in the graph is $O(p^2)$ where $p = |V|$.

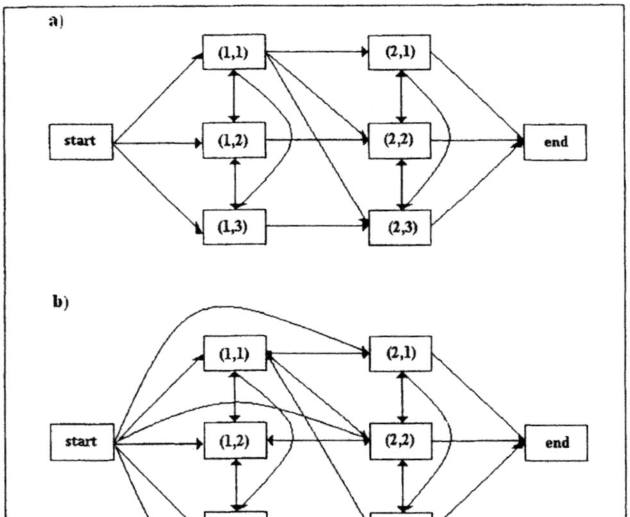

Figure 3. Two possible construction graphs for a small GAP. (r,s) denotes that job s is assigned to job r.

GRASP can also use this system in the following way. Assume that a solution to an optimisation problem can be expressed as $S = \{s_0, s_1, s_2, \ldots, s_t\}$ where $t = |S|$. Given that s_a $(a<t)$ is the most recently added element, the arcs leading from s_a can be arranged in decreasing order of cost to be considered as s_{a+1}. An RCL is then used to retain the top y% and using an equal

probability for each candidate, s_{a+1} can be calculated and added to the solution.

3.1.2 Linked List Representation Scheme

A representation technique based on dynamic lists has been used for iterative meta-heuristics by Randall (1999) and Randall and Abramson (2001b). The technique has been shown to represent a broad cross section of practical COPs. Like the 0-1 ILP system (which is commonly used to represent COPs for branch and bound based solvers), it requires some ingenuity on the part of the practitioner/researcher to represent a COP. In addition, like its counterpart, a problem may have various representations.

In this section, two common problems (GAP and the TSP) are described in terms of the list representation system. Moreover, a method of using the linked list representation with constructive meta-heuristics is described.

3.1.2.1 Example Problems

In a list representation, solutions to the GAP can be represented using a double nested list of agents, each of which contains a list of associated jobs. Thus, in the example, jobs 1, 3, 5 and 2 are assigned to agent 1 (Figure 4).

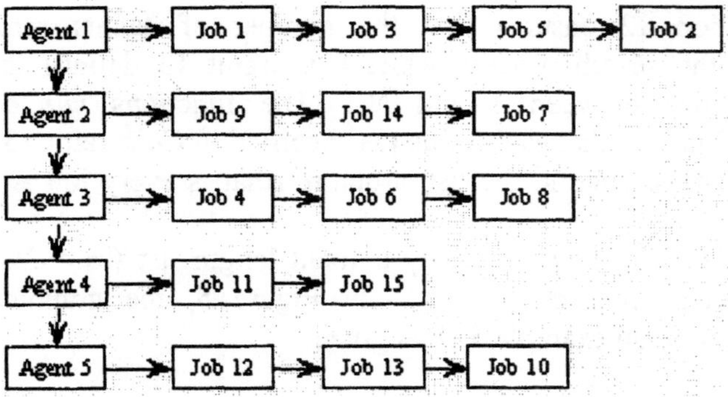

Figure 4. A 5 job, 4 agent GAP represented as a linked list.

In linked list notation, the objective of the GAP can be expressed as:

$$Minimise \sum_{i=1}^{M} \sum_{j=1}^{|x(i)|} C(i, x(i,j)) \tag{10}$$

Where:

x is the solution list. $x(i,j)$ is the jth job performed by agent i.
C is the cost matrix. $C(i,j)$ is the cost of assigning job j to agent i.
M is the number of agents.

Whilst this list notation appears similar to conventional vector form used in standard linear programmes, each sub-list contains a variable number of elements. Thus, the second summation sign in 10 requires a bound which varies depending on the length of each sub-list (that is $|x(i)|$) and changes according to the current solution state. Similarly, the constraints concerning the capacity of each agent are formed across list space. In this example, Equations 12 - 15 are the *list constraints* while Equation 11 represents the *problem constraints*.

$$\sum_{j=1}^{|x(i)|} a(i,x(i,j)) \le b(i) \quad \forall i \quad 1 \le i \le M \tag{11}$$

$$|x| = M \tag{12}$$

$$1 \le x(i,j) \le N \quad \forall i \quad 1 \le i \le M, \forall j \quad 1 \le j \le |x(i)| \tag{13}$$

$$min\_count(x) = 1 \tag{14}$$

$$max\_count(x) = 1 \tag{15}$$

Where:

a is the resource matrix. $a(i,j)$ is the resource required by agent i to perform job j.
b is the capacity vector. $b(i)$ is the capacity of agent i.
N is the number of jobs.

Equation 12 ensures that the number of agents remains static throughout the search process while Constraint 13 defines the range of element values between 1 and N. The functions *min_count*() and *max_count*() ensure that each value (as specified in Constraint 13) appears at least and at most once in the list structure (that is every job is represented once).

For the TSP, however, only a fixed length list ($|x| = N$) is needed. Each element represents a city. In Figure 5, the graphical and list representation of an example TSP is given.

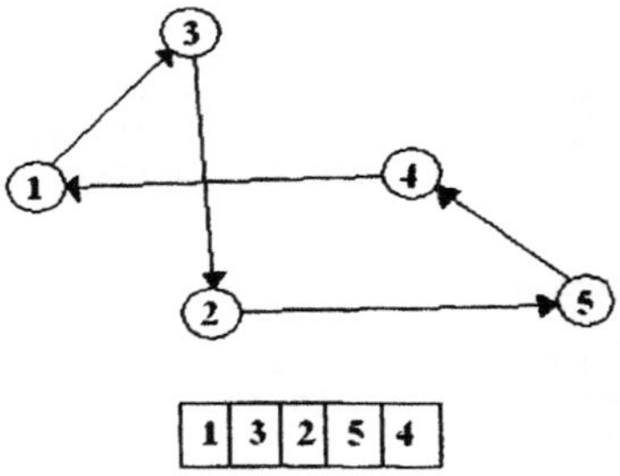

Figure 5. Graphical and list representation of a TSP.

The linked list representation of the TSP is given in Equations 16-20.

$$\textit{Minimise} \quad \sum_{i=1}^{N} d(x(i), x(pred(i,1,N,1))) \tag{16}$$

Subject to:

$$|x| = N \tag{17}$$

$$1 \leq x(i) \leq N \quad \forall i \;\; 1 \leq i \leq N \tag{18}$$

$$min\_count(x) = 1 \tag{19}$$

$$max\_count(x) = 1 \tag{20}$$

Where:

$x(i)$ is the ith city visited on the tour.
$d(i,j)$ is the distance between city i and j.
N is the number of cities.
$pred(i,j,k,l)$ returns i-1 unless i-1<j at which k is returned.

3.1.2.2 Application to Constructive Meta-heuristics

The initial solution generation technique described by Randall (1999) can be used for arbitrary COPs and consists of two phases. In the first phase, a general constructive heuristic is used to place elements throughout the list structure while the second phase applies a meta-heuristic to attain problem feasibility. A modified form of the first phase can be used with a constructive meta-heuristic in order to solve arbitrary COPs expressed in terms of linked lists. The first stage consists of selecting a set of elements to place on the list and then distributing these elements throughout the list structure. This is presented in Figure 6.

```
/* Create the vector of element values to place in the solution
list */
Case ((min_count(x)=1) AND (max_count(x)=1))
      For i=list_value_lowerbound, list_value_upperbound
            elements_to_place(i-list_value_lowerbound + 1)=i;
      End For;
Case (max_count(x)=1)
      j=list_value_lowerbound;
      For i=list_value_lowerbound, list_value_upperbound
            If (randomly_select()=true)
                  elements_to_place(j -list_value_lowerbound + 1)=i;
                  j=j+1;
            End If;
      End For;
Case ((min_count(x)=1) OR (unspecified))
      For i=list_value_lowerbound, list_value_upperbound
            elements_to_place(i - list_value_lowerbound + 1)=i;
      End For;
      j=list_value_lowerbound;
      For i=list_value_lowerbound, list_value_upperbound
            If (randomly_select()=true)
                  elements_to_place(j -list_value_lowerbound + 1)=i;
                  j=j+1;
            End If;
      End For;
/* Place the elements */
Randomly rearrange the elements contained in the elements_to_place
vector;
For i=1, length(elements_to_place)
     sublist=succ(x, sublist);
     position=|x(sublist)|;
     While (x(sublist, position) does not violate the list
     length bounds)
          sublist=succ(x, sublist);
          position=|x(sublist)|;
          If (each sublist has been tried without success) exit and
report failure;
     End While;
     x(sublist, position)=elements_to_place(i);
End For;
End.
```

Where:

x is the solution list.

elements_to_place is a vector containing the element values to place on the solution list.

randomly_select() is a function that generates a random state from the set {true, false}.

length(i) returns the length of vector i.

succ(x,i) returns i+1 unless i =|x| at which i=1.

Figure 6. The initial solution generation technique for list models taken from Randall (1999).

Rather than just taking the ith element of elements_to_place, the constructive meta-heuristic can be used to generate the element from the set elements_to_place. This is expressed in the pseudocode of Figure 7.

```
elements_to_place = Determine the set of elements on the solution list
based on the count constraints;
While(iteration not finished);
  (sublist,position) = Determine a list position that does not violate
  the list constraints;
  element = Select an element from elements_to_place based on the
  specific meta-heuristic rules;
  x(sublist,position) = element;
End While;
End.
```

Figure 7. The pseudocode for integrating the list initial solution generation technique and a constructive meta-heuristic.

3.2 Constraint Processing

As constructive meta-heuristics, by and large, do not operate on complete solutions, it may be difficult to evaluate some constraints. In general, there are three broad ways of dealing with constraints within a meta-heuristic framework[2]:

– *Feasibility Maintenance.* This method ensures that at each step of the algorithm, the solution (or group of, if the technique is population based) is feasible. The advantage is that at the end of the iteration, a feasible solution will be produced. In addition, feasibility maintenance is generally less computationally intensive than feasibility restoration (Abramson and Randall, 1999) (discussed next).

– *Feasibility Restoration.* In this method, an element is added without regard to feasibility. If the solution becomes infeasible as a result, a heuristic is invoked that changes other elements of the solution in order to produce a feasible solution again. The disadvantage is that the feasibility restoration process can be very computationally intensive and problem specific. However, as infeasible space is traversed, solutions that may be unreachable by other means, are potentially explored. It remains an open question as to whether an efficient general-purpose feasibility restoration technique can be produced for constructive meta-heuristics.

– *Disregard Feasibility.* In this method, the feasibility of a particular element (as it is added) is not considered. Whilst this approach is very efficient, many complete solutions will be infeasible (depending on the tightness and the number of constraints).

In this section, ideas for implementing a feasibility maintenance scheme will be described. Generic feasibility restoration is a difficult topic and some preliminary work is being done by Randall (2001a).

It can be more difficult to apply a feasibility maintenance approach to a constructive heuristic (such as ACO) than an iterative heuristic (such as SA and TS). This is because it may be difficult to know when (at what step) to terminate the construction process. However, termination is dependent on the type of constraints present in the problem model. There are two types of constraints:

1. *Constraints that must be satisfied at each step.* An example of such a constraint is a knapsack capacity constraint of the form

$$\sum_{i=1}^{N} c_i x_i < b \quad x_i \in \{0,1\}.$$

Regardless of the number of steps that have elapsed within an iteration, it is always possible to satisfy such a constraint.

2. *Constraints that cannot be satisfied at each step of the algorithm.*

Consider the case where a constraint of the form

$$\sum_{i=1}^{N} c_i x_i > b$$

is present in the problem model. Clearly in the beginning stages of the construction, the length of the solution vector is small and it is unlikely that this constraint will be satisfied. Hence, these types of constraints must be treated as special in the design and development of constructive solvers. Only once these types of constraints have been satisfied can the algorithm consider the termination of the solution augmentation process. Thus, once all such constraints have been satisfied, the agent (for instance, the ant in the case of ACO) can be terminated. This defines the generic iteration termination condition.

3.3 Solution Augmentation

Unlike iterative techniques that can use a variety of standard local search operators (such as swap and move), constructive heuristics have one broad operator: to add an element to the solution. The way to calculate which element to add depends on element cost, constraints and meta-heuristic specific mechanics. For both ACO and GRASP, this function produces a probability for each element. As each element must be examined, it can be computationally expensive to complete an iteration of the search process. Candidate set strategies are used to limit the set of elements examined at each step. Randall (2001c) proposes some generic candidate set strategies.

For constructive meta-heuristics there are three general ways to calculate the utility of an element.

1. *Element Cost.* This approach simply calculates the cost of an element and is suitable for problems such as TSP where the cost function consists of adding a set of separate costs together (Equation 21).

$$\Delta C_e = f(e) \tag{21}$$

Where:

ΔC_e is the incremental cost of adding element *e* to the solution,

$f(e)$ is a problem specific function specifying the incremental cost of element *e*.

2. *Element Cost + Constraint Costs*. In addition to the above point, the cost of an element is offset by the amount it violates or potentially violates the constraints. Using the system outlined in Randall and Abramson (2001b), this degree can be calculated. Constraint violation can be calculated according to the relational operators present in the constraints. For instance, if the sign of a constraint is ≤ and the left-hand side is larger than the right-hand side, the net difference is the amount of constraint violation. This is shown in Equation 22. The constraint violations (Equations 23 - 27) for the other signs are calculated in a similar matter. The total element cost is given in Equation 28. The operand in this equation is + if the objective function is to be minimised, - for maximisation.

$$(\leq) \quad cv_i = MAX(0, lhs_i - rhs_i) \tag{22}$$

$$(<) \quad cv_i = MAX(0, lhs_i - rhs_i + 1) \tag{23}$$

$$(\geq) \quad cv_i = MAX(0, rhs_i - lhs_i) \tag{24}$$

$$(>) \quad cv_i = MAX(0, rhs_i - lhs_i + 1) \tag{25}$$

$$(=) \quad cv_i = | lhs_i - rhs_i | \tag{26}$$

$$cv_i = \begin{cases} \text{if } (rhs_i = lhs_i), 1 \\ 0, \text{otherwise} \end{cases} \tag{27}$$

$$\Delta C_e = f(e) \pm \sum_{i=1}^{R} cv_i \tag{28}$$

Where:

R is the number of constraints,

cv_i is the constraint violation of constraint *I*,

lhs_i is the evaluation of the left-hand side of constraint *I*,

rhs_i is the evaluation of the right-hand side of constraint *i*.

3. *Approximation Function*. An approximation function would be used if either it is not possible to calculate the impact of the element without re-evaluating the entire solution, or it would be too difficult or computationally intensive to calculate the effect of an individual element. The latter represents the case of the QAP, where calculating the cost of individual elements is $O(n^2)$ and calculating the cost of the entire solution is also $O(n^2)$. In such cases, an approximation function would be a function with a lower worst case complexity that estimates the cost of the element. The actual solution cost can be calculated at the end of the iteration. Equation 29 gives the general form of the approximation function.

$$\Delta C_e = ap(e) \quad (29)$$

Where:

ap() is the problem specific approximation function.

For some complex problems that have difficult cost functions and constraints, a mixture of methods 2 and 3 may be appropriate. If the linked list representation system is used, its GAMS like language could be modified to accommodate the type of incremental cost function required.

Using the above notions, it may be possible to broaden the definition of the term "element". Instead of adding an individual element, a group of elements could be added at each step of the algorithm. A set of possible groups would need to be predefined. An element group consists of one or more individual element that work well together (that is have a collective high utility). This approach has been used by Randall, McMahon and Sugden (2001d) to solve network design and synthesis problems. In this approach, a subset of network routes are developed before the running of the SA meta-heuristic. This subset consists of routes that produce low cost solutions and help to maintain problem feasibility. Using the candidate grouping technique, a constructive meta-heuristic needs fewer steps per iteration in order to construct a solution. This of course reduces computational time.

4. CONCLUSION

Practitioners and researchers alike build special purpose meta-heuristic codes in order to solve combinatorial optimisation problems. This is time consuming and does not allow for easy comparison between implementations. However, a uniform framework as proposed in this paper can be used in order to build a general solver. Such a solver would allow a model developer to rapidly prototype problems and solve them using relatively new constructive meta-heuristic techniques.

The issues that have been addressed in this paper are: representation approaches including constructive graphs and linked list representation; constraint processing and solution augmentation. As shown, these issues can be treated in a general way, whereas most practitioners/researchers treat these issues differently for each problem they solve. As such they are applicable for a general solver as well as tailored solvers. We[3] are in the initial stages of building a general constructive meta-heuristic solver. This will be used to implement, test and extend the ideas contained herein. An empirical analysis of the performance of such a system can then be undertaken. Given the results of Randall and Abramson (2001b) for general iterative meta-heuristics, it is conceivable that such a system would be competitive with customised meta-heuristic and heuristic programmes.

5. ACKNOWLEDGEMENTS

The author wishes to acknowledge the financial support of the Australian Research Council.

Notes

1. The best solution can either be defined as the *global-best* (the best solution found in all iterations) or the *iteration-best* (the best solution found in the current iteration).
2. For other methods, refer to Michalewicz and Fogel (2000).

References

Abramson, D. and Randall, M. A Simulated Annealing Code for General Integer Linear Programs. Annals of Operations Research 1999; 86: 3-21.

Brooke, A., Kendrick, D., Meeraus, A. and Raman, R. *GAMS Language Guide*, GAMS Development Corporation, 1997.

Burkard, R. Quadratic Assignment Problems. European Journal of Operational Research 1984; 15: 283-289.

Dorigo, M. and Gambardella, L. Ant Colony System: A Cooperative Learning Approach to the Travelling Salesman Problem. IEEE Transactions on Evolutionary Computation 1997; 1: 53-66.

Dorigo, M., Maniezzo, V. and Colorni, A. The Ant System: Optimization by a Colony of Cooperating Agents. IEEE Transactions on Systems, Man and Cybernetics - Part B 1997; 26: 29-41.

Feo, T. and Resende, M. Greedy Randomized Adaptive Search Procedures. Journal of Global Optimization 1995; 51: 109-133.

Fourer, R., Gay, D. and Kernighan, B. *AMPL: A Modeling Language for Mathematical Programming*. California: Duxbury Press, 1993.

Glover, F. and Laguna, M. *Tabu Search*. Boston: Kluwer Academic, 1997.

Gutjahr, W. A Graph-based Ant System and its Convergence. Future Generation Computer Systems 2000; 16: 873-888.

Lawler, E., Lenstra, J., Rinnoy Kan, A. and Shmoys, D. *The Traveling Salesman Problem: A Guided Tour of Combinatorial Optimization*. Chichester: Wiley, 1985.

Mavridou, T., Pardalos, P., Pitsoulis, L. and Resende, M. (1995) "Parallel Search for Combinatorial Optimization: Genetic Algorithms, Simulated Annealing, Tabu Search and GRASP", Lecture Notes In Computer Science, 980, Ferreira, A. and Rolim, J. ed. Berlin: Springer-Verlag, 1995.

Michalewicz, Z. and Fogel, D. *How to Solve it: Modern Heuristic*, Berlin: Springer Verlag, 2000.

Ramalhinho, H. and Serra, D. Adaptive Approach Heuristics for the Generalized Assignment Problem, Working Paper Series, No. 288, UPF, 1998.

Randall, M. "A General Modelling Systems and Meta-heuristic based Solver for Combinatorial Optimisation Problems", PhD thesis, Faculty of Environmental Science, Griffith University, 1999.

Randall, M. "The Use of Feasibility Restoration in Iterative Meta-heuristics for Solving Combinatorial Optimisation Problems". Working Paper, 2001.

Randall, M. and Abramson, D. (2001) A General Meta-heuristic Based Solver for Combinatorial Optimisation Problems. Journal of Computational Optimisation and Applications 2001; 20 (to appear).

Randall, M. and Montgomery, J. "Candidate Set Strategies for Ant Colony Optimisation". Working paper, 2001.

Randall, M., McMahon, G. and Sugden, S. A Simulated Annealing Approach to Communication Network Design. Journal of Combinatorial Optimization (2001 to appear).

THE IMMUNE DISTRIBUTED COMPETITIVE PROBLEM SOLVER USING MAJOR HISTOCOMPATIBILITY COMPLEX AND IMMUNE NETWORK

Naruaki Toma[1], Satoshi Endo[2], Koji Yamada[2] and Hayao Miyagi[2]

1 Graduate School of Science and Engineering, University of the Ryukyus, 1 Senbaru, Nishihara, Okinawa, 903-0213, Japan.

2 Department of Information Engineering, Faculty of Engineering, University of the Ryukyus, 1 Senbaru, Nishihara, Okinawa, 903-0213, Japan.
tnal@eva.ie.u-ryukyu.ac.jp, {endo,koji,miyagi}@ie.u-ryukyu.ac.jp

Abstract The purpose of this paper is to propose an extended immune optimization algorithm using division as well as integration processing based on immune cell-cooperation and to investigate its validity by computer simulations. In the biological immune system, the immune cell-cooperation is a framework including MHC and immune network, the function of which is to eliminate unknown vast antigens. Our algorithm solves the division-of-labor problems for each agent's work domain inside the multi-agent system (MAS) through interactions between two agents, and those of between agents and environment through the work of immune functions. There are three functions in our algorithm: the division as well as integration processing and the co-evolutionary-like approach. The division as well as integration processing optimizes the work domain, and the co-evolutionary approach realizes equal divisions. In order to investigate the validity of the proposed method, this algorithm is applied to the "Nth agent's Travelling Salesmen Problem (called the n-TSP)" as a typical problem of multi-agent system. The property that is believed to function as solution driver for MAS shall be clarified using several simulations.

Keywords: Optimization, MHC and immune network, immune cell-cooperation, multi-agent system, division-of-labor problems.

1. INTRODUCTION

Adaptive problem solving techniques, such as neural networks and genetic algorithms, are based on information processing in biological organisms and are applied in many kinds of optimization problems, see

Thomas (1997), David et al (1997), Dasgupta et al. (1997), Goldberg (1989), and Holland (1992). A biological immune system is one of the adaptive systems and the studies of the said system are making progress these days, see Dasgupta (1999), Farmer et al (1986), Forrest et al (1990), Ishida et al (1998), and Toma et al (2000-a). In the biological immune system, the following two mechanisms are considered as important functions in eliminating invaded antigens (Janeway et al, 1997). First, major histocompatibility complex (MHC) is used to distinguish a "self" from the other □nonself□ when the nonself invades the self. Second, the immune network comprises immune cells and their connections, and then specific antibodies are produced through modification of their immune cells. Furthermore, a framework, with double functions, that sets out to eliminate the unknown vast antigens in parallel, is called `immune cell-cooperation'. From the engineering system's point of view, the immune cell-cooperation is considered a parallel-distributed system with role differentiation.

In this paper, we propose an extended immune optimization algorithm, which is based on the immune cell-cooperation, and to solve the division-of-labor problems for each agent's work domain in multi-agent system (MAS). In previous work (Toma et al, 2000-b), the immune distributed competitive problem solver had two problems; we were not able to compute even work domain and the divergence. The previous algorithm has been improved so that the immune cell-cooperation, which is considered to be a framework in a broad sense of eliminating antigens, may be applied rather than applying the local functions directly to the algorithm.

The extended algorithm solves the division-of-labor problems through interactions between the two agents, and between agents and environment by immune functions. There are three functions in our algorithm: the division as well as integration processing and the co-evolutionary-like approach. The division as well as integration processing optimizes the work domain, and the co-evolutionary approach realizes equal divisions. Through implementation of such functions, we could construct an adaptive algorithm that solves division-of-labor problems. Thereupon, we applied the proposed algorithm to the nth agent's travelling salesman problem (called the n-TSP), which is considered a typical case problem in MAS.

Some computer simulations are designed to clarify the basic performances as well as the characteristics and features of the proposed immune algorithm.

2. ISSUES OF THE TARGET PROBLEMS

2.1 Division-of-Labor Problems

Multi-agent system, which is a study in the field of distributed artificial intelligence, is an information processing technique in which autonomous agents solve problems by interactions among the agents (Russell et al, 1995). The key subject of division-of-labor problems is to focus on the issues of distribution of work for agents. The solvers of the problems are defined as follows: the domain that each agent covers is defined as work domain (WD), and the aggregate total of the work domains is defined as problem domain (PD). There are two issues to the division-of-labor problems. They are: (a) each WDi that is assigned by agent i must be divided evenly and, (b) the system performs effective division-of-labor by optimizing each WDi.

2.2 Nth Agent's TSP

As a typical case problem of the division-of-labor, we dealt with the nth agent's Travelling Salesman Problem to investigate the adaptability of our model. Travelling salesman problem (TSP) is one of the most typical combinatorial optimization problems. The objective is to find a minimal roundtrip route visiting each node exactly once. The number of salesmen in the n-TSP was increased from singular to plural number. The objective is to find the minimal tour route by division among salesmen (see Figure 2). Note that two conditions were set. First, the number of salesmen is fixed while the system runs to solve. Second, all salesmen use the same city as the starting point.

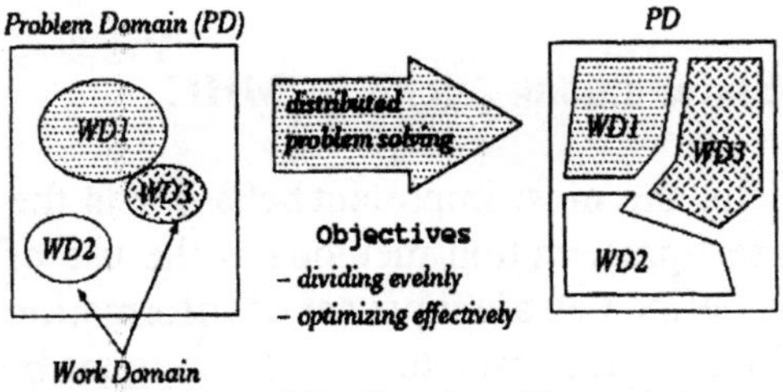

Figure 1. Aims of distributed problem solving.

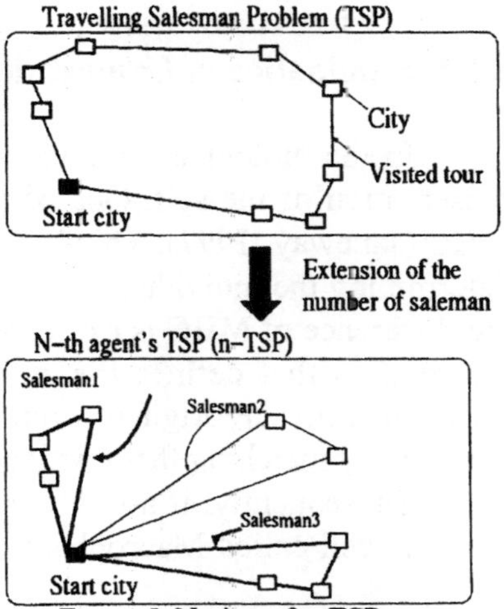

Figure 2. Notion of n-TSP.

3. PREVIOUS WORK

3.1 Modeling of an Biological Immune System

In order to construct an immune optimization for the division-of-labor problems, we developed the models of immune functions in the first instance. Our objective models are shown in Figure 3 and the details are described below in italics.

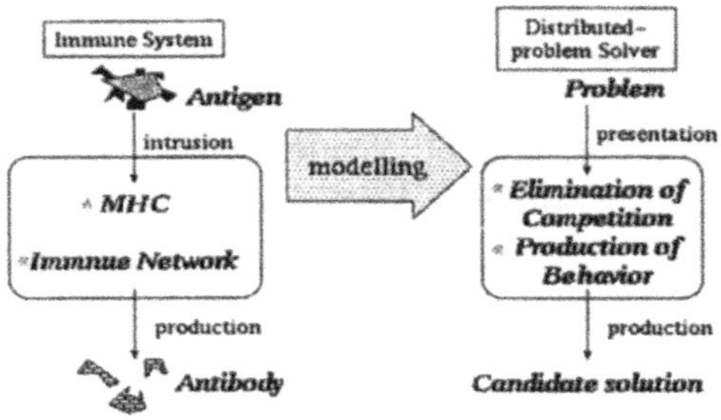

Figure 3. Modeling of biological immune system for distributed problem solver.

3.1.1 Antigens and Antibodies

Antigens are defined as problems fed into the systems or environments in which the agents exist. Antibodies are defined as solutions against the problems or ways to adapt to the environments. In accordance with these definitions, it is possible to develop the problem solver, which is adapted to the framework of a biological immune system that performs self-preservation by searching specific antibodies that react against unknown vast antigens.

3.1.2 Elimination of Competition-States among Agents by MHC

In the biological immune system, the most important behavior is the preservation of the self. One of the self-preservation methods is the use of MHC (Janeway, 1997). This MHC is defined as a unique set of information determining the individual himself. MHC that is used to identify antigens by the difference of MHC is expressed as `MHC with peptide', which is a set of information that defines the antigen. If the infected MHC on the cells is different from the original (uninfected) MHC, the individual can recognize antigens by matching them with a T cell receptor (called TcR).

Subsequently, if nonself, such as antigens, invade the self-system, the immune cell called Macrophage processes and displays invasion of nonself by differences in MHC. While the cells with different MHC exist, the self-system continues to eliminate the nonself system as shown in Figure 4.

The elimination of nonself means removal of the nonself system from the self-system, expressed in the lower part of Figure 4. By regarding the self and nonself as agents in MAS, the invaded-state considers that some agents compete with each other's work domain. In such a case, if the ability of each agent is equal, the overlapping of the respective work domain places it in an ineffectual state. The MHC is used to eliminate such states of competition.

3.1.3 Production of Behaviors by Immune Network

In the biological immune system, the Immune Network is composed of immune cells and their connections, and the antibodies are produced through modification of their immune cells. The one piece of information transmitted through the network is MHC. The procedures of the immune system producing a specific antibody are described below.

First, Macrophage recognizes antigens and transmits MHC (with the information about the antigens) to T cell. Next, the T cell that received MHC eliminates the infected cells and then it activates specific B cell that has the capability to recognize the antigens. The activated B cell produces specific antibodies. Finally, the antibodies thus produced begin the elimination of the invading antigens.

To apply such procedures as a production model of behavior of agents, the following procedures were constructed.

In the production model, first, the agent acquires the information about the environment and internal-state in the agent. Second, if the states of competition exist in the internal-state, the agent eliminates these states of competition. Third, the agent searches for a set of candidate behaviors in order to increase the fitness of internal-state in the environment. Finally, the agent extracts superior solution in the candidate behaviors.

The extracted solution is applied to the environment, and then the agent continues to process their procedures (see Figure 5). Our algorithm solves the division-of-labor problems through the elimination of competition by MHC and production of behaviors by Immune Network.

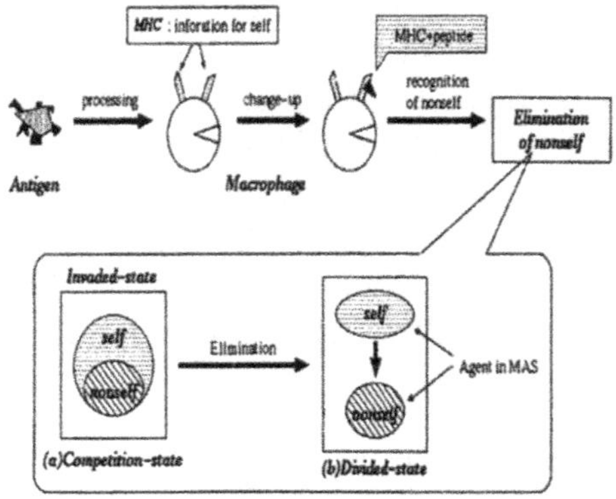

Figure 4. Elimination of Competition-States is based on the Elimination of Nonself using MHC.

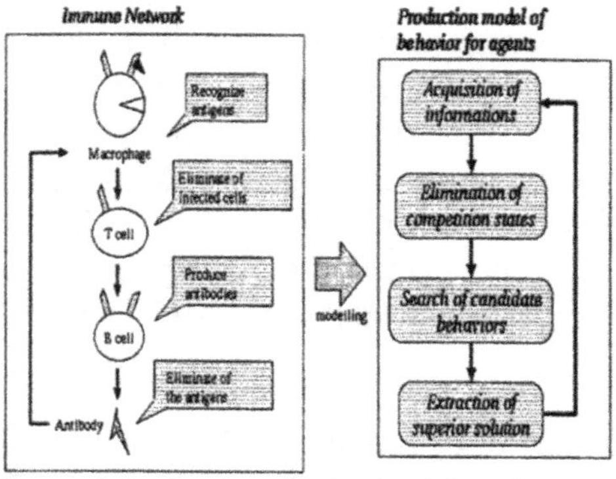

Figure 5. Production of behavior is based on production of an antibody using Inmune Network.

3.2 The Immune Distributed Competitive Problem Solver

In our method, specific antibodies are produced by transmission of the MHC through the Immune Network and re-construction of the components. The MHC is an information carrier capable of maintaining information of both the self and nonself by transforming itself into MHC with peptide (a piece of information of nonself). The situation in which MHC acquires information of nonself is referred to as the states of competition. In order to eliminate the states, we implemented the algorithm for computer simulations as described in Section 3.1. The details of our method are described below, and Figure 6 illustrates the implementation of n-TSP.

Step 1: Definition of antigens and MHC

The problem area where agents exist is defined as antigen and the internal state (for example, solution, behavior) of each agent is defined as MHC.

Step 2: Preparation of initial MHC and immune cells

All the initial MHC and immune cells are prepared at random for initialization.

Step 3: Acquisition of information

Macrophage acquires both the external and internal sets of information for calculation of the fitness and recognition of competition-states. The pieces of external information are described as MHC with internal information and the MHC is transmitted to T cell.

Step 4: Elimination of the States of Competition

If the states of competition exist in the transmitted MHC, T cell eliminates these states by means of competition processing. (The details are described in the following section).

Step 5: Production of candidate solutions

On the basis of MHC that comprise both the external and internal information that DOES NOT include the states of competition, B cell produces N number of candidate (similar) solutions that are approximated to the MHC. The design of the similarity and the way of production of similar solutions depend on the problem.

Step 6: Extraction of a superior solution

Each agent extracts a superior solution from the above similar solutions and executes the solution. If the terminal condition is not satisfactory, go to Step 3.

Four steps, from the third to the sixth steps, are one-time step. This algorithm is adapted by repeating from step 3 to step 6 until it comes to the terminal condition.

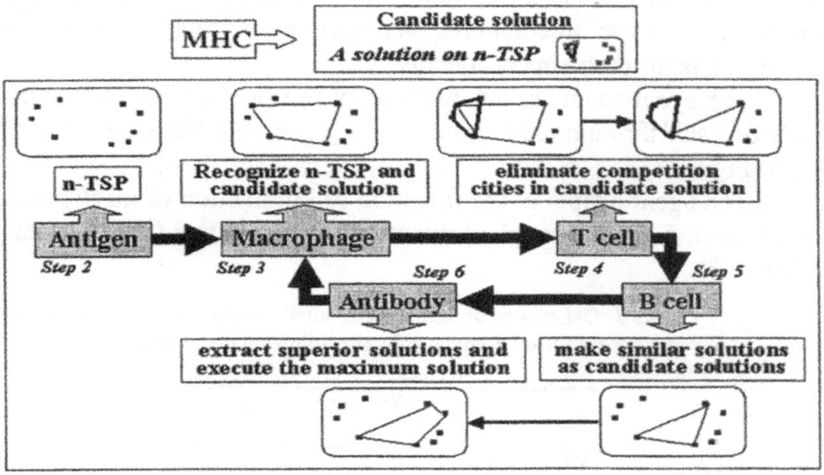

Figure 6. A Behavior example of the Immune Distributed Competitive Problem Solver.

3.3 Competition Processing

The competition processing is executed at the fourth Step in the above algorithm. A mechanism eliminates states of competition among competing agents by comparing respective fitness. The definition of terms that is used here to explain this mechanism is described in Table 1. Figure 7 illustrates a situation, which is void of competition-states based on the following two comparisons.

Step 1: Comparison of fitness in self-agent

In the competition processing, a comparison is made between f (A) and f (A'). If the f (A') without the states of competition is high, the self-agent modifies the fitness from f (A) to f (A') and the MHC from A_{MHC} to A'_{MHC} by turning over the states of competition to agent B. Otherwise go to Step2.

Step 2: Comparison of fitness between self and nonself agents

A comparison is made between f (A) and f (B). If f (A) is large, the states of competition are eliminated from agent B. Otherwise; the states are eliminated from agent A.

3.4 Concepts and Aspects

The proposed method consists of (1) elimination of competition using MHC and (2) production of behaviors using Immune Network. Based on the construction, this method has antithetic adaptations (also, see Figure 8). According to these interactions, our method is capable of solving the division-of-labor problems efficiently.

1. Each agent adapts in the direction of raising the level of fitness on the inside of itself through production utilizing Immune Network (see left-hand side of Figure 8).
2. Each agent adapts in the direction of elimination of inefficient region in each work domain through elimination of competition (see right-hand side of Figure 8).

Table 1. Definitions of terms: Here `after' denotes the situation that the states of competition are eliminated by the following two comparisons.

	Self agent: A	Nonself agent: B
Work domain, now	A_{MHC}	B_{MHC}
Work domain, after	A'_{MHC}	B'_{MHC}
Fitness, now	F(A)	f(B)
Fitness, after	F(A')	f(B')

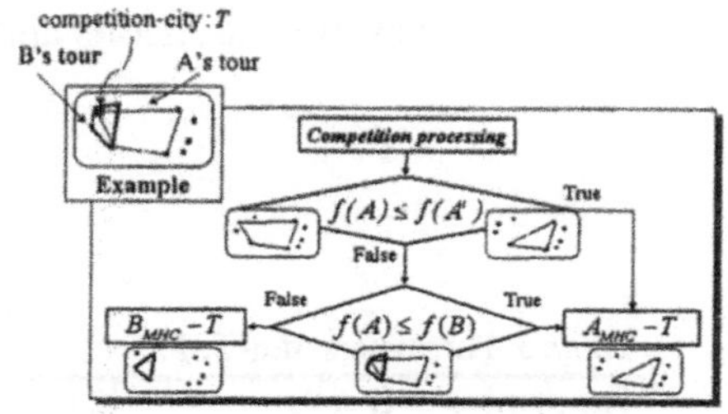

Figure 7. An example of competition processing.

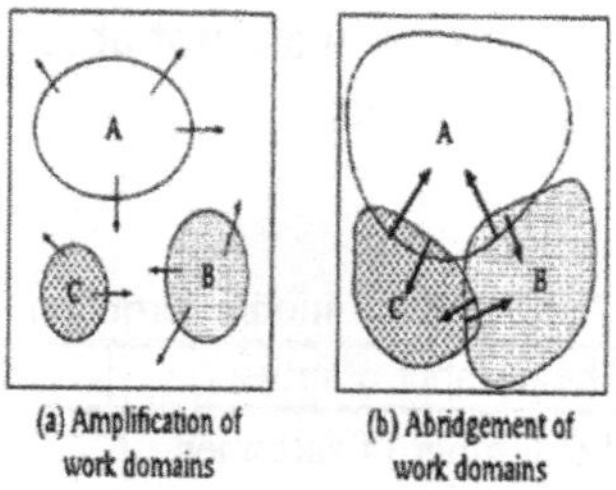

Figure 8. Basic concept of the proposed method.

4. EXPERIMENT FOR IMMUNE DISTRIBUTED COMPETITIVE PROBLEM SOLVER

4.1 Design of Simulation

In order to investigate the basic performance of our method, we applied this algorithm to the following n-TSP. The definition of problem and parameters of proposed method are set forth in Tables 2 and 3. The fitness i, the fitness of agent i, is defined as equation 1.

$$fitness_i = Cities(MHC(i)/Cost(MHC(i))+1/Cost(MHC(i)) \quad (1)$$

- *MHC(i)*: the tour of agent i.
- *Cities(MHC(i))*: the number of cities in *MHC(i)*
- *Cost(MHC(i))*: the length of *MHC(i)*

4.2 Results and Problems

Figure 9-(A) shows the acquired solution at step 50. The solution is considered as optimum solution in its own work domain. In case of simple n-TSP with a single circle, it is clear that (a) the number of cities visited by each agent is 8 and the even division of work domain is computed and (b) the solution of each agent is considered as the optimum solution. However, in case of the dual circle even division of work domain is not computed.

In addition, this method has a critical problem of divergence. Figure 9-(B) shows the transition of the fitness and the number of the cities. As shown in the figure, the fitness decreases whenever the number of cities changes, although it is improving up to step 50. The reason for these results is that the interactions are strictly based on the value of fitness. We adjusted the weights and proportion of the formula to improve it through trial and error, but we were not able to obtain the optimum that has even divisions and minimum cost.

Table 2. Definition of problem.

The number of cities	25
The number of salesmen	3
The arrangement of cities	Dual circle
Start city	Center of circle
Radius	Out=0.4, In=0.384
Terminal steps	2000

Table 3. Parameters of n-TSP.

The number of B cells	25
The number of activated B cells	25
Add city rate	0.5
Swap mutation rate	0.5

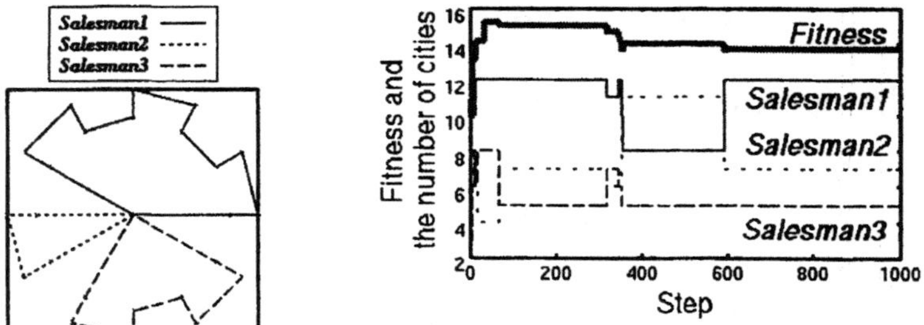

(A) An obtained solution at step 50.

(B) Transition of Fitness and the number of cities

Figure 9. Results.

4.3 Summary of the Experiment

According to the above results and Toma et al (2000-b), our method has two advantages (1) modification of the work-domain is changed while maintaining the fitness, and (2) the average costs are very small compared with GA. By contrast, the downside is, if the effect of maintaining is too strong, dividing evenly becomes difficult. According to these results, improvement for the equality in the task of division-of-labor problems is required. In this instance, our algorithm has been improved so that the immune cell-cooperation which is a framework including the MHC and the immune network may be applied rather than having it applied directly to the local functions.

5. EXTENDED ALGORITHM

In this section, we propose an extended immune optimization algorithm, which is based on the immune cell-cooperation, to solve the division-of-labor problems. In previous work for MAS (Toma et al, 2000-b), the immune distributed competitive problem solver had two problems; the even work domain and divergence could not be computed as mentioned in Section 4.3. The previous algorithm has been improved so that the immune cell-cooperation, which is a framework in a broad sense of elimination of antigens, may be applied rather than having it applied directly to the local functions.

The extended algorithm solves the division-of-labor problems through the interactions between the agents, and between agents and environment by immune functions. There are three functions in our algorithm: division as well as integration processing and co-evolutionary-like approach. The division as well as integration processing optimizes the work domain, and the co-evolutionary approach realizes equal divisions. Through implementation of such functions, it was possible to construct an adaptive algorithm that solves the division-of-labor problems. Thereupon, the proposed algorithm was applied to n-TSP. Some computer simulations are designed to clarify the basic performances as well as the characteristics and features of the extended algorithm.

5.1 Modeling of Immune Cell-Cooperation

5.1.1 Biological Immune Cell-Cooperation

In the biological immune system, the two mechanisms (MHC and immune network) are considered important in eliminating invaded antigens (see Janeway et al, 1997). Thereupon, a framework, which together with the two functions eliminates the unknown vast antigens in parallel, is called the 'immune cell-cooperation'. From the engineering system's point of view, the immune cell-cooperation is considered a parallel-distributed system with role differentiation. The roles in this system are (1) fragmentation and presentation of antigens, (2) activation of producing specific antibodies, (3) elimination of the antigens by specific antibodies, and (4) control of their functions (see Figure 10). Note that, the cell-cooperation has a set of rules not only as 'function' (fragmentation, activation) but also as 'work domain' (as defined in Section 2.1). The object of the model is to implement and to apply them to the multi-agent system.

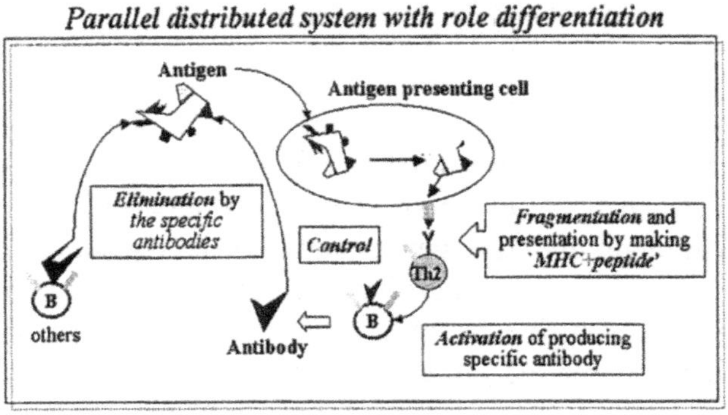

Figure 10. Concept of immune cell-cooperation.

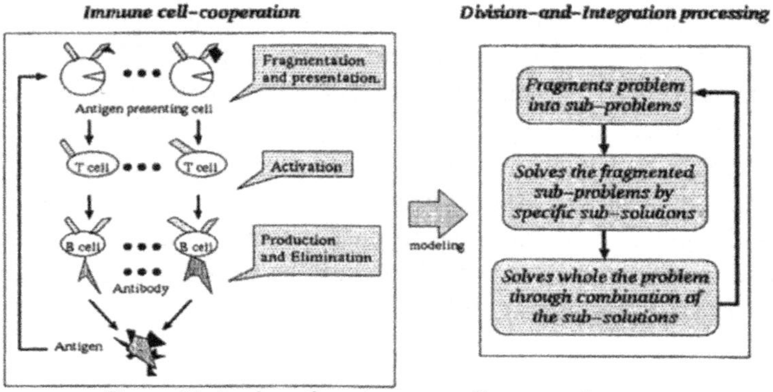

Figure 11. Analogy from immune cell-cooperation.

The fragmentation, the activation and the elimination functions have been especially chosen in this issue because their functions are minimum components to act as problem-solvers.

5.1.2 Analogy

As described in the beginning of Section 5.1.1, the biological immune cell-cooperation has four roles: fragmentation, activation, elimination and control of these functions. In order to apply the system to the division-of-labor problems, the three functions were chosen in this issue, namely the fragmentation, activation and elimination, because their functions are the minimum components necessary to perform the function as the problem-solvers. An optimization algorithm was constructed based on the concept that (1) fragments a problem, (2) solves the fragmented sub-problems by the specific sub-solutions, and (3) solves the whole problem through combination of these sub-solutions. We call these procedures division-and-integration processing (Figure 11).

In addition, we introduced a co-evolutionary concept that the immune system evolves against the virus, which also evolves to escape from the elimination of the system. In order to apply this concept as a searching method in MAS, we constructed a procedure in the likeness of the escaping virus from the changing environments (Figure 12).

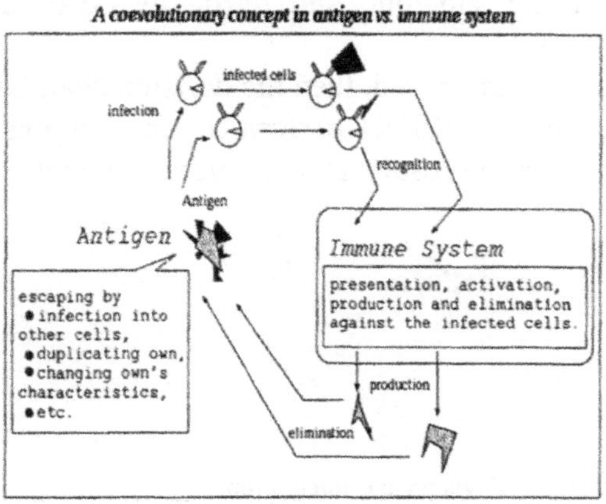

Figure 12. Illustration of a co-evolutionary like concept in the case of antigen vs. immune system.

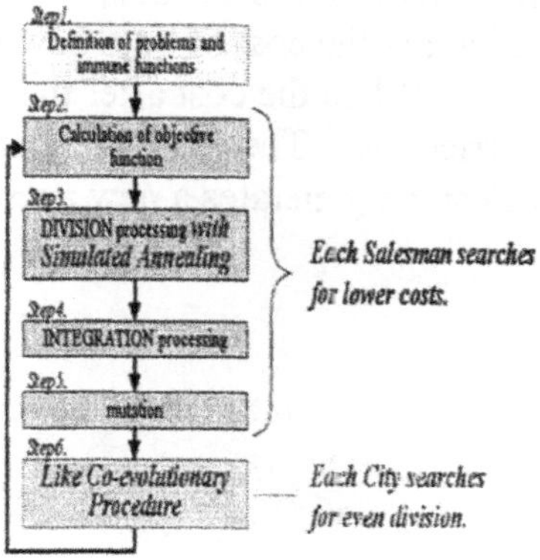

Figure 13. Procedures of modified algorithm.

5.2 Extended Algorithm

As shown in Figure 13, the algorithm solves the problems through the two searching methods, (1) division-and-integration processing by agents and (2) co-evolutionary like procedure by components in the environment. The procedures of the extended algorithm against n-TSP are as described below. In the procedures of proposed method, Step 1 is processed only once by the System for initialization.

Each agent processes are repeated from Step 2 to Step 6 for the optimization of the problem.

Step 1: Definition of problems and immune functions.

Cities and Salesmen as the problems must be defined. Salesman's unique ID and division as well as integration processing for tours of each salesman must be defined. At this point, each city has the ID as the expression of salesman visiting the city.

Step 2: Calculation of objective function.

The cost of salesman is calculated by following function.

$$Cost(S_i) = \sum distance(tour_{S_i}) \quad (2)$$

- S_i: Salesman i.
- $tour_{Si}$: the tour of Salesman i.

Step 3: Division processing with Simulated Annealing.

One salesman tries to divide his own tour into two sub tours in search of lower costs according to these steps (see also Figure 14-(A)). First, we need to decide on one sub $tour_j$ for new $salesman_j$. Second, we need to map a new $tour_i$ for $salesman_i$ except for the sub $tour_j$ and third, we need to calculate the cost of both new sub tours.

When the cost after the division is improved, then the division shall be carried out. Therefore, in the case of DIVISION performed, the original salesman generates a new agent for attaining results that are more efficient.

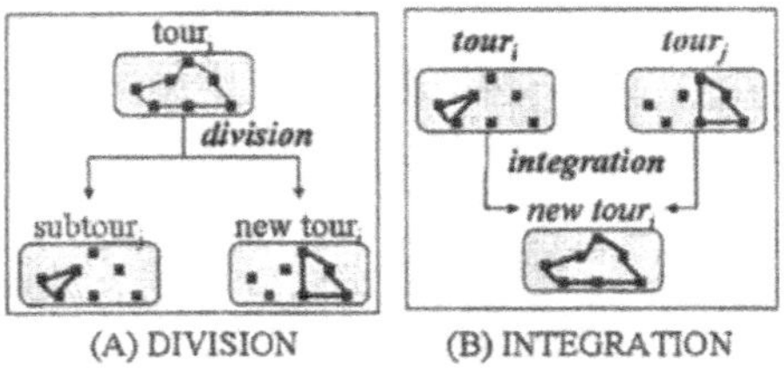

Figure 14. Examples of Division and Integration.

Step 4: Integration processing.

The two salesmen try to integrate the tours in search of lower costs by following the steps (see also Figure 14-(B)). First, we need to decide on any two tours and integration point ptr has to be made. Second, routing of new tour crossing the ptr has to be scheduled. Third, the costs of the new tour have to be computed. If the cost, after integration, is improved, the integration will be carried out. Therefore, in the case of INTEGRATION performed, the original salesman kills the other salesman for making an efficient result.

Step 5: Mutation.

Swap any two cities. In the event the costs after swapping are improved, the swap will be carried out.

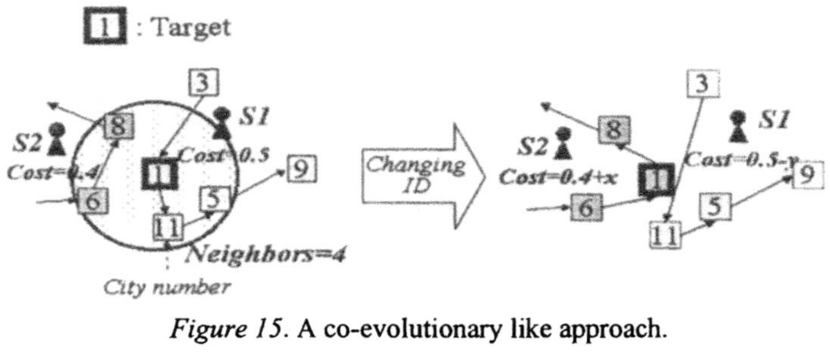

Figure 15. A co-evolutionary like approach.

Step 6: Co-evolutionary like Procedure.

This approach changes the ID of a given city depending on the costs to and from the neighboring city and the costs to process the following steps (see also Figure 15). First, check the costs of salesmen that visited neighboring cities of the target city. The neighbors are defined as N-cities close to the target city. In this example, the neighbors are visited by S1 and S2. Second, if the other salesman's cost is lower, then the city changes its ID into that of the other. Consequently, in this example, the ID of the target city is changed to S2, and then, the target city is visited by S2 as shown in the right side of Figure 15.

In this manner, each agent searches for lower costs, and each component in the environment searches for even divisions. Note that, in order to increase the number of execution times of division processing, `Simulated Annealing' is implemented at Step 3.

6. EXPERIMENT FOR EXTENDED ALGORITHM

6.1 Design of Simulations

Two computer simulations are performed to investigate the basic performance of the extended algorithm. In the first simulation, we applied the algorithm to the problem in Table 2 and then we considered the results. In the second simulation, we applied the Simple Genetic Algorithm (SGA) to the same problem and the two performances were compared. The parameters of the extended algorithm are defined in Table 4.

6.2 Experiment 1: System Behavior

Figure 16 shows an acquired solution. Because of the extension of our method, the best solution led to equal divisions and optimized the costs. The transition of the number of visited cities for each agent is shown in Table 5. The number of executions of the division processing improved owing to the Simulated Annealing. As a result, the integration processing is performed frequently and effectively.

Table 4. Parameters of the extended algorithm.

The number of neighbours	4
The number of initial agents	Under 24
Terminal steps	2000

Table 5. Distribution of the number of searching solutions.

	Trials	Times of Execution
DIVISION	6,242	211(6)
INTEGRATION	546	225(16)
Mutation	6,037	50
CITY operation	47,876	47,876
Total	60,701	48,372

In order to verify the behaviors of our method, the transition of the cost, the sum of the whole agent's costs, and the number of visited cities (dividing-number) are shown in Figure 17. The cost improves dramatically up to Step 50. A new combinational situation that alters the division-of-labor from multiple agents visiting multiple cities to three to five agents visiting approximately eight cities occurred. This occurrence is caused under the primary influence of the equalizing divisions in Step 6 of the extended algorithm. Here, a rise in the average costs for all agents is observed. In the subsequent steps, the optimization is performed repeatedly while maintaining the approximately even divisions. Finally, the transition of fitness and dividing-number converges in the optimum solution, which is a combination of eight cities with three agents in Figure 16.

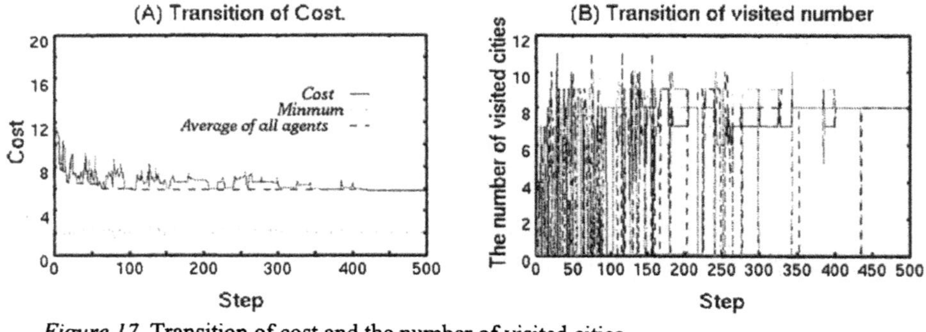

Figure 17. Transition of cost and the number of visited cities.

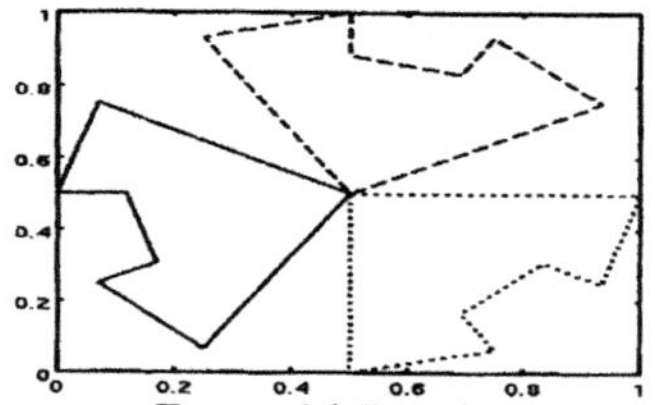

Figure 16. Results.

6.3 Experiment 2: Comparison with Simple GA

6.3.1 Design of Simple GA

The results were compared with both methods and SGA. The parameters of SGA are set in Table 6. The $fitness_i$, which is the fitness of agent i is expressed in equation 3.

The candidate solutions must be encoded for it to be applied to GA in n-TSP. We adopted the pass representation method originally proposed by Yamamura et al (1992).

Table 6. Parameters of SGA.

Population	100
Crossover rate	1.0
Mutation rate	0.01
Terminal generations	5000

$$fitness_i = 1/(Cost + Penalty) \quad (3)$$

- *Cost*: Length of the tour
- *Penalty*: Penalty for the equality

6.3.2 Results

The searching performances are compared through simulated results of 20 trials. Table 7 shows, (1) the number of the best solutions obtained, (2) the minimum, average, and the maximum cost, (3) the number of searched solutions (for example, a sum of individuals in the case of SGA), and (4) run time. In all terms, our method had better results with smaller run time than SGA. In particular, this algorithm is capable of searching for superior solution quickly on the average.

Table 7. Simulation results of 20 trials.

		SGA	Extended Algorithm
Best	Times	3	15
Cost	Min(best)	5.845231	5.845231
	Ave	6.403267	5.858212
	Max(worst)	6.949701	5.858212
Tne number of searched solutions	Min(best)	74,100	1,616
	Ave	167,000	8,488
	Max(worst)	227,100	16,580
Run tme		50sec	10sec

7. CONCLUSION

An extended immune optimization using division as well as integration processing based on the immune cell-cooperation is proposed, and the target problem has to be the optimization of the division-of-labor problems.

This method solves the problems through combination of division, integration and co-evolutionary approach. These functions are based on local interactions between the agents, and between agents and environment. In MAS, clarification of the objective function considering all agents and components in the environment is too hard a problem but it is important in the optimizing of the whole problem by using the local interactions.

Since this algorithm is capable of optimizing the division-of-labor problems, it could be expected to function effectively as the optimization algorithm in MAS. Future work would (a) analyze behaviors, adaptabilities and performances of our method, (b) define applicable problems and required conditions for application, (c) compare with similar and conventional methods, and (d) investigate its application to a more specific problem as a suitable subject for this algorithm. In addition, further work will (e) implement the other functions in the immune cell-cooperation as described in the beginning.

8. ACKNOWLEDGEMENTS

The authors would like to acknowledge Prof. Y. Itoh (Lymphocyte Biology Section, Laboratory of Immunology, National Institute of Allergy and Infectious Diseases, National Institutes of Health, Bethesda MD 20892) and members of Laboratory of Harmonious Systems Engineering, Research Group of Complex Systems Engineering, Division of Systems and Information Engineering, Graduate School of Engineering, Hokkaido University.

The first author acknowledges the Grant-in-Aid for Scientific Research (Grant-in-Aid for JSPS Fellows).

References

Back, Thomas. Proceedings of The Seventh International Conference on Genetic Algorithms. 1997; Michigan State University, East Lansing.

Dipankar Dasgupta and Zbigniew Michalewicz, Evolutionary Algorithms in Engineering Applications, Springer-Verlag, 1997.

Dipankar Dasgupta. Artificial Immune Systems and Their Applications. Springer, 1999.

David Corne, Marco Dorigo, and Fred Glover. New ideas in optimisation. 1997; McGraw-

Hill Publishing Company.
J.D. Farmer, N.H. Packard, and A.A. Perelson. The Immune system adaptation, and machine learning. Physica 22D, 1986; 187-204.
S. Forrest, and A.A. Perelson. Genetic algorithm and the Immune system. Proc. of 1st Workshop on Parallel Problem Solving from Nature 1990; 320-325.
D.E. Goldberg. Genetic algorithm, search optimization and machine learning. 1989; Addison Wesley.
Holland, J.H.. Adaptation in Natural and Artificial Systems: An Introductory Analysis With Applications to Biology. 1992; MIT Press.
Ishida, Y., Hirayama, H., Fujita, H., Ishiguro, A. and Mori, K. Immunity-Based Systems and Its Applications. 1998; CORONA.
Charles A. Janeway, Jr., Paul Travers ; with assistance of Simon Hunt, Mark Walport. Immunobiology : The Immune System in Health And Disease. 1997; Garland Pub.
S. Russell and P. Norvig. Artificial Intelligence: A Modern Approach. 1995; Prentice-Hall.
Toma, N., Endo, S. and Yamada, K. The Proposal and Evaluation of an Adaptive Memorizing Immune Algorithm with Two Memory Mechanisms. Journal of Japanese Society for Artificial Intelligence. 2000; Vol.15, No.6, pp.1097-1106.
Toma, N., and Endo, S., Yamada, K. and Miyagi, H. The Immune Distributed Competitive Problem Solver with MHC and Immune Network. Intelligent Engineering Systems through Artificial Neural Networks 2000; vol.10 (editor C.H. Dagli et al.), ASME PRESS SERIES (ISBN 0-7918-0161-6), pp.317-322
Yamamura, M., Ono, T., and Kobayashi, S. Character-Preserving genetic algorithms for Travering salesman problem. Journal of JSAI 1992; Vol.7 No.6, pp.1049-1059.

SEARCH ALGORITHMS FOR NURSE SCHEDULING WITH GENETIC ALGORITHMS

Ahmad Jan, Masahito Yamamoto and Azuma Ohuchi

Laboratory of Harmonious Systems Engineering, Research Group of Complex Systems Engineering, Division of Systems and Information Engineering, Graduate School of Engineering, Hokkaido University, North 13, West 8, Kita-ku, Sapporo 060-8628, Japan.
{jan, masahito, ohuchi}@complex.eng.hokudai.ac.jp

Abstract This paper investigates the use of a genetic algorithms (GA) approach to a multi-objective optimization problem, in particular the nurse scheduling problem (NSP). Because GAs are computationally intensive algorithms, there is a strong need to make it effective. We define effective as producing good results in a short time. Our efforts so far revealed that for solving of the NSPs using a GA approach, unwanted premature convergence occurs in the early stages of the search. This is due to the presence of strong constraints and limiting of the search to feasible regions only. Therefore, diversification of the solution space plays a very important role for exploration of the potentially unexplored regions of the solution space. A mutation operator, or some kind of niching method is supposed to overcome this problem. The authors' task is to enhance optimization of the search performance of GA for NSPs. First a simplified and later a more useful version of the problem are examined. The existence of the global optimum for the simplified version of the problem is known. However, the existence of the global optimum for the later version of the problem is unknown. Therefore, for the later version of the problem the objective is to acquire solutions as close to the Pareto efficient front as possible. So far the best results have been acquired by applying a simple and cost effective block-wise mutation operator that we call an escape operator. Throughout computer simulations, the aforementioned difficulties are analyzed and the efficiency of the so called *escape* operator is confirmed.

Keywords: Nurse Scheduling Problem, Genetic Algorithms, Decision Making, Pareto Ranking Scheme.

1. INTRODUCTION

There are many different ways of solving manpower scheduling and in particular nurse scheduling problems. For the purpose of this paper, nurse scheduling solution methods are classified into four types following the recommendations of Bradley and Martin (1990): exact cyclical, heuristic cyclical, exact non-cyclical and heuristic non-cyclical. The problem in hand is of a heuristic cyclical nature. Although heuristic algorithms do not guarantee to find the optimal solution, they tend to find very good solutions in a short time. Concerning heuristic cyclical approaches for solution of NSPs, there have been only a few papers published so far. Yamamura et al. (1995) proposed a Cooperative Genetic Algorithm (CGA) and recently Yamamoto et al. (1998) proposed a collective multi-agent approach using pheromone style communication of real ants for solution of NSP. The search procedure in the multi agent approach begins with an initial infeasible schedule. However, this approach does not guarantee generation of the feasible schedules, which is a serious drawback for real world scheduling. In this paper, the approach proposed by Yamamura et al. (1995) is taken into consideration.

CGA initializes its search with an initial feasible schedule and continues searching in feasible search regions only. CGA successfully generates feasible schedules. However, due to presence of strong constraints and restricting the algorithms to search in the feasible search region only, premature convergence occurs in the early stages of search. In CGA new candidate solutions are created through the application of an extended two-point crossover. Newly created candidate solutions are evaluated and updated using Pareto ranking scheme. Only Pareto efficient or non-dominated solutions (see Figure 1) are allowed to survive for further competition and are kept in the Pareto efficient front. From Pareto efficient front of all non-dominated candidate solutions, only one solution is randomly selected and replaced with the previous schedule. Also, in each generation an archive of all Pareto efficient solutions are updated and kept for the decision maker. The search continues until the pre-defined objectives are met or the time set for the search is over. Finally, the acquired archived set of the Pareto efficient solutions are presented for the decision maker.

The organization of this paper is as follows. In Section 2, the nurse scheduling problem is briefly explained. In Section 3 hard and soft constraints, fitness calculation for the individual nurses and also the objective functions for the entire schedule is explained. In Section 4 algorithmic flow of cooperative genetic algorithms for the problem and also the concept of Pareto optimality are described. In Section 5 results of the computer simulations are discussed. Finally, discussion and conclusions are given in Section 6.

2. NURSE SCHEDULING PROBLEM (NSP)

According to Bagchi (1999), scheduling is an optimization process by which limited resources are allocated over time among parallel and sequential activities. Such situations develop routinely in factories, publishing houses, shipping, universities, hospitals, airports, etc. This paper deals with the solution of a nurse scheduling problem using genetic algorithms (GAs). GAs have proven their usefulness in various domains. The roots of genetic algorithms lie in Charles Darwin's discoveries in the 19 *th* century: the means of natural selection, survival of the fittest, and theories of evolution. The reader seeking more information on genetic algorithms can begin with Goldberg's book, *Genetic Algorithms* in Search, Optimization and Machine Learning (1989).

NSP as a scheduling task consists of the assignment of shifts and holidays to nurses for each day on the time horizon, taking into consideration a variety of conflicting interests or objectives between the hospitals and individual nurses. Given a number of nurses with specific skills and working agreements, a contract may consist of general constraints as there are restrictions on the number of nurses for each shift; the maximum number of shifts in a week, a month, etc. Moreover, a number of personal wishes or desires representing nurses' preferences are allowed. For instance, a demand for the desired day off, demand for doing a certain shift on a certain day with a certain nurse, etc. Conventionally, every nurse works on three shifts; day shift, night shift, and late night shift and has some holidays.

Throughout this paper, shifts and day offs are denoted as a single byte as follows: ***'d'*** - day shift (8:00 - 16:00), ***'n'*** - night shift (16:00 - 24:00), ***'l'*** - late-night shift (0:00 - 08:00), and ***'h'*** - day offs. The schedules have to satisfy a given number of nurses for each shift (hard constraints). Moreover, it is desired to satisfy as many wishes of the individual nurses as possible (soft constraints). In this paper, the maximum number of nurses *(MAXN)* and days *(MAXD)* to be scheduled is set to 15 nurses and 30 days. Also a 15-day working history is considered to ensure that unpopular shifts are evenly distributed.

3. HARD AND SOFT CONSTRAINTS

Table 1 shows the hard constraints considered in this paper. Violation of the hard constraints is not allowed. It is desired to satisfy soft constraints as much as possible; however, their violation does not lead to generation of infeasible schedules.

3.1 Fitness Calculation

A fitness function denoted by F_i is calculated for each nurse. All nurses are assigned the same fitness function. F_i consists of the following three factors:

F_{ip} Fitness of the night shift pattern in respect to its order and length.

F_{ic} The number of consecutive night shifts in respect to its length.

F_{id} The interval between night shifts.

Each factor of F_i represents a predefined requirement. If it is violated, corresponding penalties are assigned. Detailed explanation of each factor of F_i is given in Jan et al. (1999). Overall fitness of individual nurses F_i is calculated according to equation 1.

$$F_i = \alpha.F_{ip} + \beta.F_{ic} + \gamma.F_{id} \tag{1}$$

α, β, γ are parameters, representing specific criterion of the hospitals, chief nurses, etc. The objective function for the entire schedule is formalized as:

Maximize $$avg = \frac{1}{N}\sum_{i=1}^{MAXN} F_i \tag{2}$$

Maximize $$dev = \sqrt{\frac{1}{MAXN}\sum_{i=1}^{MAXN} F_i^2 - \frac{1}{MAXN^2}\left(\sum_{i=1}^{MAXN} F_i\right)^2} \tag{3}$$

Subject To Hard constraints (see Table 1)

Table 1. Hard constraints and nurses' preference considered in this paper

Hard Constraints:

1. Each nurse can work only one shift in a day.
2. Number of day shift nurses must be $\geq$ the required number of nurses for day shift
3. Number of night shift nurses must be = the required number of night shift nurses.
4. Number of late shift nurses must be = the required number of late shift nurses.

Nurses' Preferences:

The right to select their days off as they wish.

It is desirable to attain the maximization of the average of the fitness of all nurses, and minimization of the variance of the fitness of all nurses. A schedule with *avg=0* and *dev=0* (see Figure 6) represents the global optimum or the desired absolute solution to the problem. Moreover, for individual nurses $F_i = 0$ (see Figure 5) represents the best fitness value.

The existence of the global optimum for the later model of NSPs considered in this paper is unknown. Thus it is desirable to attain the values of *avg* and *dev* as close to the global optimum (*zero*) as possible.

4. ALGORITMIC FLOW OF CGA FOR NSPs

4.1 Recollection of the Concept of the Pareto Optimality

According to Bagchi (1999) solutions to a multi-objective optimization problem are mathematically expressed in terms of non-dominated or "*superior*" points. The non-domination property of a solution may be explained as follows. In a minimization problem, a solution vector is partially dominated by another
vector $x^{(2)}$ (written as $x^{(1)} \succ x^{(2)}$), when no component value of $x^{(2)}$ is less

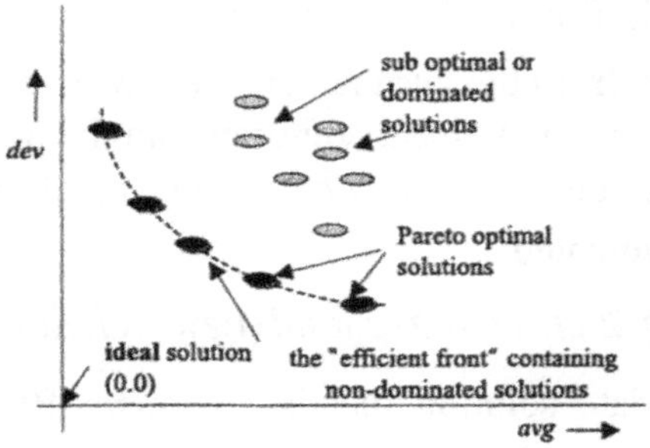

Figure 1. Graphical illustration of the Pareto efficient front for

than $x^{(1)}$ and at least one component of $x^{(2)}$ is strictly greater than $x^{(1)}$. In a minimization problem if $x^{(1)}$ is partially less than $x^{(2)}$, we say that solution $x^{(1)}$ dominates $x^{(2)}$ or the solution $x^{(2)}$ is inferior to $x^{(1)}$. Any member of such vectors that is non-dominated by any other member is said to be *non-dominated* or *non-inferior*. In multi-objective optimization problems, there usually exist many solutions that are optimal in the *Pareto* sense. Owing to such plurality of optimal decisions, the most desirable decision may be selected after one has generated the *Pareto optimal* or non-dominated solutions. The final solution thus selected as the most desirable, or at least the best-compromised solution, is called the *preferred* solution.

4.2 Preference Handling and Decision Making for NSPs

As indicated in Horn (1997) and Coello (2000), preferences can also be a *priori*, a *posteriori*, or in an *interactive* way when using evolutionary algorithms (EAs). If preferences are expressed a *priori*, the decision maker (DM) has to define his/her preferences in advance (before actually performing the search). In the second case, we search first, and decide later. This is the category where most evolutionary multi-objective optimization (EMOO) approaches fall into, including this paper (i.e. EMOO techniques based on Pareto ranking). In this case we use an EA to search the alternatives, where "best possible" normally means non-dominated or Pareto efficient solutions. The third case is the search of EA using preferences from the DM, but assuming that such preferences could change over time.

4.3 Cooperative GA for NSPs

To begin with, let us briefly describe the search procedure of CGA for NSPs. After the initialization of the initial feasible schedule and 15 days of the working history of the previous month, all nurses are evaluated according to their fitness F_i. To create a new schedule, two nurses are randomly selected from the current schedule. Moreover, two crossing points are selected so that only legal children can be generated. As crossover progresses new solutions new_ch are generated and updated if they are explicitly non-dominated:

$$if((new\_ch.avg \geq current.avg) and (new\_ch.dev \leq current.dev)) \quad (4)$$

Any candidate solution that is no longer Pareto efficient is simply discarded from the list of the candidate solutions. These potential candidate solutions represent the "Pareto efficient solutions". From the set of Pareto efficient solutions, only one solution is randomly selected and replaced with its inferior parents. At this point one generation is completed. Also a separate list of the set of observed Pareto efficient solutions are archived and updated in each generation. The search continues while for all nurses $F_i \neq 0$ or $time \leq MAXTIME$.

When one of the above mentioned termination criteria are met, the acquired set of solutions (schedules) is presented for the decision maker.

4.4 Design of the Mutation Operator

Because GAs are computationally intensive algorithms, there is a strong need to make it effective. We define effective as producing good results in a short time. As mentioned in section 1, CGA initializes its search with an initial feasible schedule and continues searching in feasible search regions only. This approach guarantees successful generation of feasible schedules, i.e. it preserves hard constraints throughout the search process of CGA.

It is implied by our experiments that the unwanted premature convergence occurs in the early stages of the search. Traditional mutation fails to divert search towards potentially unexplored regions in the search space (see Figure 7 left and Figure 8 left). It is due to the presence of strong constraints and restricting CGA to search in the feasible region only. To emphasize the role of mutation operator, we will rather call it the *"escape"* operator. The basic idea beyond the design of the escape operator is illustrated in Figure 2.

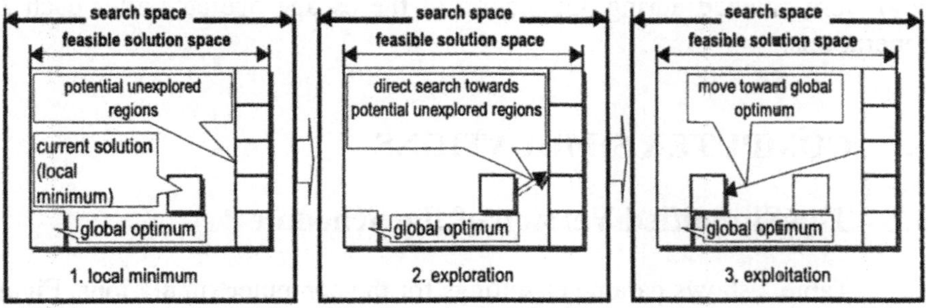

Figure 2. Exploration and exploitation of the potential unexplored regions in the search space.

This operator is supposed to:

- divert the search from the local minimum;
- direct the search towards potential unexplored regions in the search space;
- enable GAs to generate solutions as close to the global optimum as possible; and
- desirably be simple and cost effective.

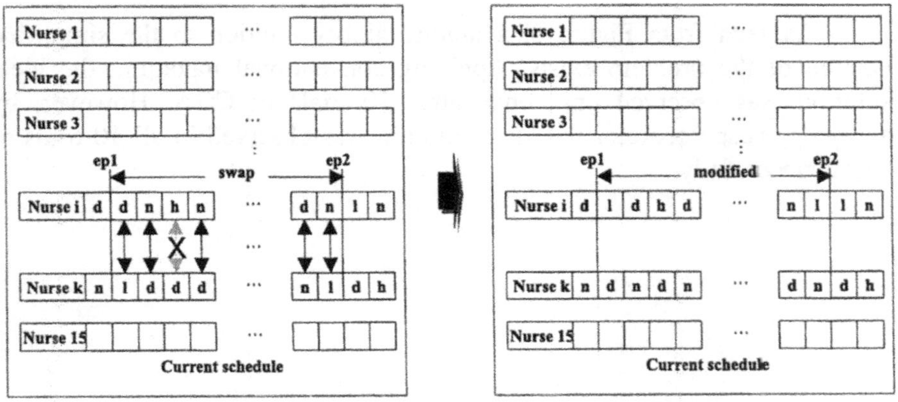

Figure 3. Illustration of the application of the escape operator to NSPs.

Figure 3 illustrates an example of the implementation of the escape operator for NSPs. During application of the escape operator, two different nurses (nurse[i], nurse[k]) and two different escaping points *(ep1, ep2)* are randomly selected. If nurse[i].shift[j] and nurse[k].shift[j] fall between escaping points *ep1* and *ep2*, their corresponding shifts are simply swapped (see figure 3). Since the days off are considered as the nurses' preferences, escape operator is designed so that it does not disrupt them.

This simple operator does not require any other additional computational effort. Only the determination of the frequency of application of the escape operator is sufficient. For the current model of NSPs presented in this paper, the best observed values of the objective functions (avg and dev) were attained during application of the escape operator after each 11 generations.

5. COMPUTER SIMULATIONS

5.1 The Simplified Version of the Schedule

Table 2 shows parameter settings for the computer simulations. Figure 4 shows the initial schedule and Figure 5 shows the acquired schedule for the simplified version of the problem (with escape operator being applied).

Table 2. Parameter settings for the experiments

Time horizon (MAXD)	30 days
Working history of the previous month (MAXH)	15 days
Max number of nurses (MAXN)	15 nurses
Max time (MAXTIME)	5 minutes
CPU	Celleron 300 MHz
$\alpha=\beta=\gamma$	1.0

As seen from Figure 6, a unique global solution to the simplified version of the problem exists. Applying conventional mutation, the ideal solution was observed only once after 10 trials of CGA. However, by applying escape operator the ideal solution was observed in all 10 trials of CGA successfully.

	1	2	3	4	5	6	7	8	9	10	11	12	13	14	15	1	2	3	4	5	6	7	8	9	10	11	12	13	14	15	16	17	18	19	20	21	22	23	24	25	26	27	28	29	30	fitnes	
Nurse[01]	l	l	n	n	d	d	d	d	d	d	d	d	d	d	d	l	l	l	l	l	l	l	l	l	l	l	l	l	l	l	l	l	l	l	l	l	l	l	l	l	l	l	l	l	l	F:	-2720
Nurse[02]	d	l	l	n	n	d	d	d	d	d	d	d	d	d	d	l	l	l	l	l	l	l	l	l	l	l	l	l	l	l	l	l	l	l	l	l	l	l	l	l	l	l	l	l	l	F:	-2721
Nurse[03]	d	d	l	l	n	n	d	d	d	d	d	d	d	d	d	n	n	n	n	n	n	n	n	n	n	n	n	n	n	n	n	n	n	n	n	n	n	n	n	n	n	n	n	n	n	F:	-2724
Nurse[04]	d	d	d	l	l	n	n	d	d	d	d	d	d	d	d	n	n	n	n	n	n	n	n	n	n	n	n	n	n	n	n	n	n	n	n	n	n	n	n	n	n	n	n	n	n	F:	-2729
Nurse[05]	d	d	d	d	l	l	n	n	d	d	d	d	d	d	d	d	d	d	d	d	d	d	d	d	d	d	d	d	d	d	d	d	d	d	d	d	d	d	d	d	d	d	d	d	d	F:	-676
Nurse[06]	d	d	d	d	d	l	l	n	n	d	d	d	d	d	d	d	d	d	d	d	d	d	d	d	d	d	d	d	d	d	d	d	d	d	d	d	d	d	d	d	d	d	d	d	d	F:	-625
Nurse[07]	d	d	d	d	d	d	l	l	n	n	d	d	d	d	d	d	d	d	d	d	d	d	d	d	d	d	d	d	d	d	d	d	d	d	d	d	d	d	d	d	d	d	d	d	d	F:	-576
Nurse[08]	d	d	d	d	d	d	d	l	l	n	n	d	d	d	d	d	d	d	d	d	d	d	d	d	d	d	d	d	d	d	d	d	d	d	d	d	d	d	d	d	d	d	d	d	d	F:	-529
Nurse[09]	d	d	d	d	d	d	d	d	l	l	n	n	d	d	d	d	d	d	d	d	d	d	d	d	d	d	d	d	d	d	d	d	d	d	d	d	d	d	d	d	d	d	d	d	d	F:	-484
Nurse[10]	d	d	d	d	d	d	d	d	d	l	l	n	n	d	d	d	d	d	d	d	d	d	d	d	d	d	d	d	d	d	d	d	d	d	d	d	d	d	d	d	d	d	d	d	d	F:	-441
Nurse[11]	d	d	d	d	d	d	d	d	d	d	l	l	n	n	d	d	d	d	d	d	d	d	d	d	d	d	d	d	d	d	d	d	d	d	d	d	d	d	d	d	d	d	d	d	d	F:	-400
Nurse[12]	d	d	d	d	d	d	d	d	d	d	d	l	l	n	n	d	d	d	d	d	d	d	d	d	d	d	d	d	d	d	d	d	d	d	d	d	d	d	d	d	d	d	d	d	d	F:	-361
Nurse[13]	n	d	d	d	d	d	d	d	d	d	d	d	l	l	n	d	d	d	d	d	d	d	d	d	d	d	d	d	d	d	d	d	d	d	d	d	d	d	d	d	d	d	d	d	d	F:	-441
Nurse[14]	n	n	d	d	d	d	d	d	d	d	d	d	d	l	l	d	d	d	d	d	d	d	d	d	d	d	d	d	d	d	d	d	d	d	d	d	d	d	d	d	d	d	d	d	d	F:	-601
Nurse[15]	l	n	n	d	d	d	d	d	d	d	d	d	d	d	l	d	d	d	d	d	d	d	d	d	d	d	d	d	d	d	d	d	d	d	d	d	d	d	d	d	d	d	d	d	d	F:	-601
																																														avg	-1109
																																														dev	977.4

history of the previous month ◄—► current month

Figure 4. The simplified initial schedule.

	1	2	3	4	5	6	7	8	9	10	11	12	13	14	15	1	2	3	4	5	6	7	8	9	10	11	12	13	14	15	16	17	18	19	20	21	22	23	24	25	26	27	28	29	30	fitnes	
Nurse[01]	l	l	n	n	d	d	d	d	d	d	d	d	d	d	d	l	l	n	n	d	d	d	d	d	d	d	d	d	d	d	l	l	n	n	d	d	d	d	d	d	d	d	d	d	d	F:	0
Nurse[02]	d	l	l	n	n	d	d	d	d	d	d	d	d	d	d	d	l	l	n	n	d	d	d	d	d	d	d	d	d	d	d	l	l	n	n	d	d	d	d	d	d	d	d	d	d	F:	0
Nurse[03]	d	d	l	l	n	n	d	d	d	d	d	d	d	d	d	d	d	l	l	n	n	d	d	d	d	d	d	d	d	d	d	d	l	l	n	n	d	d	d	d	d	d	d	d	d	F:	0
Nurse[04]	d	d	d	l	l	n	n	d	d	d	d	d	d	d	d	d	d	d	l	l	n	n	d	d	d	d	d	d	d	d	d	d	d	l	l	n	n	d	d	d	d	d	d	d	d	F:	0
Nurse[05]	d	d	d	d	l	l	n	n	d	d	d	d	d	d	d	d	d	d	d	l	l	n	n	d	d	d	d	d	d	d	d	d	d	d	l	l	n	n	d	d	d	d	d	d	d	F:	0
Nurse[06]	d	d	d	d	d	l	l	n	n	d	d	d	d	d	d	d	d	d	d	d	l	l	n	n	d	d	d	d	d	d	d	d	d	d	d	l	l	n	n	d	d	d	d	d	d	F:	0
Nurse[07]	d	d	d	d	d	d	l	l	n	n	d	d	d	d	d	d	d	d	d	d	d	l	l	n	n	d	d	d	d	d	d	d	d	d	d	d	l	l	n	n	d	d	d	d	d	F:	0
Nurse[08]	d	d	d	d	d	d	d	l	l	n	n	d	d	d	d	d	d	d	d	d	d	d	l	l	n	n	d	d	d	d	d	d	d	d	d	d	d	l	l	n	n	d	d	d	d	F:	0
Nurse[09]	d	d	d	d	d	d	d	d	l	l	n	n	d	d	d	d	d	d	d	d	d	d	d	l	l	n	n	d	d	d	d	d	d	d	d	d	d	d	l	l	n	n	d	d	d	F:	0
Nurse[10]	d	d	d	d	d	d	d	d	d	l	l	n	n	d	d	d	d	d	d	d	d	d	d	d	l	l	n	n	d	d	d	d	d	d	d	d	d	d	d	l	l	n	n	d	d	F:	0
Nurse[11]	d	d	d	d	d	d	d	d	d	d	l	l	n	n	d	d	d	d	d	d	d	d	d	d	d	l	l	n	n	d	d	d	d	d	d	d	d	d	d	d	l	l	n	n	d	F:	0
Nurse[12]	d	d	d	d	d	d	d	d	d	d	d	l	l	n	n	d	d	d	d	d	d	d	d	d	d	d	l	l	n	n	d	d	d	d	d	d	d	d	d	d	d	l	l	n	n	F:	0
Nurse[13]	n	d	d	d	d	d	d	d	d	d	d	d	l	l	n	n	d	d	d	d	d	d	d	d	d	d	d	l	l	n	n	d	d	d	d	d	d	d	d	d	d	d	l	l	n	F:	0
Nurse[14]	n	n	d	d	d	d	d	d	d	d	d	d	d	l	l	n	n	d	d	d	d	d	d	d	d	d	d	d	l	l	n	n	d	d	d	d	d	d	d	d	d	d	d	l	l	F:	0
Nurse[15]	l	n	n	d	d	d	d	d	d	d	d	d	d	d	l	l	n	n	d	d	d	d	d	d	d	d	d	d	d	l	l	n	n	d	d	d	d	d	d	d	d	d	d	d	l	F:	0
																																														avg	0
																																														dev	0

history of the previous month ◄—► current month

Figure 5. The acquired schedule.

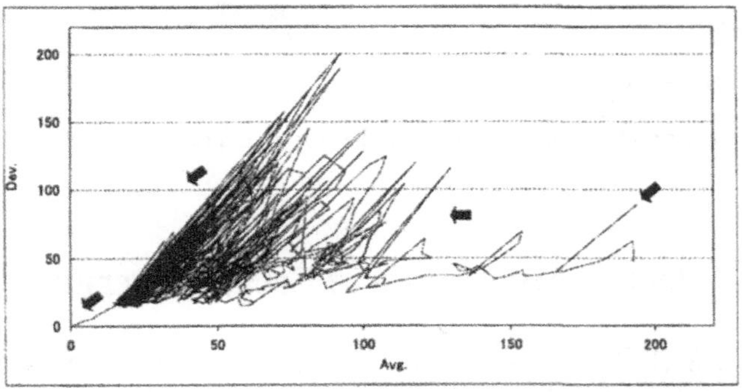

Figure 6. Transition of the avg. and dev. for the simplified version of the schedule.

5.2 The Schedule With Days off Being Added

The existence of the global optimum for this version of the problem is unknown. Therefore, it is desirable to attain the acquired solutions as close to the global optimum as possible. As seen from Figure 7 (left), applying conventional mutation premature convergence occurs in the relatively early stages of the search

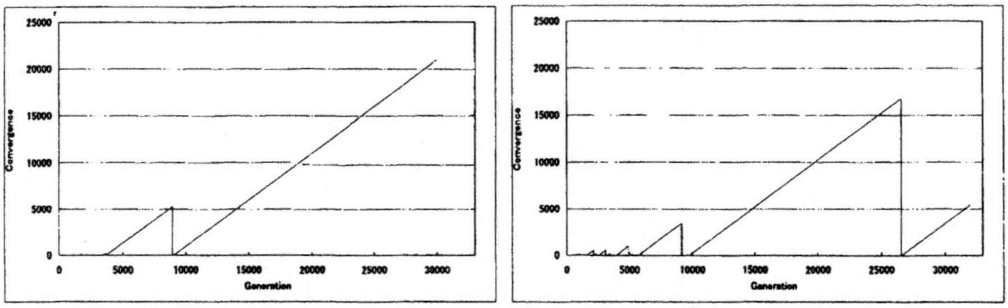

Figure 7. Convergence to the Pareto front: left-with mutation. and right with escape operator being applied.

Figure 8 left and right show distribution of the solution space prior to and after the application of the escape operator, respectively. As seen from Figure 8, applying escape operator, the search space has been dramatically diversified resulting in generations of superior solutions (see Figure 10 right). Figure 11 upper and below show the initial and a randomly selected acquired schedule (from the set of Pareto efficient solutions) at *MAXTIME*, respectively.

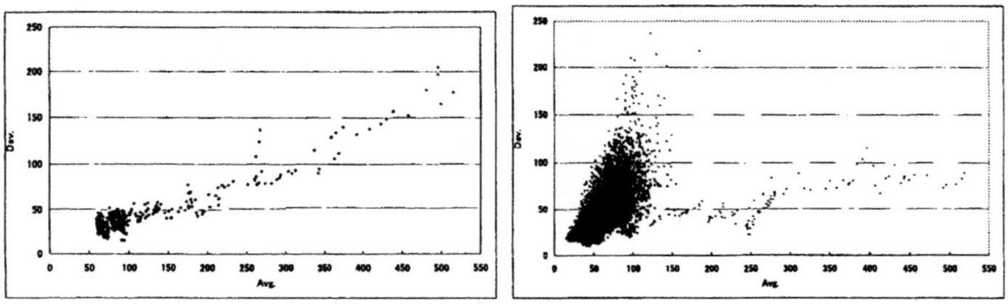

Figure 8. Distribution of the solution space: left-prior to and right-after the application of the escape operator.

5.3 The Effect of Application of Escape Operator

Figure 8 (left) shows distribution of the solution space prior to and (right) after the application of the escape operator. As seen from Figure 6 (right) this operator successfully diversifies the solution space, resulting in generation of superior solutions (see Figure 10 right).

5.4 The Acquired Number of Non-dominated Solutions in MAXTIME

Figure 9 compares the number of non-dominated solutions in each generation with conventional mutation and escape operator being applied. It represents the archived set of (all Pareto non-dominated solutions) from the beginning of the search to MAXTIME (one trial). As seen from Figure 9 applying the escape operator, the number of these solutions increases from 3 to 9.

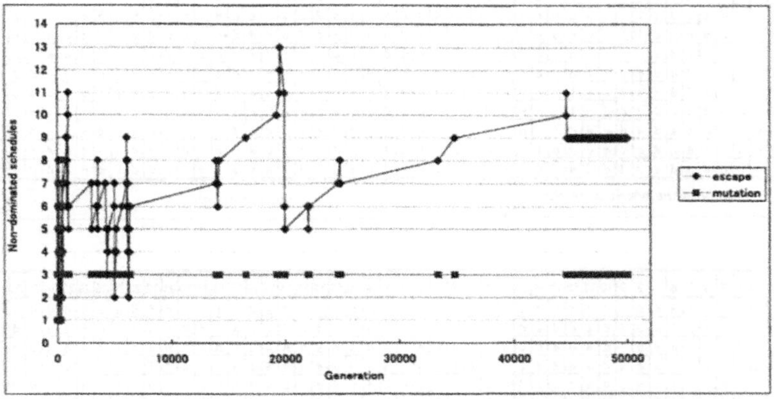

Figure 9. The number of non-dominated solutions per generation with days off being added.

Also Figure 10 (right) shows that the acquired solutions are more adjacent to the global optimum.

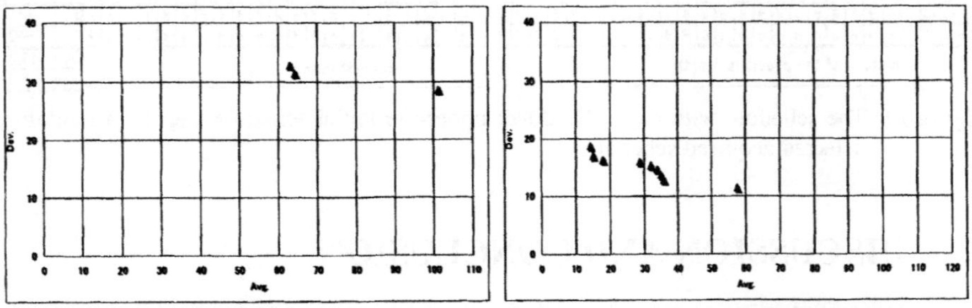

Figure 10. The acquired set of Pareto non-dominated schedules in MAXTIME: left-prior to and right-after the application of the escape operator with days off being added.

	1	2	3	4	5	6	7	8	9	10	11	12	13	14	15	1	2	3	4	5	6	7	8	9	10	11	12	13	14	15	16	17	18	19	20	21	22	23	24	25	26	27	28	29	30	fitnes	
Nurse[01	l	l	n	n	d	d	d	d	d	d	d	d	d	d	d	l	l	l	l	h	h	l	l	l	l	l	l	l	l	l	l	l	l	l	l	l	l	l	h	l	l	l	l	l	l	F:	-2041
Nurse[02	d	l	l	n	n	d	d	d	d	d	d	d	d	d	d	l	l	l	l	l	l	l	h	h	l	l	l	l	l	l	l	l	l	l	l	l	l	l	l	l	l	l	l	l	l	F:	-2322
Nurse[03	d	d	l	l	n	n	d	d	d	d	d	d	d	d	d	n	n	n	n	l	l	n	l	l	n	n	n	h	h	n	n	n	n	n	n	n	n	n	l	n	n	n	n	n	n	F:	-2325
Nurse[04	d	d	d	l	l	n	n	d	d	d	d	d	d	d	d	n	n	n	n	n	n	n	h	n	n	n	n	n	n	n	n	n	n	n	n	n	n	n	n	n	n	n	n	n	n	F:	-2449
Nurse[05	d	d	d	d	l	l	n	n	d	d	d	d	d	d	d	d	d	d	d	n	n	d	n	n	d	d	d	n	n	d	d	d	h	h	d	d	d	d	n	d	d	d	d	d	d	F:	-768
Nurse[06	d	d	d	d	d	l	l	n	n	d	d	d	d	d	d	d	d	d	d	d	d	d	n	d	d	d	d	d	d	d	d	d	d	d	d	d	d	d	h	h	d	d	d	d	d	F:	-365
Nurse[07	d	d	d	d	d	d	l	l	n	n	d	d	d	d	d	d	d	d	d	d	d	d	d	d	d	d	d	d	d	d	d	d	d	d	d	d	d	h	d	h	d	d	d	d	d	F:	-576
Nurse[08	d	d	d	d	d	d	d	l	l	n	n	d	d	d	d	d	d	d	d	d	d	d	d	d	d	h	d	d	d	d	d	d	d	d	d	d	d	d	d	d	d	d	d	d	d	F:	-529
Nurse[09	d	d	d	d	d	d	d	d	l	l	n	n	d	d	d	d	d	d	d	d	d	d	d	d	d	d	d	d	d	d	d	d	d	d	d	d	h	h	d	d	d	d	d	d	d	F:	-484
Nurse[10	d	d	d	d	d	d	d	d	d	l	l	n	n	d	d	d	d	d	d	d	d	d	d	d	d	d	d	d	h	d	d	d	d	d	h	d	d	d	d	d	d	d	d	d	d	F:	-441
Nurse[11	d	d	d	d	d	d	d	d	d	d	l	l	n	n	d	d	d	d	d	d	d	d	d	h	h	d	d	d	d	d	d	d	d	d	d	d	d	d	d	d	d	d	d	d	d	F:	-400
Nurse[12	d	d	d	d	d	d	d	d	d	d	d	l	l	n	n	d	d	d	d	d	d	d	d	d	d	d	d	d	d	h	d	d	d	d	d	d	d	d	d	d	d	d	d	d	d	F:	-361
Nurse[13	n	d	d	d	d	d	d	d	d	d	d	d	l	l	n	d	d	d	d	d	d	d	d	d	d	d	d	d	d	d	d	d	h	h	d	d	d	d	d	d	d	d	d	d	d	F:	-441
Nurse[14	n	n	d	d	d	d	d	d	d	d	d	d	d	l	l	d	d	d	d	d	d	d	d	h	h	d	d	d	d	d	d	d	d	d	d	d	d	d	d	d	d	d	d	d	d	F:	-601
Nurse[15	l	n	n	d	d	d	d	d	d	d	d	d	d	d	l	d	d	h	d	d	d	d	d	d	d	d	d	d	d	d	d	d	d	d	d	d	d	d	d	d	d	d	d	d	d	F:	-601
	history of the previous month															current month																														avg	-980
																																														dev	796.6

	1	2	3	4	5	6	7	8	9	10	11	12	13	14	15	1	2	3	4	5	6	7	8	9	10	11	12	13	14	15	16	17	18	19	20	21	22	23	24	25	26	27	28	29	30	fitnes	
Nurse[01	l	l	n	n	d	d	d	d	d	d	d	d	d	d	d	l	l	n	n	h	h	d	d	d	d	d	d	d	d	d	d	l	l	n	n	d	d	d	h	d	d	d	d	d	d	F:	-1
Nurse[02	d	l	l	n	n	d	d	d	d	d	d	d	d	d	d	l	n	n	d	d	d	d	h	h	d	d	d	d	d	l	l	n	n	d	d	d	d	d	d	d	d	d	d	d	d	F:	-42
Nurse[03	d	d	l	l	n	n	d	d	d	d	d	d	d	d	d	d	l	l	n	n	d	d	d	d	d	d	d	h	h	d	d	d	d	l	l	n	n	d	d	d	d	d	d	d	d	F:	-5
Nurse[04	d	d	d	l	l	n	n	d	d	d	d	d	d	d	d	d	d	d	l	l	n	n	h	d	d	d	d	d	d	d	d	d	d	d	d	l	l	n	n	d	d	d	d	d	d	F:	-4
Nurse[05	d	d	d	d	l	l	n	n	d	d	d	d	d	d	d	d	d	d	d	l	l	n	n	d	d	d	d	d	d	d	d	d	h	h	d	d	d	d	l	l	n	n	d	d	d	F:	-16
Nurse[06	d	d	d	d	d	l	l	n	n	d	d	d	d	d	d	d	d	d	d	d	d	l	l	n	n	d	d	d	d	d	d	d	d	d	l	l	n	n	h	h	d	d	d	d	d	F:	-5
Nurse[07	d	d	d	d	d	d	l	l	n	n	d	d	d	d	d	d	d	l	l	n	n	d	d	d	d	d	d	d	d	d	l	l	n	n	d	d	d	h	d	h	d	d	d	d	l	F:	-21
Nurse[08	d	d	d	d	d	d	d	l	l	n	n	d	d	d	d	d	d	d	d	d	l	l	n	n	d	h	d	d	d	d	d	d	l	l	n	n	d	d	d	d	d	d	d	d	d	F:	-13
Nurse[09	d	d	d	d	d	d	d	d	l	l	n	n	d	d	d	d	d	d	d	d	d	d	d	d	d	l	l	n	n	d	d	d	d	d	d	d	h	h	d	d	d	d	l	l	n	F:	-12
Nurse[10	d	d	d	d	d	d	d	d	d	l	l	n	n	d	d	d	d	d	d	d	d	d	l	l	n	n	d	d	h	d	d	d	d	d	h	d	d	l	l	n	n	d	d	d	d	F:	-4
Nurse[11	d	d	d	d	d	d	d	d	d	d	l	l	n	n	d	d	d	d	d	d	d	d	d	h	h	d	l	l	n	n	d	d	d	d	d	d	d	d	d	l	l	n	n	d	d	F:	-5
Nurse[12	d	d	d	d	d	d	d	d	d	d	d	l	l	n	n	d	d	d	d	d	d	d	d	l	l	n	n	d	d	h	d	d	d	d	d	d	l	l	n	n	d	d	d	d	d	F:	-13
Nurse[13	n	d	d	d	d	d	d	d	d	d	d	d	l	l	n	d	d	d	d	d	d	d	d	d	l	l	n	n	d	d	d	d	h	h	d	d	d	d	d	d	l	l	n	n	d	F:	-45
Nurse[14	n	n	d	d	d	d	d	d	d	d	d	d	d	l	l	n	d	d	d	d	d	d	d	h	h	d	d	l	l	n	n	d	d	d	d	d	d	d	d	d	d	l	l	n	n	F:	-41
Nurse[15	l	n	n	d	d	d	d	d	d	d	d	d	d	d	l	n	n	h	d	d	d	d	d	d	d	d	d	d	l	l	n	n	d	d	d	d	d	d	d	d	d	d	d	l	l	F:	-40
	history of the previous month															current month																														avg	-17.8
																																														dev	15.5

Figure 11. The schedule with days off added: upper-the initial schedule, below- a randomly selected acquired schedule.

6. DISCUSSION AND CONCLUSION

Generally speaking, to solve NSPs using either conventional or EC approaches, we would face the following problems: firstly, how to design an adaptive model for NSPs and secondly, how to propose an adaptive algorithm for the solution of the proposed model of the problem. The design of the model of NSPs is a very difficult task because objectives of the hospitals are not fixed (varying from hospital to hospital). Consideration of a real model of NSPs would very likely result in limitation of the design of the proposed search algorithms to the given model of the problem. Consideration of an artificial model of the problem would limit our ability to evaluate the acquired solution(s) because of the multi-objective nature of the problem. Therefore, assuming that the model of the problem is well defined, generation of a set of non-dominated solutions that are slightly different from each other (that is Pareto efficient) seems to be a good strategy. Because of the multi-objective nature of the problem, instead of a single solution, CGA presents the set of all non-dominated solutions for the decision maker. Only the decision maker is able to decide which acquired schedule meets the criteria as the best solution. Let us summarize our observations regarding the solutions of NSPs with a GA approach. Because NSPs represent a real-world problem, it is desired to generate feasible schedules. To generate feasible schedules it is easier to start the search with an initial feasible schedule. To do so, the search would get restricted only to the feasible portion of the search space leading to the premature convergence. To overcome the above mentioned problem the so called escape operator was proposed. As the computer simulations showed, the proposed operator successfully directed the search towards potentially unexplored regions of the solution space. This simple technique enabled CGA to generate quantitatively and qualitatively better solutions. In order to confirm the effectiveness of the CGA for the solution of NSPs, in our future work we would like to address a more complex model of the problem.

References

Yamamura M., Yamagishi M., Ase H. Genetic algorithm, Sangyou Tosho, 1995; 3: 89-126.

Jan A., Yamamoto M., Ohuchi A., Operations Research from Theory to Real Life. Proceedings of the 15 th National Conference for Operation Research Inc. ASOR Queensland Branch and ORSJ Hokkaido Chapter Joint Workshop 1999; 1: 615-628.

Yamamoto M., Kawamura H., Ohuchi A. Collective Approach to Optimization Problems. Proceedings of ITC-CSCC '98, 1998; 1479-1482.

Bradley D. and Martin J. Continuous Personnel Scheduling Algorithms: A Literature Review. Journal of the Society for Health Systems 1990; 2: 8-23.

Coello C. Handling. Preferences in Evolutionary Multiobjective Optimization: A Survey. Congress on Evolutionary Computation 2000, IEEE Piscataway, La Jolla California, USA 2000; 1: 30-37.

Goldberg, D. Genetic Algorithm in Search, Optimization and Machine Learning. Addison-Wesley, Reading, MA 1989.

Richard K. Belew and Michael D. Vose. Foundations of Genetic Algorithms. Morgan Kaufman Publishers Inc 1997.

Tapan P. Bagchi. Multiobjective Scheduling by Genetic Algorithms. Kluwer Academic Publishers 1999.

PARALLEL AGENT PASSING TABU SEARCH ALGORITHM FOR GRAPH PARTITIONING PROBLEM

Taichi Kaji

Otaru University of Commerce
3-5-21, Midori, Otaru, 047-8501, Japan
kaji@res.otaru-uc.ac.jp

Abstract This paper implements standard Tabu Search (TS) and the Simulated Annealing (SA) algorithm, and examine the performance of the basic type of TS and SA for a multi-graph-partitioning problem. The results show that there is a tendency for the solutions obtained using standard TS and SA to be trapped in bad local optimum and TS is too computationally expensive in its pure form. In this respect, the increasing availability of parallel machines offers an interesting opportunity to explore the possibility of parallel TS algorithms, in order to reduce computational time for TS and to obtain better solutions. However, the parallelism for TS generally induces a loss of quality of the best solution found, because it significantly limits the movement possibilities between different subproblems. Hence, we present how the difficulty caused in parallel algorithm for this problem has been overcome.

Key words: Parallel algorithm, Graph-partitioning problem, Tabu search, Agent, Simulated annealing

1. INTRODUCTION

There are several types of graph partitioning problems depending on the number of partitions and their condition of partitions. Such problems arise in the very large scale integration (VLSI) design (Jepsen and Gelatt, 1983), layout problems (Lee et al., 1979), and scheduling problems (Sarkar, 1989). In particular, researchers have addressed the bi-partitioning problem using meta-heuristics (Fujisawa et al 1994; Johnson, 1989)]. Many of the practical applications have been in the complicated problem domain. Now, we consider a useful multi-graph partitioning problem. This multi-graph partitioning problem is more difficult to solve using heuristics than the bi-graph partitioning problem. It is hard to compute a good approximate solution (Bui and Jones, 1992). Hence this paper considers TS and SA, which have been successfully applied to bi-graph partitioning, for the multi-graph partitioning problem.

First, the standard TS and SA are implemented for the multi-graph-partitioning problem, and the performance of both meta-heuristic algorithms are examined. The solutions obtained by both algorithms were approximately similar, and the TS algorithm in pure form required more computational time than SA. However, there is a tendency for the solutions obtained using the basic type of TS and SA to be trapped in bad local optimums. The purpose of this paper is to explore the possibility of the use of the parallel TS algorithm, to reduce the computational time of TS and to obtain better solutions, under the circumstance of the increase in availability of parallel machines.

In iterative search techniques including TS, various different types of parallelism may be distinguished (Mohr, 1988). We consider here only two general classes according to the method used for the parallelizing move. In the first type of parallelism, we consider the classes in which the search for the next move can be parallelized. This will generally require the partition of the set of feasible moves into p subsets. Each of these subsets is assigned to one process which computes its best member move. The best overall of these p moves is then determined and applied to the current solution (Mohr, 1988; Taillard, 1989). The second type of parallelism treats the problem of performing several moves concurrently, called multi-move parallelism. This can be done by partitioning the problem itself into several independent subproblems. The complete solution is then obtained by combining the subsolutions. This method needs no communication or synchronization except for its initialization and for consolidating the subsolutions at the end. The difficulty with this type of parallelism is that it strongly limits the movement possibilities between different subproblems and thus generally induces a loss of quality of the best solution found (Fiechter, 1994). Following is a description of how the difficulty of this problem can be overcome.

The above-mentioned two types of classes are adapted to the multi-graph-partitioning problem, and new ideas to overcome the difficulty of developing a parallel algorithm for this problem are presented. One approach has been taken to parallelize neighbouring moves and has reduced the computational time significantly. Another main approach, which considers the second type of parallelism, for obtaining high-quality solutions has also been proposed. This parallelism has been implemented using a framework in which a number of independent moves operate on each different partial solution space, to improve the quality of solutions. However, the problem with this kind of parallelism by iterative search is that it strongly limits the movement possibilities between different subproblems and thus generally induces a loss of quality of the best solution found. We overcome this difficulty by attractive searching based on TS, the extension of searching by super node move, and the diversification of producing subproblems by generator agents. This produces cooperatively better solutions from partial solutions obtained by parallel optimizing with reduced computational time. Finally, as the algorithms are well suited for parallel computation, an implementation for parallel computation is described. Numerical results obtained show the efficiency of the parallel algorithm in terms of the quality of the solution and computational time. The results show that the proposed algorithm yields better solutions than SA and standard TS, and reduces the computational time significantly.

2. MULTI-GRAPH PARTITIONING PROBLEM

In this paper, the multi-partitioning problem is relevant to the extent that the total cost of the edges in the cut set is minimized. This problem is complex but multipurpose in comparison with the bi-partitioning problem examined previously. Let $D(V,E)$ be a graph where V is a set of n nodes and E is a set of edges. Given $V_1, V_2, \cdots, V_k$, where any subset V_i is called a block, $V_1 \cup V_2 \cup \cdots \cup V_k = V$, and $V_i \cap V_j = \phi$, for $i \neq j$, a graph partition is defined as a set partition $\{V_1, V_2, \ldots, V_k\}$ such that the block size, which is the number of nodes for each block, is of a fixed size. The multi-graph-partitioning problem can be formulated mathematically as follows.

$$\begin{aligned} &\min \sum_{i\in V_k, j\in V_l, \forall k<l \in\{1,2,\cdots,k\}} e_{i,j} \\ &\text{subject to } |V_i| = M_i, \quad i = 1,\cdots,k, \\ &\sum_{i\in\{1,\cdots,k\}} M_i = n, \qquad (1) \end{aligned}$$

where $e_{i,j}$ is the edge cost and M_i is a block size that is given a positive number. The number of subsets k is the size of the partition and is a fixed number.

3. SEQUENTIAL BASIC META-HEURISTIC FOUNDATIONS

3.1 Tabu Search Algorithm

The TS (Glover and Laguna, 1997) is a meta-heuristic that guides a local heuristic search procedure to explore the solution space beyond local optimality. Widespread successes in practical applications of optimization have spurred a rapid growth of TS in the past few years. The basis for the TS may be described as follows. TS begins in the same way as an ordinary local search, proceeding iteratively from one point (solution) to another until a chosen termination criterion is satisfied. Each $x \in X$ has an associated neighbourhood $N(x) \subset X$, where X is the solution space of a given problem, and each solution $x' \in N(x)$ is reached from x by an operation called a *move*. The process in which the TS method seeks to transcend local optimality is based on an evaluation function that chooses the highest evaluation move in terms of the objective function and a tabu restriction. The tabu restriction employs a strategy of modifying $N(x)$ as the search progresses, effectively replacing it by another neighbourhood $N(\alpha,x)=N(x)-\alpha$, where α is called a *tabu list* consisting of s elements of solutions visited during the recent past, to avoid cycling. However, the tabu list memory is designed to exert a more subtle effect on the search through the use of attributive memory, which records information about solution attributes that change in moving from one solution to another. A key aspect of TS is the use of tabu list structures and neighbourhood structures to organize the way in which the space is explored.

We show standard TS algorithm by representing the neighbourhood structure and tabu memory structure for this problem. First we explain the neighbourhood structure for the multi-graph partitioning problem, that is based on the exchange of nodes. We consider a solution $x=(V_1,V_2,\ldots,V_k)$, where V_i is the subset of partitioned nodes of the graph. Each solution x has the following neighbourhood $N(x)$.

$$N(x) = \{x' = (V_1, V_2, \cdots, V_s', \cdots, V_w', \cdots, V_k) : V_s' = (V_s \setminus \{i\}) \cup \{j\}, \\ V_w' = (V_w \setminus \{j\}) \cup \{i\}, i \in V_s, j \in V_w, s \neq w]\}. \quad (2)$$

Next, we consider a tabu list structure. A data structure for the tabu list will be used to store a partial range of solution attributes rather than complete visited solutions. For this problem, the attribute can consist of nodes that are shifted by moves executed. When a node v of block V_i is shifted to another block V_j, we store v as the attribute in the tabu list to avoid returning v to V_i. In general, the TS become significantly stronger by including longer term memory. In this paper longer term memory is also used. When node v is shifted, we increase the longer term memory element LTM[v] by 1. The evaluation function for exchanging the nodes v_i and v_j is modified using longer term memory, that is, incremental cost + (LTM[v_i] +LTM[v_j])□*BIAS* / 2, where *BIAS* is a parameter.

3.2 Simulated Annealing Algorithm

The SA algorithm (Aarts and Korst, 1989) also provides an iterated hill-climbing mechanism as TS and local search do. Moves that degrade the current solution by an amount □*E*, which is a difference in cost, that is, cost of current solution – cost of selected neighbor solution, are accepted with probability $e^{-\Delta E/T}$. T is a control parameter analogous to temperature in the annealing of a physical system. In an optimization framework, we anneal a solution by performing standard iterative improvement steps, but now accept some degraded solutions according to this temperature rule. Central to any annealing algorithm is the need to evaluate many moves. During initial annealing, T is large; in this hot regime, most uphill moves are accepted. As iterative improvement proceeds, the temperature T is slowly reduced. Near the end of annealing, T is small; in this cold regime, few uphill moves are accepted. The purpose of the relatively high-temperature phase of the search is to discover the gross features of the search space. Successively lower temperatures identify more and more details while solutions become more and more localized. Ultimately, SA yields a very good solution as a result of a search process.

For this problem, we use the same neighborhood structure as TS in the SA algorithm. A new solution can be generated by choosing two nodes $i \in V_s$, $l \in V_t$ for all s and t (ʻs), and exchanging these two nodes between block V_s and V_t. The difference in cost $\delta(i;l \mid V_s, V_t)$, which is the same as □$E$, can be calculated incrementally from the following expression,

$$\delta(i;l \mid V_s, V_t) = \Big(\sum_{k \in V_s} e_{i,k} + \sum_{k \in V_t} e_{l,k}\Big) - \Big(\sum_{k \in V_t, k \neq l} e_{i,k} + \sum_{k \in V_s, k \neq i} e_{l,k}\Big). \tag{3}$$

The decision to accept new solutions is based on the above-mentioned basic acceptance criterion.

3.3 Computational Results for Basic Sequential Meta-heuristic Algorithms

In this section, the performance of standard TS and SA algorithms for the multi-graph-partitioning problem are evaluated. Relative error cannot obtained for this problem because it is a NP-complete problem. Consequently obtained local optimum values are evaluated in the common computational environment. We take average data for a random algorithm applied ten times. The algorithms used here were coded in C++ and implemented on an Intel Celeron processor/400MHz. There are several parameters in the algorithm that have to be adjusted. First, let us examine the effect for the most important parameter, the tabu list size *tabulength*, for TS. The way to obtain a good tabu list size for a given problem is simply to observe the quality of solutions when size is varied. We therefore solved randomly generated test problems for graphs with 500 nodes. We found that better solutions are obtained when *tabulength*=20, and therefore propose that *tabulength* be set to value 20. Similarly, the *BIAS* used in longer term memory is 0.20, by which better solutions are obtained. Next, consider the parametric values of SA in order to obtain good solutions in a reasonable amount of computational time. The initial *tmp* (T) is 15 and the stop *tmp* (T) is 0.01, since, in spite of extending the above range of *tmp*, we did not obtain a significant improvement in the quality of the solutions. *phi*=0.97 is found which is the cooling schedule parameter, gives an acceptable balance between solution quality and computational speed. It appears that initial *r*=100 and *tau*=1.03, where *r* is the initial length of Marcov chain and *tau* is the control length of Marcov chain, yield good compromises between the quality of the solutions obtained and the time required.

Assessing the performance of standard TS and SA using the above-mentioned parameters, eight partitioning test problems for graphs were solved with numbers of nodes ranging from 100 to 1000, which are randomly generated. The number of edges for the graphs is 10% of the number of nodes. Table 1 shows the cost and CPU time for TS and the average, best and worst costs, and the CPU time obtained after ten repetitions of the SA. The same quality of solutions were obtained in both algorithms. However, it was found that the time required by the TS algorithm is longer than that required by the SA algorithm.

Table 1. Computational results for basic tabu search and simulated annealing

Problem	Tabu Search		Simulated Annealing			
Size	cost	time(s)	mean	best	worst	time(s)
100	3812	7.25	3704.8	3687	3734	29.49
200	16335	93.66	16269.6	16238	16342	75.41
300	37710	423.36	37794.0	37744	37879	147.51
400	68843	1357.88	68968.8	68770	69104	230.50
500	108055	2210.26	108425.6	108302	108526	307.12
600	157788	4456.19	157643.2	157549	157880	385.80
700	216037	5797.71	216348.2	216155	216730	464.98
800	284427	7385.28	284325.4	284153	284828	548.88
900	362989	9205.00	362609.4	362369	363012	628.17
1000	448997	10832.89	447511.0	447367	447840	700.38

4. PARALLEL TABU SEARCH

In TS techniques, different types of parallelism may be distinguished. Here only two general classes are considered according to the method used for parallelism moves. The first type of parallelism we consider consists of searching the next move concurrently. The second type can be considered by partitioning the problem itself into several independent subproblems. These algorithms are very well suited for implementation on a Multiple Instruction Stream, Multiple Data Stream (MIMD) (Kumar et al., 1994) parallel mechanism. In MIMD computers, each processor is capable of executing a different program independent of the other processors. Algorithms for problems requiring control parallelism usually map well onto MIMD parallel computers because control parallelism requires multiple instruction streams. The parallel algorithms have been implemented on a network by The Message-Passing Interface (MPI) library (Gropp et al., 1999). MPI specification is a widely used significant scientific and engineering method on parallel computers. MPI is a library, not a language. It is an attempt to collect the best features of many message-passing systems. The message-passing model posits a set of processes that have only local memory but are able to communicate with other processes by sending and receiving messages. It is a defining feature of the message-passing model that data transfer from the local memory of one process to the local memory of another requires operations to be performed by both processes.

4.1 Parallel Neighbour Move Algorithm

First, the search of the next move to be performed can be parallelized. This will generally require the partition of the set of feasible moves in P subsets. Each of these subsets is assigned to one process which computes its best move. The overall best of these P moves is then determined and applied to the current solution. That is, we divide the set of neighbourhood $N(x)$ into $N_1(x), N_2(x), \cdots, N_p(x)$, such that $N_1(x) \cup N_2(x) \cup \cdots \cup N_p(x) = N(x)$, and $N_i(x) \cap N_j(x) = \emptyset, \textit{ for } i \neq j$, where P is the number of processors. Each partitioned neighbourhood is assigned to each processor, which computes the best solution, x_i for $N_i(x)$ for $i = 1, 2, \ldots, P$. The obtained best neighbour solution of each processor is mutually communicated between all processors together. Then we determine a best solution in $N(x)$ that is $min\{x_1, x_2, \ldots, x_p\}$, and apply it to the current solution. The synchronization is required at each iteration step. Normally, this technique requires extensive communication since the synchronization is required at each step. However, standard TS for this problem has a broad neighbourhood space, and this search for the neighbourhood space is very time-consuming. Therefore, it is only worth applying to problems in which the search for the best move is relatively complex and time-consuming. It is estimated that this proposed parallel algorithm, called the parallel neighbour move algorithm, will allow reduction of the computational time considerably.

4.2 Parallel Agent Passing Tabu Search Algorithm

As discussed in Section 4.3, the parallel neighbour move algorithm allows us to reduce the computational time considerably. However, the quality of solutions obtained by this algorithm is the same as that of those obtained by the standard algorithm. Therefore, multi-move parallelism has been implemented using a framework in which a number of independent moves operate on each different partial solution space, to improve the quality of solutions. However, the problem with this type of parallelism in iterative search is that it significantly limits the movement possibilities between different subproblems and thus generally induces a loss of quality of the best solution found. Ideas are presented that overcome this difficulty by attractive searching based on TS, by extending the searching using super node move, and by diversification of subproblems produced using generator agents. In particular, the agent on computers contributes to providing better solutions. It is simple, generates subproblems, which can be moved from one processor to another, and has learning strategies and learns the state of each solution assigned on the processors. The parallel algorithm based on these concepts is called the parallel agent passing TS algorithm.

Intensification Strategy

First, consider the parallel attractive searching move using TS. The type of parallelism consists of performing several moves concurrently. This can be achieved by partitioning the problem itself into several independent subproblems. Each of the processes can then apply moves to its assigned subproblem, independently of those made in other subproblems. The strategy used here is to divide the current solution in P independent partial solutions, $x_i = (V_l, V_m)$, which are constructed by two mutually prime blocks where $V_l \cup V_m = V_{S_1}, V_l \cap V_m = \phi$, such that $V_{S_1} \cup V_{S_2} \cup \cdots \cup V_{S_p} = V$, $V_{S_i} \cap V_{S_j} = \phi$. The partition between these two prime blocks is then optimized independently for each subproblem. We apply the TS to this optimized process to intensify optimization of the subproblem, which converges rapidly as the characteristic of TS. It is clear that the cost thus optimized can only be less than that of the original solution.

Finally, the entire procedure is clearly suited for parallel computing, as optimizations are performed independently for each procedure, and can thus be done concurrently. The complete solution is then obtained by combining the subsolutions. This method needs no communication or synchronization except for its initialization and for the grouping of the subsolutions at the end of each move.

Super Node Move

Furthermore, searching is extended by the super node move, adding this strategy at the end of each move. The super node move corresponds to the displacement of a complete portion in the block from one solution to another. This choice is motivated by noting that different good solutions of the same instance generally contain a great proportion of identical part of solution. We consider enlarging the search, which cannot progress solely by swapping of two nodes, using this super node move. To explain this super node move in detail, we consider two blocks V_s, V_t, such that, node $i, j \in V_s$ and $l, m \in V_t$. Let $\delta(i; l \mid V_s, V_t)$ be the difference in cost when nodes i and l are exchanged, which is verified by equation (3). The difference in cost when super nodes (*i,j*) and (*l,m*) are exchanged is represented as $\delta(i,j;l,m\mid V_s,V_t)$, which is defined as follows:

$$\delta(i,j;l,m\mid V_s,V_t)=(\sum_{k\in V_s,k\neq j} e_{i,k}+\sum_{k\in V_s,k\neq i} e_{j,k}+\sum_{k\in V_t,k\neq m} e_{l,k}+\sum_{k\in V_t,k\neq l} e_{m,k}) \\ -(\sum_{k\in V_t,k\neq l,m} e_{i,k}+\sum_{k\in V_t,k\neq l,m} e_{j,k}+\sum_{k\in V_s,k\neq i,j} e_{l,k}+\sum_{k\in V_s,k\neq i,j} e_{m,k}). \tag{4}$$

We have a disadvantage in the two node exchange, which is given as follows.

Theorem 1.

If $\forall l\in V_t,\exists j\in V_s; e_{i,j}>2\cdot\max(\sum_{u\in V_t,u\neq l} e_{i,u},\ \sum_{u\in V_s,u\neq i} e_{l,u})$, then $\delta(i;l\mid V_s,V_t)$>0.

Proof.

The precondition of the theorem implies that $\sum_{k\in V_s} e_{i,k}\geq e_{i,j}$ and $\sum_{k\in V_t,k\neq l} e_{i,k}+\sum_{k\in V_s,k\neq i} e_{l,k}<e_{i,j}$.

Hence,

$$\delta(i;l\mid V_s,V_t) \\ =(\sum_{k\in V_s} e_{i,k}-(\sum_{k\in V_t,k\neq l} e_{i,k}+\sum_{k\in V_s,k\neq i} e_{l,k}))+\sum_{k\in V_t} e_{l,k}>(e_{i,j}-e_{i,j})+\sum_{k\in V_t} e_{l,k}\geq 0.$$

Therefore, there is a small probability of adaptation of this neighbouring move. However, if the super node move is used as in the case of Theorem 2, we have an advantage.

Theorem 2.

We have $\delta(i,j;l,m\mid V_s,V_t)$<0, if it satisfies the following equations:

$$\sum_{u\in V_s,u\neq j} e_{i,u}+\sum_{u\in V_s,u\neq i} e_{j,u}<\sum_{u\in V_t,u\neq l} e_{i,u}+\sum_{u\in V_t,u\neq m,l} e_{j,u}, \tag{5}$$

$$\sum_{u\in V_t,u\neq m} e_{l,u}+\sum_{u\in V_t,u\neq l} e_{m,u}<\sum_{u\in V_s u\neq i} e_{l,u}+\sum_{u\in V_s,u\neq i,j} e_{m,u}, \tag{6}$$

$$e_{i,m}=e_{l,j}=0. \tag{7}$$

Proof.

$$\delta(i,j;l,m\mid V_s,V_t) \\ =\{(\sum_{k\in V_s,k\neq j} e_{i,k}+\sum_{k\in V_s,k\neq i} e_{j,k})-(\sum_{k\in V_t,k\neq l,m} e_{i,k}+\sum_{k\in V_t,k\neq l,m} e_{j,k})\} \\ +\{(\sum_{k\notin V_t,k\neq m} e_{l,k}+\sum_{k\in V_t,k\neq l} e_{m,k})-(\sum_{k\notin V_s,k\neq i,j} e_{l,k}+\sum_{k\in V_s,k\neq i,j} e_{m,k})\}$$

$$=\{(\sum_{k\in V_s,k\neq j} e_{i,k}+\sum_{k\in V_s,k\neq i} e_{j,k})-(\sum_{k\in V_t,k\neq l} e_{i,k}+\sum_{k\in V_t,k\neq m,l} e_{j,k})\} \\ +\{(\sum_{k\in V_t,k\neq m} e_{l,k}+\sum_{k\in V_t,k\neq l} e_{m,k})-(\sum_{k\in V_s,k\neq i} e_{l,k}+\sum_{k\in V_s,k\neq j,i} e_{m,k})\}+(e_{i,m}+e_{l,j})<0.$$

Then, it is possible to take advantage of this neighbouring move.

Generator Agent

However, the real ability of parallel TS algorithm cannot be achieved solely by using the above-mentioned strategies, because it significantly limits the movement possibilities between different subproblems. Therefore, an agent is introduced which manages producing subproblems which can be diversified. The subproblems are generated by a simple agent, called a generator agent, with the following simple characteristics:

- It visits all processors and obtains the information regarding the partial solution in each processor. (This agent has some memory).
- It produces independent partial solutions mutually considering the partial problems of other processors, which can be combined without any contradiction.
- It can have various strategies for generating partial problems and for movement between each processor.

This agent, with simple ability, manages producing subproblems passing through processors. The agent has a strategy which produces various partial problems randomly, to give the system significant diversity. That is, there are a number of combinations of partial solutions, and the limitations on the movement possibilities between different subproblems can be overcome. Moreover, using this agent, a parallel computing system environment can be constructed, which does not depend on each computer's ability (CPU speed). It can implement the non synchronized parallel computing algorithm easily, which is the most flexible parallel algorithm.

4.3 Computational Results for Parallel Tabu Search Algorithm

The performance of the parallel neighbour move algorithm is evaluated, using eight partitioning test problems with the number of nodes ranging from 100 to 1000. Figure1 illustrates the trace of CPU time required for two processors and four processors using the parallel neighbour move algorithm and the sequential standard TS algorithm. The results show that the parallel neighbour move algorithm reduces the computational time in proportion to the number of processors. It should be noted that this type of parallelism for TS, in spite of using a simple parallel algorithm, allows considerably reduced CPU time.

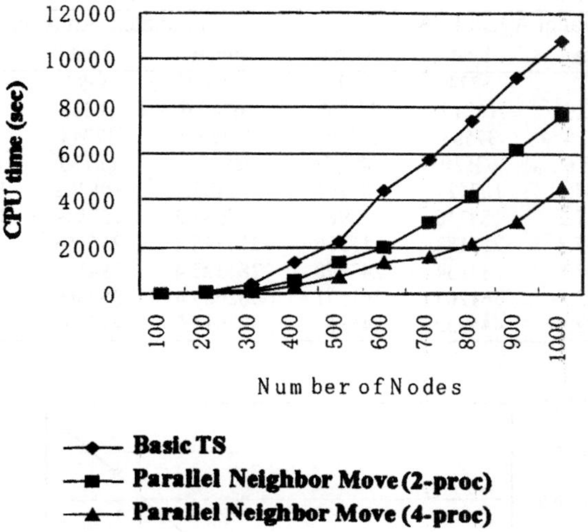

Figure 1. CPU time for parallel neighbour move algorithm

Next, the parallel agent passing TS algorithm was tested with four processors for the eight partitioning test problems ranging from 100 to 1000 nodes. For each size and configuration, the procedure was applied ten times. In tests for this research, *inner loop*=5 was always taken, which is the step of the move, and *tabulenth*=2, which seems to yield good solutions. Table 2 shows the average cost, the best cost, and the worst cost obtained after ten repetitions of the parallel agent passing TS algorithm and SA method and the cost for the standard TS method. The computational results show that the parallel agent passing TS algorithm gives a better solution than the SA algorithm or standard TS, as a matter of course. In Figure 2, it is observed that the computational time required for the parallel agent passing TS algorithm is approximately 1/3 of that required for the standard TS algorithm. This parallelism is efficient in the improvement of the quality of solutions and for a reduction in the computational time required.

Table 2. Computational results for parallel agent passing tabu search in comparison with basic meta-heuristics

Problem	Parallel Agent Passing TS			Simulated Annealing			Basic
Size	mean	best	worst	mean	best	worst	TS
100	3787.2	3774	3801	3704.8	3687	3734	3812
200	16270.0	16219	16334	16269.6	16238	16342	16335
300	37645.0	37583	37706	37794.0	37744	37879	37710
400	68653.6	68520	68761	68968.8	68770	69104	68843
500	107607.6	107427	107761	108425.6	108302	108526	108055
600	156887.2	156746	156988	157643.2	157549	157880	157788
700	215295.8	215183	215356	216348.2	216155	216730	216037
800	283407.8	283134	283778	284325.4	284153	284828	284427
900	361363.8	361204	361571	362609.4	362369	363012	362989
1000	446172.0	445922	446351	447511.0	447367	447840	448997

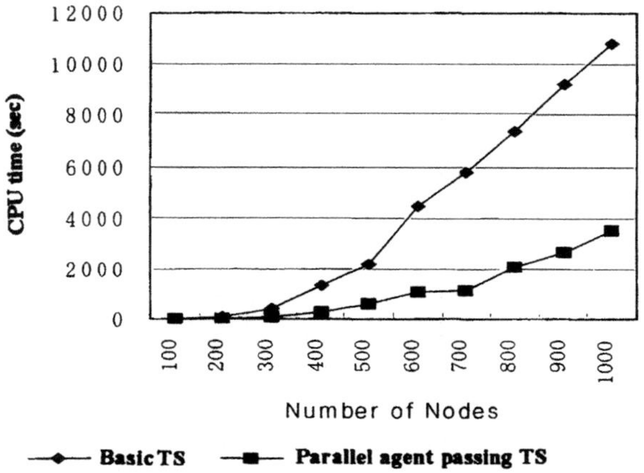

Figure 2. CPU time for parallel agent passing TS using four processors.

5. CONCLUSION

In this paper, approximation algorithms based on the parallel framework to solve multi-graph-partitioning problems have been proposed. The first approach of the parallel neighbour move algorithm reduced the computational time significantly in spite of being based on a simple concept, and confirmed the effectiveness of parallel algorithms. Another parallel agent passing algorithm, which is a multi-move type algorithm using an agent, gives better solutions than the SA and standard TS and reduces the computational time considerably. Ideas are presented that overcome the difficulty caused in parallel algorithms by attractive searching based on TS, the extension of searching by the super node move, and diversification of producing subproblems using generator agents. The parallel agent passing algorithm can be optimized well by partitioning the problem itself into several independent subproblems. It should be emphasized that a simple agent for diversification in the production of subproblems has been used, which is optimized independently and this revealed the importance of the diversification concept of TS. We have further shown the importance of the extension of searching by the super node move, which is the displacement of a complete portion of the node from one solution to another. We illustrated here how TS can be used to guide an optimization process at a high level rather than at every basic step of the search. The results for the multi-graph-partitioning problem revealed that this method can efficiently be used to obtain near optimal solutions for a wide variety of partitioning problems.

In the foreseeable future, parallel computers will certainly advance rapidly in terms of power and will become an interesting means for solving of *NP*-hard problems. High-level TS appears to offer a good framework for the development of parallel heuristic search procedures.

References

Aarts, E. and Korst, J. Simulated Annealing and Boltzmann Machines, John Wiley & Sons, 1989.

Bui, T. N. and Jones, C. Finding Good Approximate Vertex and Edge Partitions is NP-Hard, Inform. Proc Lett. 1992: 42: 153-159

Fiechter, C.-N. A Parallel Tabu Search Algorithm for Large Travelling Salesman Problems, Discrete Applied Mathematics 1994: 51: 243-267

Fujisawa, K., Kubo, M. and Morito, S. An Application of Tabu Search to the Graph Partitioning Problem, The Transactions of the Institute of Electrical Engineers of Japan 1994: 114-C / 4: 430-437,

Glover, F. and Laguna, M. Tabu Search, Boston: Kluwer Academic Publishers 1997

Gropp, W., Lusk, E. and Skjellum, A. Using MPI, Cambridge: The MIT Press 1999

Jepsen, D. S., and Gelatt, C. D. Macro Placement by Monte Carlo Annealing. Proc. International Conf. On Computer Design 1983: 495-498

Johnson, D., Aragon, C., Mcgeoch, L. and Schevon, C. Optimization by Simulated Annealing: An Experimental Evaluation; Part I, Graph partitioning, Operations Research 1989: 37 / 6: 865 –892

Kumar, V., Grama, A., Gupta, A. and Karypis, G. Introduction to Parallel Computing, The Benjamin/Cummings Publishing Company, Inc. 1994

Lee, J. G., Vogt, W. G. and Mickle, M. H. Optimal Decomposition of Large-Scale Networks, IEEE Trans., System Man Cybernet. 1979: 9: 369-375

Mohr, T. Parallel Tabu Search for the Graph Coloring Problem, ORWP 88/11. DMA, Ecole Polytechnique Federale de Lausanne, 1988

Sarkar, V. Partitioning and Scheduling parallel Program for Multiprocessors, MIT Press, Cambridge: 1989

Taillard, E. Parallel Taboo Search Technique for the Jobshop Scheduling Problem, ORWP 89/11. DMA, Ecole Polytechnique Federale de Lausanne, 1989

HEURISTIC TECHNIQUES FOR SCHEDULING RAILWAY TRACK MAINTENANCE

M. Lake[1], L. Ferreira[2] and E. Kozan[3]

[1&2] School of Civil Engineering

[3]School of Mathematical Sciences
Queensland University of Technology
GPO Box 2434 Qld 4001 Australia
{m.lake, l.ferreira, e.kozan@qut.edu.au}

Abstract The maintenance of a rail network significantly contributes to the total cost of operating the network. Therefore substantial savings in railway expenditure can be achieved if track maintenance is conducted in the most cost-effective manner. Railways, however, operate under the conflicting objectives of minimising the infrastructure expenses while continuing to provide adequate service. This paper reports on the development of a model for the short-term scheduling of track maintenance activities, after the train schedule and maintenance activities have been planned. The model schedules the track maintenance activities, including assigning the maintenance crews to the activities, with the objective of minimising the total maintenance costs. This maintenance scheduling model has been solved using a two stage heuristic technique, with the first stage generating a feasible solution and the second stage implementing one of simulated annealing, local search or tabu search. The paper presents the results of applying the model to a problem that is indicative of the size experienced in the railway industry, with twenty track segments and a seven-day scheduling period. The three heuristic methods are then compared for the solution values obtained and the time frame required.

Keywords: Maintenance, scheduling, heuristics.

1. BACKGROUND

The maintenance of a rail network significantly contributes to the total cost of operating the network. It has been estimated that for freight operations in Australia, 25-35 percent of total train operating expenses are related to track maintenance (Ferreira and Higgins, 1998). The maintenance of existing networks therefore needs to be undertaken according to a plan that maximises the overall net benefit to the rail operator.

Railways operate under the conflicting objectives of minimising the infrastructure expenses while continuing to provide adequate service. This conflict between maintenance and train operations has intensified over recent years due to a number of factors. Increased traffic on railways and the reduction of route and track kilometres have raised traffic densities, reducing the time available to conduct maintenance, while also increasing track maintenance requirements. In addition, the railway industry in Australia is undergoing very significant changes with respect to ownership and management models, as well as operational practices. There is an increasing move towards the separation of the track ownership and the track infrastructure management, making the need to schedule track maintenance activities extremely important.

2. TRACK MAINTENANCE SCHEDULING

The majority of the research into track maintenance has focused on the planning of the maintenance, that is, determining what maintenance should be conducted. A relatively small amount of research has been carried out on reducing the conflict between the trains and track maintenance by the short term scheduling of the maintenance and the required resources. One such study (Ruffing, *et al*, 1994), which schedules the maintenance by altering the train schedule, does not assign the maintenance crews to activities. When the number of crews for a maintenance scheduling region is small, however, the maintenance schedule should not necessarily be constructed with the assumption that enough crews are available to conduct the activities. The schedule may then need to be substantially altered to accommodate the availability of the crews.

Research was also conducted by Higgins (1998), with the weighted objective of minimising the maintenance activity finishing time and the expected interference delays due to delayed trains and maintenance. The objective of minimising the maintenance activity finishing times may not be the most appropriate, however, as work place agreements may not permit intensive work for a certain time, then no work for the remainder of the schedule. The train schedule is treated as a fixed input, and assigns the maintenance crews to the activities, however, these crews are assigned to an activity permanently, so a different crew cannot conduct the maintenance activity at a later time.

The finishing time of the maintenance activity was minimised in Higgins (1998) without considering individual set-up and take-down times, only using one length of time if the work is conducted in one consecutive block or another if the maintenance is done in more than one block. The opportunity therefore exists to develop a model which optimises the short-term scheduling of track maintenance activities and the crews required, considering the factors discussed above.

3. MODEL FORMULATION

The problem of scheduling the short-term track maintenance from an existing maintenance plan and train timetable has been formulated as a binary integer non-linear programming problem. Included in the formulation is the assignment of crews to maintenance activities and the activities, which may or may not require track closure, to track segments for given time periods. The track segment term used is a length of railway track in which only one train or activity that requires track closure may be at any one time. The objective is the minimisation of the cost of conducting the track maintenance.

3.1 Parameters and Variables

The parameters and variables used in the formulation are given below, with the decision variable first.

$$x_{ijkt} = \begin{cases} 1 & \text{if crew } i \text{ conducting activity } j \text{ on track segment } k \text{ in time period } t \\ 0 & \text{otherwise} \end{cases}$$

p_{jk} Number of time periods of productive time required for maintenance activity j on track segment k

$sutd_j$ Number of time periods required for set up and take down of maintenance activity j

$$a_{kt} = \begin{cases} 1 & \text{if track segment } k \text{ is available in time period } t \\ 0 & \text{otherwise} \end{cases}$$

$$v_{it} = \begin{cases} 1 & \text{if crew } i \text{ available in time period } t \\ 0 & \text{otherwise} \end{cases}$$

c_{ijt} The cost for crew i to be carrying out activity j in time period t

$$g_{ik} = \begin{cases} 1 & \text{if crew } i \text{ allowed to conduct maintenance on track segment } k \\ 0 & \text{otherwise} \end{cases}$$

$$h_{ij} = \begin{cases} 1 & \text{if crew } i \text{ allowed to conduct activity } j \\ 0 & \text{otherwise} \end{cases}$$

$$m_{j_1 j_2} = \begin{cases} 1 & \text{if activity } j_1 \text{ must be conducted prior to activity } j_2 \\ 0 & \text{otherwise} \end{cases}$$

$$tc_j = \begin{cases} 1 & \text{if activity } j \text{ requires track closure} \\ 0 & \text{otherwise} \end{cases}$$

$$sf_{jt} = \begin{cases} 1 & \text{if activity } j \text{ can be conducted in time period } t \\ 0 & \text{otherwise} \end{cases}$$

t_{min} The minimum (first) time period

t_{max} The maximum (last) time period

3.2 Objective Function and Constraints

The objective function (1) consists solely of the cost of the crews to conduct the track maintenance. The track constraint (2) enforces that track segment k is available in time period t before an activity can be scheduled on it for the maintenance activities that require track closure. This constraint also limits the number of activities that require track closure on each segment k at each time period t to one. The amount of maintenance constraint (3) enforces that the total number of time periods for activity j on track segment k is the production time plus the number of time periods for the set up and take down. The number of set up and take downs required depends on the number of consecutive time periods or blocks the work is conducted in.

The crew constraint (4) enforces that track maintenance crew i is available in time period t before the crew can be assigned to an activity. This constraint also limits the number of tasks that crew i can be assigned to in a time period t to one. The activity constraint (5) ensures that an activityis allowed to be conducted at a certain time period t to be assigned to that time period. The crew on track and crew on activity constraints (6 and 7) ensure a crew i must be allowed to conduct maintenance activity j or work on track segment k respectively. The ordering constraint (8) enforces any sequence requirements for the activities on each track segment k. The formulation allows different crews to be assigned to the same activity at different times or different track segments.

$$\text{Min } z = \sum_i \sum_j \sum_t (c_{ijt} \times \sum_k x_{ijkt}) \quad (1)$$

$$\sum_i \sum_{j \text{ if } tc_j > 0} x_{ijkt} \le a_{kt} \qquad \forall k,t \quad (2)$$

$$\sum_i \sum_t x_{ijkt} = p_{jk} + \frac{1}{2}\Big(\sum_{t>1} [abs(\sum_i x_{ijkt} - \sum_i x_{ijkt-1})] + \sum_i x_{ijkt_{\min}} + \sum_i x_{ijkt_{\max}}\Big) \times (sutd_j) \quad \forall j,k \quad (3)$$

$$\sum_j \sum_k x_{ijkt} \le v_{it} \qquad \forall i,t \quad (4)$$

$$\sum_i x_{ijkt} \le sf_{jt} \qquad \forall j,k,t \quad (5)$$

$$\sum_j x_{ijkt} \le g_{ik} \qquad \forall i,k,t \quad (6)$$

$$\sum_k x_{ijkt} \le h_{ij} \qquad \forall i,j,t \quad (7)$$

$$\max_t (t \times \sum_i x_{ij_1kt}) < \min_{t \text{ if } \sum_i x_{ij_2kt} > 0} (t \times \sum_i x_{ij_2kt}) \qquad \forall j_1, j_2, k$$

$$\text{where } p_{j_1k} > 0, p_{j_2k} > 0 \text{ and } m_{j_1j_2} > 0 \quad (8)$$

4. SOLUTION METHOD

Due to the non-linearity and the large number of variables, applying optimisation techniques to create a solution for the formulation is difficult. The application of commercially available optimisation software was attempted, as discussed in the next section, however, this met with limited success. The main use of the results from the optimisation package is therefore to validate the solutions for the heuristic techniques.

The solution technique applied to the formulation consists of two stages. The first stage generates a feasible solution, that is a solution that satisfies all the constraints but is unlikely to be the optimal solution. A number of the constraints are not effected by the current solution, for example, if crew one cannot conduct activity two ($h_{12} = 0$) then all the

x_{12kt} variables must be zero, as they can never be one in a feasible solution. Therefore, a boolean variable, $change_{ijkt}$, which is one if the x_{ijkt} variable can be one or otherwise zero, can be created. The constraints that can be considered once in determining the value of $change_{ijkt}$ and then do not need to be considered when calculating the feasibility of each individual solution include (with the equation number in brackets): the activity permitted constraint (5); the crew permitted on the track constraint (6) and the crew allowed on activity constraint (7). Constraints that are considered initially, as they can reduce the amount of variables with a $change_{ijkt}$ value equal to one, but that also need to be used to determine the feasibility of each solution include: the track closure constraint (2); the amount of maintenance constraint (3) and the crew available constraint (4). For the crew availability constraint, for example, if the crew is unavailable in a time period, the $change_{ijkt}$ variable for that time and crew can be set to false. This constraint still needs to be considered for each solution, however, as a crew could be available and given two activities to conduct in a time period, causing the solution to be infeasible. The ordering constraint (8) can only be considered for each individual solution. The creation of the *change* variable reduces the amount of computation required to calculate the feasibility of a solution.

The procedure to find a feasible solution is in two parts: the first generates a starting solution; and the second adjusts this solution until feasibility is achieved or is determined not to be able to be obtained. The two possible exits to the first part of the process of finding a feasible solution, one of which creates the starting solution and continues to the next part of the procedure, which attempts to resolve the infeasibilities in the solution. The other option is that the process found that too much maintenance was required for a track segment. If no block greater than the productive maintenance required plus one set up and take down exists, the process loops, assigning the activity to the largest block available, adding a set up and take down each time. If there is maintenance still required when the available blocks are less than one set up and take down time, then there is no possible way to schedule the required maintenance on the track segment. The operator would need to adjust the inputs, for example, permitting the activity to be conducted for more time periods, permitting more crews to conduct the work, or allowing the crews that can conduct the maintenance to be available for more time.

As for the procedure that generates the initial solution, which is the starting point for the process to find a feasible solution, there are two possible exits. The first exit is the generation of a feasible solution, with the program continuing to the heuristic technique. The second is the termination of the program with a message if a feasible solution cannot be generated for the problem. This is determined in the procedures for resolving an infeasibility for each of the four constraints that need to be considered for each solution.

The second stage implements one of three heuristics: simulated annealing, local search or tabu search. All of these heuristic techniques involve identifying a neighbourhood of solutions that can be reached from a given solution, with transitions from one solution to another, called moves, preformed iteratively from the solution generated initially, in an attempt to obtain a near optimal solution. Local Search does not allow moves to a non-improving solution, while Tabu Search avoids convergence to a local optimum by allowing non-improving moves when no improving moves are available. To prevent the procedure from cycling the same solutions repeatedly, the moves recently visited are recorded and considered tabu, which prohibits the process from returning to these solutions. Further explanation of tabu search can be found in Glover (1990), while simulated annealing is described in Kirkpatrick *et al* (1983). To prevent convergence to a local optimum in Simulated Annealing, a non-improving solution can be accepted based on a probability, which depends on the value of the objective function with the previous solution, the value with the new solution and a control parameter. The algorithm for implementing tabu search is similar to the one used by Higgins (1998), with the aspiration criteria and the solution re-initialising to the best found after a certain number of iterations. Results generated from this algorithm were significantly better than for the same procedure with no re-initialisation.

The solution neighbourhood investigated for all the heuristic methods is defined by the following possible five transitions:

1) *Move Block:* A block of maintenance is moved to a different available time window, which is at least the same size. The crews assigned to the activity are not necessarily the same as the crews that were conducting the maintenance in the previous time slot.

2) *Split Block:* A block of maintenance is divided into two separate blocks of work, one of which is moved to another available time window. The set up and take down time required is added to the maintenance done, as the work is now in two separate blocks.

3) *Join Blocks:* Two blocks of maintenance are joined together in a different time window. The set up and take down time required is deducted from the maintenance done, as the work is now done in one block.

4) *Partial Move of Block:* Some of the time periods for a block of work are moved to the other side of the block, without moving the whole block of work. For example, if work is done in time periods 3-6, the maintenance could be 'shifted' along if the time periods 7 and 8 are available, resulting in the work conducted in time periods 5-8.

5) *Swap Crew:* A crew in a time period is replaced by another available crew.

There are three stopping methods available for each heuristic, stopping after: a set number of iterations; a set number of iterations since the solution value changed or, if reached first, the overall maximum iterations; and whichever is obtained first of the maximum iterations or a solution less than a certain value.

5. APPLICATION AND RESULTS

The LINGO optimisation package (LINDO Systems Inc, 1999) was applied to solve the formulated problem. Generally, LINGO could not solve the formulation given in the previous section. In 33 different data sets, which only contained up to 24 time periods and ten track segments, only seven actually returned a solution, with the rest concluding the problems were, incorrectly, infeasible.

A new formulation was therefore devised which contains two decision variables and a squared term instead of the absolute value, for which LINGO returned solutions in the majority of the small data sets attempted. The two decision variable formulation has a number of restrictions that do not exist for the other formulations. The major restriction is that an activity cannot be conducted at the same time on different track segments. The ordering constraints were also omitted as LINGO permitted only a quasi-ordering constraint of the activities.

Therefore, any solution feasible for the two variable form, which cannot have an activity conducted more than once at the same time, is feasible for the single variable form, which is the formulated solved by the heuristic methods. Despite the reformulation of the problem, however, LINGO is only able to solve extremely small problems. For problems with more than 12 time periods, LINGO could not return a feasible solution for either formulation. This is related to the number of variables, as the data sets that returned infeasible results were the larger data sets, with between 180 and 960 decision variables. The largest number of variables for which a feasible result was obtained was 384. Problems of the size experienced in the real world with scheduling periods of a week, however, can contain over sixty thousand variables.

The three heuristic techniques were applied to each problem 10 times, with the stopping criteria of 200 iterations since the solution values changed. In all but one case, a solution that was less than or equal to the best solution obtained by LINGO was found. Summary results for a number of small data sets, for which LINGO returned a feasible solution for one of the formulations, can be seen in Table 1. The heuristic methods all returned a solution within 2 seconds for most of the problems. In comparison, over half the problems required more than 30 minutes on LINGO. The heuristic with the best performance was simulated annealing, with the average percent of a solution better than or equal to LINGO of 83.7%. Tabu search was not far below this with a figure of 82.6%, however, the times required were also longer. Therefore the heuristic techniques are returning solutions comparable with those obtained from optimisation techniques.

Table 1. The results for small problems from LINGO and the heuristic techniques.

Heuristic	Average % <= Best Lingo Solution	Average Time (Secs)	Average Objective Function
Simulated Annealing	83.7	1.60	743
Tabu Search	82.6	1.96	744
Local Search	79.3	1.18	748

The heuristic methods have been applied to scheduling problems of a size typically experienced in the field. The results shown below are for a situation based on information for a section of railway track operated and maintained by Queensland Rail in Australia. The track section in which the maintenance is being scheduled for one week has 20 track segments or signalling blocks, with 4 available crews and 5 different maintenance activities required.

Investigation was conducted into the most appropriate parameters for tabu search and simulated annealing. Twelve runs of each of the three heuristic techniques for fifty thousand iterations were conducted with the chosen parameters. Figure 1 shows the solution value over the number of iterations. Tabu search improves very slowly initially, while local search is the fastest improving at first. The final results from local search and tabu search are quite variable, while the results from simulated annealing are more consistent and within a smaller range throughout the process. The objective function values for all the heuristics do not decrease much in the last twenty five thousand iterations.

Table 2 gives the average, minimum and maximum final solution and the percentage decrease of the solution value on the initial feasible solution. Simulated annealing is the best performing heuristic, as was found for the smaller problems. Local search actually gives a lower average objective function than tabu search, however the average percentage improvement is less.

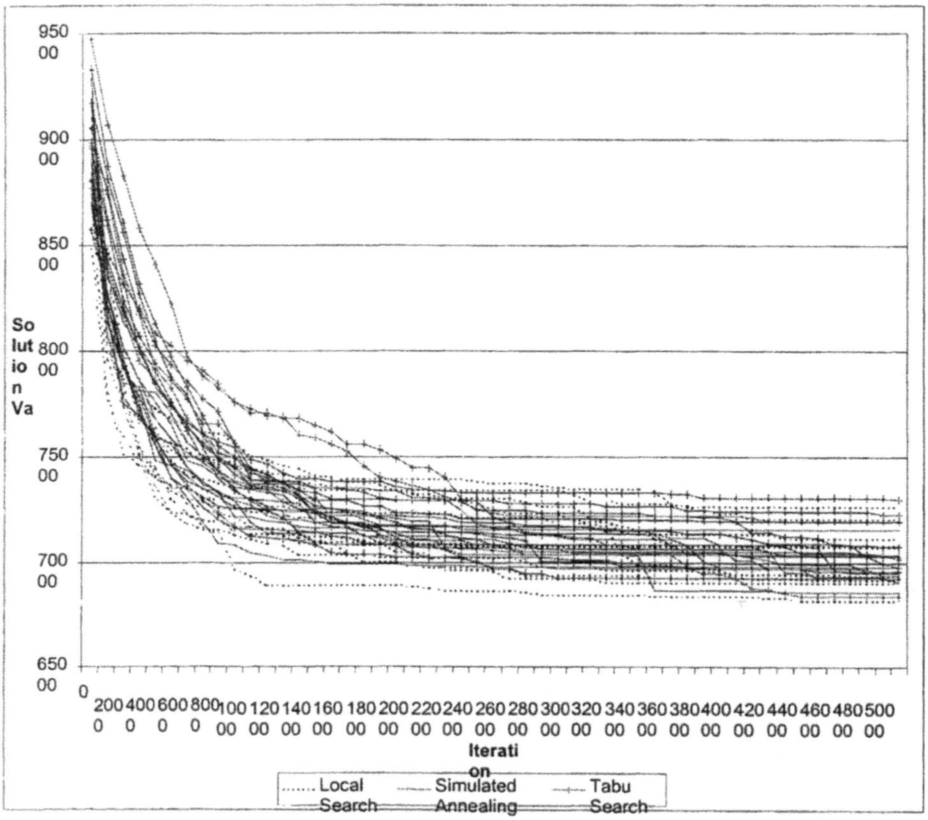

Figure 1. The solution value over the number of iterations.

Table 2. The overall values for 12 runs of the heuristic techniques.

Heuristic	Average Objective Function	Average % Improve on Initial Feasible Solution	Minimum Objective Function	Maximum Objective Function
Simulated Annealing	69833	21.7	68650	71550
Tabu Search	70646	21.5	68450	73000
Local Search	70350	20.9	68250	72650

6. CONCLUSION

A model has been developed for the short-term scheduling of track maintenance activities, after the train schedule and maintenance activities have been planned. The model schedules the track maintenance activities, including assigning the maintenance crews to the activities, with the objective of minimising the total maintenance costs. The formulation incorporates the individual set-up and take-down times required each time an activity is conducted and is solved using a two stage heuristic technique. The first stage generates a feasible solution, while the second stage implements either simulated annealing, local search or tabu search.

The results from the heuristic techniques are validated by comparison to an optimisation package. The heuristic methods are applied to a problem of the size experienced in the railway industry, with twenty track segments and a seven-day scheduling period. Simulated annealing is the best performing heuristic, with the best final results, measured by the average solution or the percentage improvement on the initial feasible solution. Simulated annealing is also preferred over tabu search as the latter improves very slowly at first. The results from local search, as would be expected, are more variable, giving the worst overall performance.

7. ACKNOWLEDGMENTS

The research presented here was funded by the Australian Research Council under a SPIRTS grant and by the Network Access Division of Queensland Rail. The authors would like to thank in particular Mr John Davey for his support in the continuing development of the model. The views expressed in this paper are solely those of the authors.

References

Ferreira, L. and Higgins, A. Scheduling Rail Track Maintenance. Proceedings of International Conference on Traffic and Transportation Studies, American Society of Engineers, Beijing, China, 1998; 820-829

Glover, F. Tabu Search: A Tutorial. Interfaces 1990; 20(4): 74-94

Higgins, A. Scheduling of Railway Track Maintenance Activities and Crews. Journal of the Operational Research Society 1998; 49(10): 1026-1033

Kirkpatrick, S., Gelatt, C. and Vecchi, M. Optimisation by Simulated Annealing. Science 1983; 220(4598): 671-680

LINDO Systems Inc. LINGO User's Guide. LINDO Systems Inc, Chicago, Illinios; 1999

Ruffing, J. A. and Marvel, B. M. An Analysis of the Scheduling of Work Windows for Railway Track Maintenance Gangs. Proceedings of the International Heavy Haul Conference, International Heavy Haul Association 1994; 1-4.

OPTIMISING A HORIZONTALLY INTEGRATED PUSH/PULL HYBRID PRODUCTION SYSTEM IN A FOUNDRY

Paul Corry and Erhan Kozan

School of Mathematical Sciences
Queensland University of Technology, Australia
{p.corry, e.kozan@qut.edu.au}

Abstract This study investigates a horizontally integrated push/pull hybrid production system (HIHPS) of a large foundry in the Asia-Pacific region. Components are produced according to a pushing philosophy and are stored in a buffer when complete. Assembled goods draw components from the buffer and are made to order. The objective is to minimise the sum of ordering, inventory and tardiness costs. Currently, the inventory decisions are made based on the experience of personnel within the foundry. Operations research provides an opportunity to improve efficiency by using computational techniques to determine good inventory decisions. Traditional inventory and lot-sizing problems determine optimal policies using mathematical analysis made possible by simplifying assumptions. These assumptions are invalid for the HIHPS problem studied in this paper so that mathematical analysis is extremely difficult if not impossible. As an alternative, simulation is combined with simulated annealing to determine a good inventory policy. Experimental results are used to evaluate the performance of this approach and to identify strategies for developing even better techniques.

Key words: Scheduling, heuristics, push/pull hybrid production system, simulated annealing

1. INTRODUCTION

Traditional inventory systems can be classified as either pushing or pulling systems. Pushing systems such as material requirements planning (MRP) calculate production orders on the basis of forecasted demand. Pulling systems such as just-in-time (JIT) (Sugimori et al 1977, Kimura and Terada 1981) calculate production orders based on actual demand. Whilst a pulling approach keeps on-hand inventories to a minimum, Browne et al (1988) mentions that a Kanban approach cannot easily respond to unexpected changes in market demand. Pushing approaches buffer against surges in demand by keeping on-hand inventories of finished goods. Studies such as Sarker and Fitzsimmons (1989) have shown that push and pull control strategies have different advantages and disadvantages. Hybrid systems integrate pushing and pulling policies to gain the advantages both have to offer.

A hybrid system can be either vertically or horizontally integrated. Vertically integrated systems apply a pushing policy at an upper planning level and a pulling policy at a lower planning level; for example, MRP at the plant wide level and a kanban policy applied locally between cells (Karmarkar, 1986). This study investigates a horizontally integrated system, which uses a pushing system to control some stages of production and a pulling system to control other stages.

Takahashi, Hiraki and Soshiroda (1994) investigate a pull/push system with initial pulling stages and final pushing stages to determine optimal production quantities and safety stock levels. Cochran and Kim (1997) use simulated annealing to determine safety stock levels and the optimal junction point between initial pushing stages and the final pulling stages. This type of system is called a horizontally integrated push/pull hybrid production system (HIHPS). The foundry problem investigated by this study can be classified as a multi-product, non-serial HIHPS.

This paper proposes an optimisation model for the horizontally integrated push/pull hybrid production system within a foundry. The junction point between pushing and pulling stages takes the form of a buffer situated between the final stage of component production (pushing) and a single assembly stage (pulling). It is required to find suitable safety stock and order quantities for each component kept in the buffer at the junction point. The objective is to minimise total production and inventory costs. There is a large body of literature that addresses this kind of problem using mathematical analysis to determine an optimal policy (see Bookbinder and Cakanyildirim (1999), Gavirneni (2001), Graves et al (1993), Winston (1993)). Unfortunately, these techniques always make analysis possible by some simplifying assumption. For example, constant demand, known lead times and single product type are common assumptions that cannot be applied to the foundry problem. Previous studies have not addressed inventory problems with stochastic demand, uncertain lead-times and multiple product types on the one production line under a HIHPS, so that new techniques must be developed. This study proposes a combination of simulation and simulated annealing to solve the complex inventory problem within the HIHPS foundry. The approach of combining simulation with an optimisation heuristic can also be applied to complex inventory problems in other industries with similar production environments.

2. A HIHPS PROBLEM

This study proposes a foundry HIHPS model consisting of two stages for component production and a single assembly stage. The first stage of component production is a single moulding line, which produces all the component castings. After a cooling period, component castings are ready for machining at stations dedicated to particular component types. It is assumed that a component station can process several component types but all components of the same type are processed at the same station. Each component type represents a family of distinct components with similar characteristics. After a batch of components has been completed at the machining stage, it enters the buffer waiting to be removed for an assembly job. The ordering policy is a periodic review (*S*, *ss*) policy, that is, when the net buffer level for a given component falls below *ss*, an order is placed so that the net buffer level is replenished to a level *S*. Whilst component production is initiated by internal orders, assembled goods are ordered by customers. If there are insufficient component stocks in the buffer to complete an assembly job, it must wait until replenishment stocks arrive. Figure 1 is a station precedence diagram describing the proposed HIHPS model.

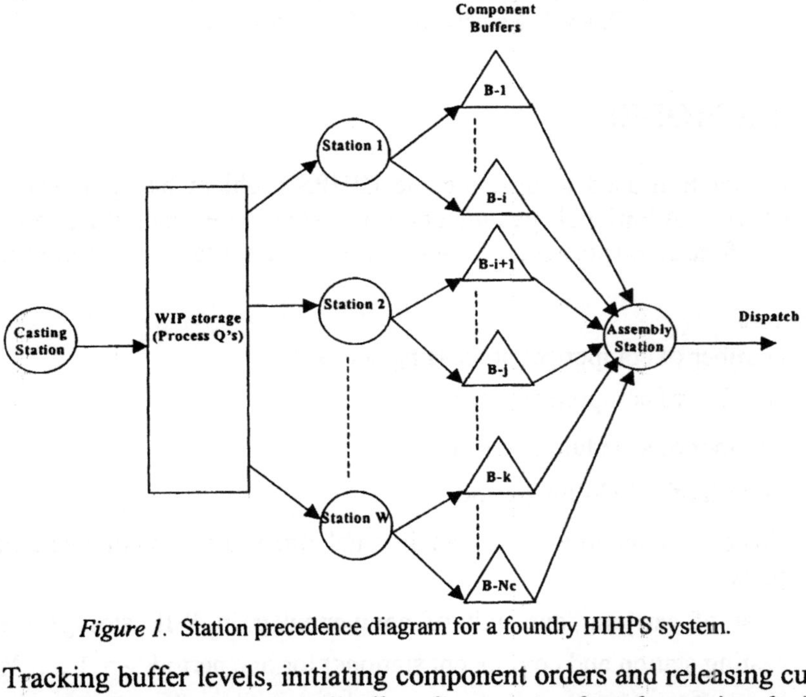

Figure 1. Station precedence diagram for a foundry HIHPS system.

Tracking buffer levels, initiating component orders and releasing customer orders to the system occurs periodically. Customer orders that arrive during any period are released to the system at the beginning of the next period. The necessary buffer replenishment orders are calculated and released at the same time as shown in Figure 2. Because replenishment orders are made internally and not through an external party, traditional economic order quantity (EOQ) methods make assumptions that over simplify the problem. For example, if lead-time demand for a product is uncertain but the lead-time is known with certainty, then a modification of the EOQ analysis could be applied. For the foundry HIHPS however, lead-time is not known with certainty and is highly dependent on the current workload. Another point is that a buffer decision can influence other parts of the system. For example, lot-sizing decisions influence the number of set-up operations and queue lengths, which in turn affect the lead-time of component orders. For these reasons a model is formulated to combine the effects of buffer decisions with scheduling in the system. By conducting a number of simulations over a long enough period of time, the model can give a good indication of system performance under a proposed set of buffer decisions.

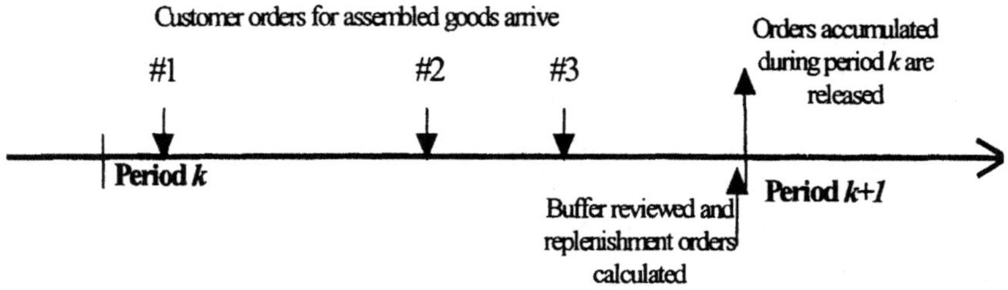

Figure 2. Periodic review of net buffer levels.

3. THE MODEL

The notation used to describe the HIHPS problem being investigated is given below. Additional parameters not listed here, including probability distribution functions (pdf) describing demand, are discussed in a later section.

Parameters

W Number of component processing stations

N_C Number of component types

N_P Number of scheduling periods

L_P Length of scheduling period

c^{ORD} Fixed cost incurred when an internal order for a component batch is placed

c_g^{WIP} Cost of carrying one unit of component g in WIP storage (between casting station and component stations) for one period, $g = 1, ..., N_C$

c_g^{BUF} Cost of carrying one unit of component g in buffer storage (between component stations and assembly station) for one period, $g = 1, ..., N_C$

$B_g(0)$ Initial buffer stock level for component g, $g = 1, ..., N_C$

Decision variables

S_g buffer replenishment level of component, $g = 1, \ldots, N_C$

ss_g buffer reorder point (safety stock) for component $g = 1, \ldots, N_C$

Simulation variables and parameters

Related to Inventory Costs $g = 1, \ldots, N_C$:

$B_g(t)$ Number of component g in buffer at time t

WIP_g^k Set of replenishment jobs for component g in-process carried over duration of period k

t_0^k Time at which period k begins, $t_0^k = (k-1)L_P$

c_q^{TRD} Cost charged on job q for each late period, $q \in J$

C_q The next period beginning after q is completed, $q \in J$

d_q Job q is due to be completed by the beginning of period d_q (due-date) $q \in J$

BOM_q Set of components required for assembly job q

J_k Set of assembly jobs arriving in period k-1 $, J = \bigcup_{k=1}^{N_P} J_k$

$J1_k$ Set of externally ordered casting jobs arrived in k-1, $J1 = \bigcup_{k=1}^{N_P} J1_k$

$J2_k$ Set of externally ordered component jobs arrived in k-1, $J2 = \bigcup_{k=1}^{N_P} J2_k$

JC_k Set of replenishment jobs released in period k, $JC = \bigcup_{k=1}^{N_P} JC_k$

j_g^k Job index of replenishment job for component g placed in period k,

X_j Order quantity of job $j \in J \cup J1 \cup J2 \cup JC$

$B_g^{\text{net}}(k)$ is the net buffer level of component g at beginning of period k before replenishments are ordered. It is calculated by the following expression.

$$B_g^{\text{net}}(k) = B_g(0) + \sum_{l=1}^{k-1} \left(X_{j_g^l} - \sum_{\substack{q \in J_l \\ g \in BOM_q}} X_q \right) - \sum_{\substack{q \in J_k \\ g \in BOM_q}} X_q$$

where

$$X_g^k = \begin{cases} S_g - B_g^{\text{net}}(k) & \text{if } B_g^{\text{net}}(k) \le ss_g \\ 0 & \text{otherwise} \end{cases}, \qquad j_g^k \in JC_k,\ g=1,\ldots,N_C,\ k=1,\ldots,N_P$$

The HIHPS model presented in this study is based on the following assumptions: Buffer levels are reviewed periodically according to an *(S, ss)* policy; and Buffer and WIP inventory costs are charged periodically on the number of units carried from the beginning of a period to the end. Unlike many traditional inventory problems, there is a considerable lag between placing an order to remove components, and when the components are finally removed. For this reason the *net* buffer level is used to make decisions, rather than the observed level. If the observed level were used, the ordering system would not react to a shortage until it actually occurred. Using the net level allows the ordering system to react earlier, when on-hand assembly orders indicate a future shortage.

The objective of the model is to minimise the sum of inventory costs, ordering costs and tardiness costs. Inventory holding cost consists of two components, work-in-process inventory costs and buffer holding costs. Evaluating total cost requires a simulation of the system conducted over an appropriate time horizon. This involves generating externally ordered jobs according to a given set of products and components. Generating jobs and scheduling them is performed in a simulation of the system based on the given input parameters. To generate a schedule the simulation uses a dispatching rule. Once the simulation is complete, all the data necessary to evaluate the objective, *TC*, is available.

Objective Function

TC is the total ordering, tardiness and holding costs over the planning horizon.

$$\min TC(\mathbf{S},\mathbf{ss}) = |JC| c^{\mathrm{ORD}} + \sum_{q\in J} c_q^{\mathrm{TRD}} \max\left(0, C_q - d_q\right)$$

$$+\sum_{g=1}^{N_C}\sum_{k=1}^{N_P}\left[c_g^{\mathrm{WIP}} \sum_{q\in WIP_g^k} X_q + c_g^{\mathrm{BUF}} \max\left(0, \min(B_g(t_0^k), B_g(t_0^{k+1}))\right)\right]$$

Subject to $S_g \geq ss_g + 1,\; g = 1,\dots,N_{\mathrm{C}}$

$S_g,\; ss_g \geq 0 \;\; g = 1,\dots,N_{\mathrm{C}}$

$J, JC, C_q, d_q, X_q, WIP_g^k, B_g(t)$ are parameters of a simulation.

4. A CASE STUDY

Toowoomba Foundry is a leading manufacturer in the Asia Pacific region. Wheels and wheel components for heavy transport are the highest selling products in a large range of machined and assembled iron castings. There are many different wheel specifications requiring various component types. Components are produced in two stages feeding a large in process inventory acting as a buffer before the assembly stage. Wheels are assembled in a single stage drawing components from the buffer. When the level of a component in the buffer falls below a given safety stock level, an order is placed to replenish the buffer to a given maximum level for that component. When external orders are made for assembled goods, the necessary components are drawn from the buffer and assembled to the specifications of the order. The problem is to determine for each component a safety stock and maximum buffer level to achieve minimum production and holding costs.

At Toowoomba foundry wheels for heavy transport are assembled from two main components, a *drum* and either a *spider* or *hub*. All components are cast on the moulding line and are then machined in either the drum cell or the spider/hub cell. Each cell consists of a robotic arm, which transfers semi-finished components between the surrounding ring of machines. For modelling purposes each cell can be considered a single stage in production. After machining is complete, the components enter the buffer and are assembled after a customer order has been placed. Two stages during the production of wheels require constraints to be satisfied. The casting stage is constrained by shift patterns and the grade of iron produced in each shift. The assembly stage is constrained by the availability of components. Figure 3 illustrates the product routings within the HIHPS at the foundry.

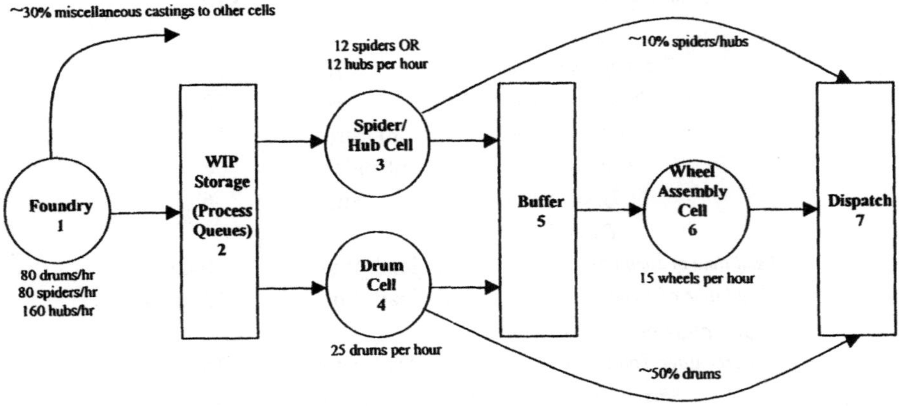

Figure 3. Process diagram for production of components and assembled goods.

Data has been collected at Toowoomba Foundry for the simulation parameters. Because the data set is quite large, it is not provided in complete detail, but a summary is given. Probability distributions related to demand were estimated based on the previous twelve months demand. External order quantities were generated from uniform distributions, and inter-arrival times from exponential distributions. Lot-sizes typically ranged from 10 or 20 units up to 50, 100 or 150 units. Average inter-arrival times range from 2 days to about 200 days. Table 1 gives a summary of other important simulation parameters. This data is representative of foundry operations and replicates the workload within the Toowoomba Foundry. It is based on actual data extracted from records of the previous year of production, and advice from expert staff. The data provides a useful tool for comparing operations research techniques with current practice at Toowoomba Foundry.

In total, there are 32 types of wheel and 24 distinct component types. A scheduling period was taken to be one day and the objective evaluated over twelve 4 week periods, 336 days, ie, N_P=336. All components pass through a casting station (1) and then through a component station. Drums are machined at station (2) whilst hubs and spiders are machined at station (3). Casting has an average cycle time of 45 seconds whilst the machining cycle times are much larger, averaging around 3 minutes. Wheels are assembled at station 4 with comparable cycle time to component machining.

Table 1. Important parameters of Toowoomba Foundry simulations

Descriptions	Parameters	Value
Time Horizon		
Number of scheduling periods (days)	N_P	336 days
Operating Costs (g=1,...,N_C)		
Holding cost in WIP1 ($/unit/day)	c_g^{WIP}	0.5c
Holding cost in buffer ($/unit/day)	c_g^{BUF}	1c
Tardiness cost ($/unit/day)	c_g^{TRD}	$1
Ordering cost ($/order)	c_g^{ORD}	$10
Products		
Types of component	N_C	24
Types of wheel assembly	N_A	32
Types of drum	nDrums	12
Types of spider	nSpiders	6
Types of hub	nHubs	6
Due-dates		
Lead time of components (days)	C_lead_time	6 days
Lead time of assemblies (days)	A_lead_time	13 days
Miscellaneous Castings		
Percentage Total Production	MC_ratio	35 %
Percentage Ductile	MC_DuctRatio	25 %
Percentage Grey	MC_GreyRatio	75 %
Min order quantity	MC_Xmin	100 units
Max order quantity	MC_Xmax	300 units
Cycle time	MC_rho	0.75 minute
Lead times	MC_LeadTime	3 days

5. SIMULATION OF DEMAND

Simulating the HIHPS at Toowoomba Foundry requires a set of randomly generated external orders, and a method of generating schedules. Schedules are generated according to the following assumptions and the critical ratio rule.

- A casting job must begin and finish within a designated shift interval.
- Only jobs of the same metal type can be cast in the same shift (*grey*: drums, *ductile*: hubs/spiders).
- A batch of components is available to the buffer immediately after the completion of machining the entire batch.
- Components used in an assembly job are removed from the buffer immediately after the start of the job.
- An assembly job can begin at time *t* if: there are sufficient component stocks in the buffer at time *t;* or by the time buffer stocks for a component are depleted to zero, replenishments are scheduled to arrive.
- Set-up times are sequence dependent.

Before schedules can be generated it is necessary to generate customer orders and determine replenishment jobs. In addition to component buffer replenishment jobs to replenish buffer stocks, there are non-buffer jobs that must be generated to realistically simulate loading of the system. This can be seen in Figure 3, which illustrates the possible product routes. There are four types of jobs that can be processed in stages involving wheel production: customer ordered assembly jobs; component replenishment jobs; customer ordered component jobs; and miscellaneous casting jobs. In addition to the parameters defined for the HIHPS model, the simulation requires parameters from which input for the HIHPS model can be randomly generated. The parameters describe the different job types and demand patterns. The following is a list of the additional parameters required for the simulation:

- Number of assembled product types and component types;
- Sequence of machines for processing each product;
- Cycle times for each product;
- Set-up matrices for products processed at each machine;
- Probability distribution functions (pdf) for inter-arrival times of external orders for each product;
- pdf's for order quantities of external orders for each product; and
- Function to determine the due-dates of jobs.

Given all these input parameters, VB Marcos in Microsoft Excel were used to generate the input data necessary for the scheduling module and objective evaluation. For scheduling, the critical ratio rule was implemented using Microsoft Visiual C++. The two applications communicate via text files.

6. SOLUTION

Good solutions to the Toowoomba Foundry problem were obtained using a simulated annealing algorithm (Kirkpatrick et al 1983). Simulated annealing has been applied to a diverse range of combinatorial optimisation problems. The algorithm attempts to escape local optima by strategically accepting non-improving neighbour solutions. A neighbour of the current solution is chosen at random and automatically accepted if TC is decreased. If the neighbour causes an increase ΔTC in the objective, then it is accepted with probability exp($-\Delta TC/T$). The temperature T, begins at a relatively high value T_0 so that most non-improving neighbours are accepted. As the algorithm progresses, T is reduced at regular intervals by $T' = T_{\text{fact}}T$, where $0 < T_{\text{fact}} < 1$. When T is reaching its lowest values towards the end of the process, non-improving neighbours are rarely accepted. The initial temperature, T_0 and cooling factor, T_{fact} are controllable parameters of the process.

The HIHPS model was used with simulated annealing and applied to the representative data-set from Toowoomba Foundry. At the beginning of a simulated annealing run, three sets of externally ordered jobs (J_k, $J1_k$ and $J2_k$) and initial buffer levels are generated, and an initial solution ($\mathbf{S_0}, \mathbf{ss_0}$) is supplied. The external job sets and initial buffer levels are not changed throughout the optimisation. Given below is the pseudo-code of the simulated annealing algorithm used in this study.

HIHPS-Simulated Annealing Algorithm:

```
read initial solution
T = T0
while iteration limit not exceeded do
        for a fixed number X iterations do
                generate random neighbour solution
                if improvement then accept neighbour for next iteration
                else accept with probability exp(-ΔTC/T)
        end for
        T = T * Tfact
end while
output best solution found
```

Neighbour solutions are generated by applying randomly sized jumps of up to 40 units on either S_i or ss_i. The jump is applied to each component with a probability of 0.3. Once **S** and **ss** are determined, the replenishment job set JC is determined and the objective evaluation begins. One objective evaluation consists of scheduling machine operations from the job sets J, $J1$, $J2$ and JC, over the 336 days. The critical ratio dispatching rule is used to generate the schedules. Simulated annealing was run for 2000 simulations with $T_0 = 65000$ and $T_{\text{fact}} = 0.9$. The initial solution used was $S_i = 150$ and $ss_i = 0$, $i = 1, ..., N_C$ with objective value $TC = \$78036.51$, Figure 4 shows total cost changing over 1 run of simulated annealing and Table 2 shows the best solution found, $TC = \$28487.09$. Overall, simulated annealing improved on the initial solution by about 63%. Total computation time was about 17 hours on a Pentium 300MHz Windows NT machine.

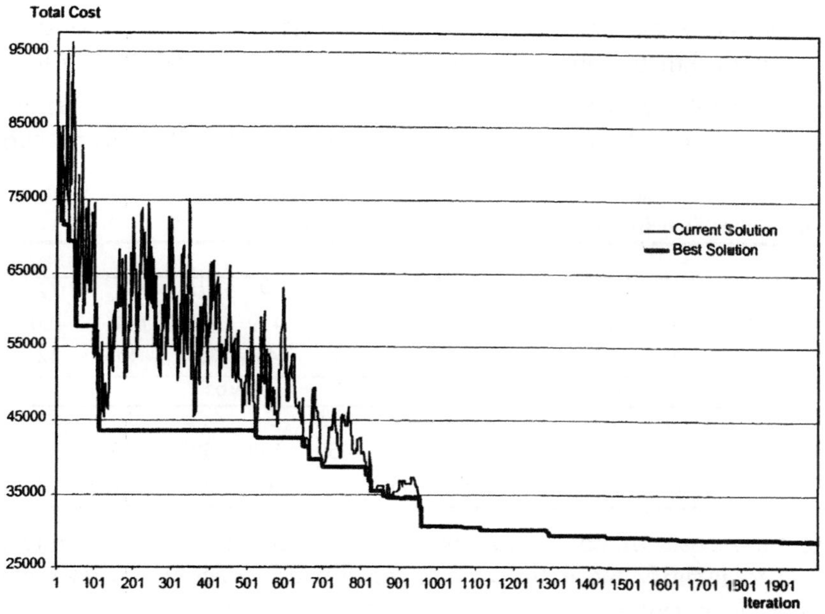

Figure 4. Improvement in solution by simulated annealing.

Table 2. Best solution found, S_i and ss_i , $i = 1, ..., N_C$

	Drums											
i	*1*	*2*	*3*	*4*	*5*	*6*	*7*	*8*	*9*	*10*	*11*	*12*
S_i	290	270	160	210	415	160	525	440	140	165	325	240
ss_i	215	65	5	0	35	0	70	15	25	80	20	65
	Spiders						**Hubs**					
i	*13*	*14*	*15*	*16*	*17*	*18*	*19*	*20*	*21*	*22*	*23*	*24*
S_i	90	225	150	255	155	235	175	145	600	255	390	355
ss_i	10	160	110	25	45	65	105	75	35	160	35	55

The same dataset used for HIHPS simulated annealing was used to evaluate current practices at Toowoomba Foundry. Maximum buffer levels (S_i) are determined from the expected annual assembled goods demand for that component. Components are separated into two categories, A = less than 4500 units per year; and B = greater than 4500 units per year. Maximum buffer levels for category A are determined based on an expected 6 week replenishment cycle ($T_A^{\text{rep}} = 6$), and a 3 week cycle for category B ($T_B^{\text{rep}} = 3$). The following calculation is used to determine the maximum buffer levels.

$$S_i \approx \frac{T_{cat}^{\text{rep}}}{52} E(D_i), \ cat = A, B.$$

Safety stock levels are simply calculated as 10% of the maximum buffer levels, that is $ss_i \approx 0.1 \times S_i$. Based on these calculations, Table 3 gives the decision variables determined using current practices at Toowoomba Foundry.

Maximum buffer levels are rounded to the nearest five units. Using these decisions, replenishment jobs are determined and the critical ratio rule was applied and the total cost was calculated, $TC^{TF} = \$54617.27$. By using the simulated annealing solution, total costs are reduced by 48% from current practices.

Table 3. Toowoomba Foundry current practice solution, S_i and ss_i, $i = 1, ..., N_C$

	Drums											
i	*1*	*2*	*3*	*4*	*5*	*6*	*7*	*8*	*9*	*10*	*11*	*12*
S_i	385	305	290	385	150	340	515	645	450	385	450	420
ss_i	39	31	29	39	15	34	52	65	45	39	45	42
	Spiders						Hubs					
i	*13*	*14*	*15*	*16*	*17*	*18*	*19*	*20*	*21*	*22*	*23*	*24*
S_i	380	645	615	385	385	355	450	450	130	595	225	225
ss_i	38	65	62	39	39	36	45	45	13	60	23	23

The total costs of both solutions were divided into their separate cost components to identify the deficiency in the foundry practice. This includes work-in-process inventory (process queues before component stations) cost, buffer inventory costs, tardiness costs and set-up costs. Table 4 gives the cost break down for the two solutions. In the TF-practice solution, tardiness costs are much higher than the HIHPS-SA solution. This results from a combination of two reasons, low safety stock levels and high maximum buffer levels. Low safety stocks result in later reactions to future shortages. Maximum buffer levels that are too high mean that processing times of replenishment batches are very large and therefore take longer to reach the buffer. Whilst larger lot-sizes result in a reduction of set-up costs, this is overwhelmed by the increase in tardiness costs.

Table 4. Cost break downs of TF practice and HIHPS-SA solutions.

Costs:	WIP	Buffer	Tardiness	Set-up	Total
TF-practice	6987.85	16293.42	28446.00	2890.00	54617.27
HIHPS-SA	2826.60	12689.49	4881.00	8090.00	28487.90

A second experiment was conducted to test the consistency of solutions over a variety of simulated datasets. The test involved the simulated annealing solution given previously, which is denoted X_1 and another unrelated solution X_2. Additionally, the given solution was perturbed slightly into two new solutions to test the sensitivity of the objective to minor changes in the decision variables. The perturbation used was to interchange the values S_i and ss_i with the values of another component S_j and ss_j. In particular, the perturbed solution X_{11} interchanged values for components 10 and 16, whilst the second perturbed solution X_{12} interchanged the values for components 7 and 21. After 100 trials of 336 day simulations, the following averages were observed for the solutions tested: $\overline{TC}(X_1) = \$66262.40$; $\overline{TC}(X_{11}) = \$63307.40$; $\overline{TC}(X_{12}) = \$65849.70$; and $\overline{TC}(X_2) = \$79160.50$.

These averages are quite large compared with the evaluation of X_1 with the original dataset, TC = \$28487.09. However, the original value is within the range observed in the 100 trials, ($\min TC(X_1) = \$26330.70$ and $\max TC(X_1) = \$195562.00$). One-sided, paired t-tests were performed to determine whether the observed means were significantly different and how many simulations were required before this could be detected. Each solution was compared against the original, X_1. Figure 5 shows the *p*-values of the t-tests evolving as more simulations are performed.

The t-tests indicated that solutions X_1 and X_{12} were not significantly different, but the p-values for X_1 compared with X_{11} and X_2 indicate a strong difference between the means. The differences were significant at the 5% level after about 15 simulations. The fact that X_{11} was definitely better than the original solution indicates that swapping values could be a useful extension to the "stepping" neighbourhood used for the HIHPS simulated annealing algorithm. That is, improvements may still be found using a "swapping" neighbourhood when the stepping neighbourhood is trapped in a local optimum.

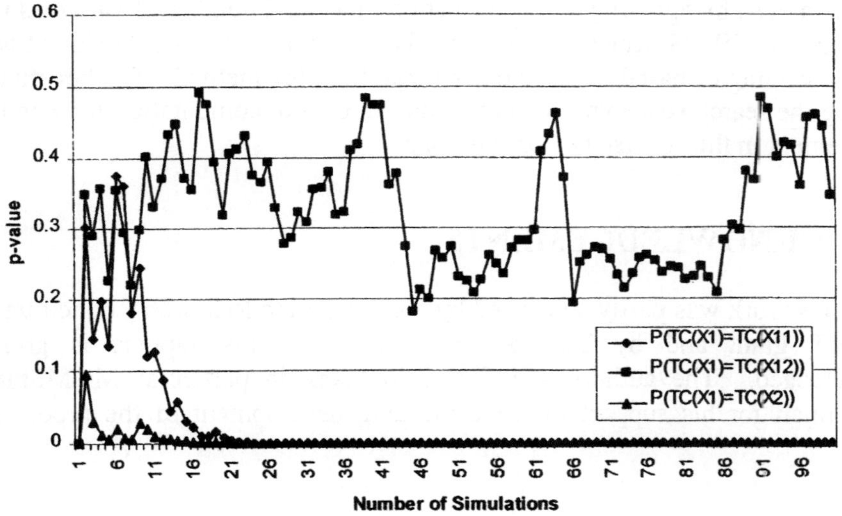

Figure 5. t-Test comparisons of different solutions.

These results also indicate that about fifteen simulation datasets would be necessary to obtain robust solutions. This would increase computation time by a factor of fifteen, which with a faster computer may be acceptable under a yearly or monthly review of inventory and lot-sizing decisions. Alternatively, the number of datasets could be reduced if a candidate set of good solutions were kept. Once simulated annealing is complete, more datasets can be introduced to conduct a more vigorous comparison of the solution candidates.

7. CONCLUSION

This study has investigated a complex inventory and lot-sizing problem originating from a push/pull hybrid production system in a foundry. To the authors knowledge, previous studies have not addressed problems with the degree of difficulty present in this system. This study has layed the groundwork for solving such difficult problems using a combination of simulation and heuristic optimisation. In particular, the simulated annealing algorithm was applied and achieved significant improvement over the initial solution provided and the current practice at Toowoomba Foundry. Although computation time is quite costly at 17 hours, inventory decisions are typically reviewed monthly or yearly so that this length of time is acceptable.

Future work on this problem will implement an optimisation heuristic with multiple simulation datasets and keep a candidate set of good solutions for further evaluation at the end. Another modification will be to reduce the number of decision variables so that the search space is a more manageable size. This can be achieved by applying a transformation of a small number of variables into the full set of HIHPS decision variables. This approach is to be combined with solution techniques based on the Nelder-Mead Simplex method. The benefits of reducing the search space will be a reduction in overall computation time and an improvement in the consistency of solutions.

8. ACKNOWLEDGEMENTS

This work was partly supported by the Australian Research Council under a SPIRT grant and by Toowoomba Foundry. This support is greatly acknowledged. The authors would like to thank in particular Mr Mithiran Kirupairajah for his support in the continuing development of the production system.

References

Bookbinder, J. H. & Cakanyildirim, M. Random lead times and expedited orders in (Q, r) inventory systems. European Journal of Operational Research, 1999: 115, 300-313.

Browne, J, Harhen, J & Shivnan, J. *Production Management Systems: A CIM Perspective.* Reading MA: Addison Wesley, 1988.

Cochran, J. K. & Kim, S. S. Optimum junction point location and inventory levels in serial hybrid push/pull production systems. International Journal of Production Research, 1997, 36.4: 1141-1155

Gavirneni, S. An efficient heuristic for inventory control when the customer is using a (s,S) policy. Operations Research Letters 28, 2001, 187-192.

Graves, S. C., Rinnoy Kan, A. H. G. & Zipkin, P. H. (Eds). Logistics of production and inventory. Handbooks in operations research and management science (4). Netherlands, Elsevier Science Publishers, 1993.

Karmarkar, U. S. Integrating MRP with kanban/pull system (Working Paper Series #QM8615). Graduate School of Management, University of Rochester 1986,

Kimura, O. & Terada, H. Design and analysis of pull system, a method of multistage production control. International Journal of Production Research, 1981, 19: 241-253.

Kirkpatrick, S., Gelatt, C. D. & Vecchi, M. P. Optimization by simulated annealing. Science, 1983, 220: 671-680

Sarker, B. R. & Fitzsimmons, J. A. The performance of push and pull systems: a simulation and comparative study. International Journal of Production Research, 1989, 27: 1715-1732.

Sugimori, Y., Kusunoki, K., Cho, F. & Uchikawa, S. Toyota production system and Kanban system materialization of just-in-time and respect-for-human system. International Journal of Production Research, 1977, 15, 553-564.

Takahashi, K., Hiraki, S. & Soshiroda, M. Pull-Push integration in production ordering systems. International Journal of Production Economics, 1994, 33: 155-161.

Winston, W. L. *Operations Research: Applications and Algorithms*. California: International Thomson Publishing, 1994.

DESIGN OF ADAPTIVE SELF-NAVIGATED AIRSHIP IN SIMULATED ENVIRONMENT

Keiko Motoyama[1], Keiji Suzuki[2], Hidenori Kawamura[1], Masahito Yamamoto[1], and Azuma Ohuchi[1]

[1] *Graduate School of Engeneering,*
Hokkaido University,
Nishi 8, Kita 13, Kita-ku, Sapporo, Hokkaido 060-8628, Japan
{keiko, kawamura, masahito, ohuchi}@complex.eng.hokudai.ac.jp

[2] *Faculty of System information Science.,*
Hakodate Future University,
116-2, Nakano, Kameda, Hakodate, Hokkaido, 041-8655, Japan
suzukj@fun.ac.jp

Abstract The final goal of this research is to realize a small airship robot that can automatically achieve a given task. The airship is subjected to strong inertial forces and air resistance. Although reinforcement learning methods could be expected to control a small airship, the unstable property of the airship prevents the learning methods from achieving control of it.
In order to design an automatically controlled airship, sensory information is especially important. We assume using like ultrasonic transducers which have been widely used as a cheap and light way to provide mobile robots with accurate range finders. This paper verifies the difference in control performance of the airship between a variety of sensory setup. We simulated a small airship with the Cerebellar Model Articulation Controller (CMAC) as a reinforcement learning method which is enabled to deal with generalization problems, on the assumption that we use ultrasonic transducers afterward.
The experimental results showed that the learning performance was not always proportional to the amount of sensory information, and different behavior was acquired according to differences in the sensory setup.

Keywords: Reinforcement Learning, Airship Control

1. INTRODUCTION

The final goal of this research is to realize a small airship robot that can automatically achieve a given task, such as an indoor observation at a disaster emergency, in an unstable real environment. Robots such as an automatically controlled airship will be of great use in various area. Especially, compared with a big airship, a small airship is easy to pass through a narrow passage, and can maneuver more effectively in indoor environment.

However, it is very difficult to control an airship automatically. For instance, the airship has great inertia and unstable dynamics, and it is greatly affected by temperature and atmospheric pressure. It is also affected by the condition of the battery used in a small airship, because the battery is so small that the supplied voltage is altered in each trial. In order to cope with such a dynamic and unknown environment, the airship should be able to learn and acquire specific patterns of behavior.

There are two main problems as barriers to accomplish in this purpose. One is to develop a learning method for a self-Navigated airship. Much interest has recently been shown in reinforcement learning as a possible method for robot learning with little or no a priori knowledge about the environment or a robot itself (Sutton, 1998, Singh, 1996, Murao et al., 1997, Tan, 1993). Such learning methods have enabled us to realize reactive and adaptive behaviors in several applications with a higher capability (Lin, 1993). However, reinforcement learning was conventionally intended for the situation that agents, which have constant characteristics, optimize their actions for a simple task in the dynamic environment based on stable change in the simulation. They did not assume the situation that the characteristics of environmental fluctuation itself changes. In this research, the airship must learn in a real dynamic environment. Though we need to consider the learning method in order to make the airship learn in such an environment, here we use CMAC which can generalize the sensor space.

Another problem is how to design the airship. In order to design an automatically controlled airship, a lot of things must be decided. For example, it is necessary to consider what kind of sensor, motor, and power supply will be appropriate for the airship. Sensory information is especially important. As the airship does not have a large buoyant force enough to be able to hold heavy sensory equipment, we assume using ultrasonic transducers. They have been widely used as a cheap and light way to provide mobile robots with accurate range finders. Sonar using ultrasonic transducers typically measure the round trip time of flight required for a pulse of emitted energy to travel from the ultrasonic emitter to a reflecting object and then echo back to the receiver.

In this study, the authors simulated a small airship control with CMAC as a reinforcement learning method which is enabled to deal with generalization problem, on the assumption that we use ultrasonic transducers afterward. We verified the difference in control performance of the airship between a variety of sensory setup, and design policy for sensory setup.

In Section 2, the problem domain concerned with the small airship in the simulation environment is described. In Section 3, the algorithm of CMAC based Q-Learning is explained. Experimental results are discussed in Section 4. The discussion focuses on the difference between a variety of sensory setup. Finally, a summary is presented in Section 5.

2. PROBLEM DOMAIN

This problem considers a simulated airship control in a continuous environment. The airship exists in a two-dimensional rectangular area surrounded by walls and divided into three rooms by partition walls as shown in Figure 1. The task for the airship is to pass through the rooms in turn without crashing into walls. This environment commonly causes a perceptual aliasing problem which reinforcement learning robots often face. This experimental environment investigates the relationship between sensory settings and perceptual aliasing.

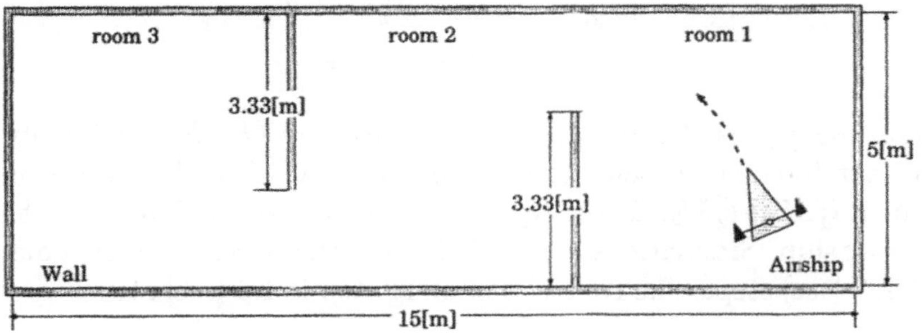

Figure 1. Experimental environment

The environment and the structure of the airship are assumed as follows: The rectangular area is assumed as 5[m] × 15[m] surrounded by the walls. The room is divided into three rooms by partition walls, of which length is 3.33[m]. In each room, it is assumed that there is no current of air in space. The airship has two propellers P_1, P_2. The distance between the airship's center of gravity M_0 and each propeller is $l = 10.0$[cm]. The mass of this airship is $M = 7.0$[g] and the mass of each propeller is $M_p = 1.0$[g]. In this simulation, we assume the airship has no volume, that is, it is a point with mass. A propeller $P_i (i = 1, 2)$ can produce a force $f_i (0 \leq |f_i| \leq F_{max} = 1.0[N])$. Driving these propellers, the airship can move in the two-dimensional space freely. The resistance coefficients for driving force and for rotating force are $D_v = 4.0$[kg/s] and

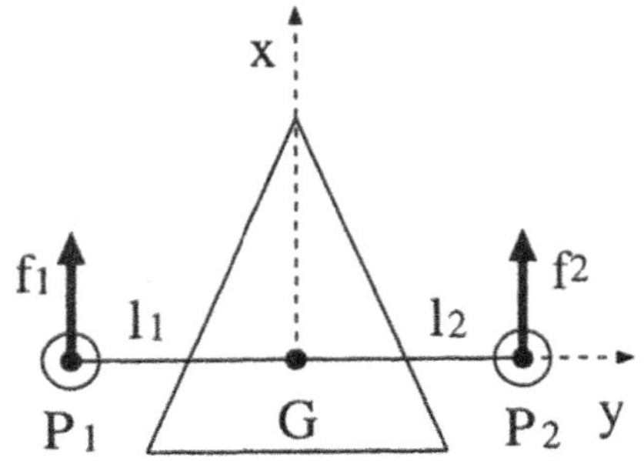

Figure 2. Structure of airship

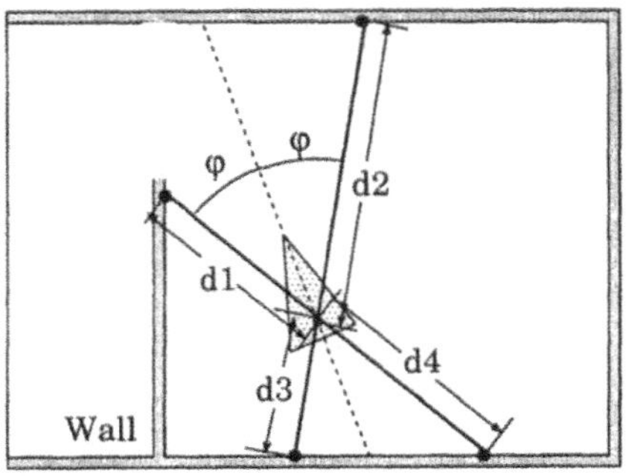

Figure 3. Sensor

$D_w = 80.0$[kg/s], respectively. Also there are secondary delays because of the effect of inertia. The airship has some sensors as mentioned in section 4, we assume, to detect a distance from walls.

The equations of motion for the airship are

$$M\ddot{x}_G + D_v\dot{x}_G + F_x = 0, \quad (1)$$

$$M\ddot{y}_G + D_v\dot{y}_G + F_y = 0, \quad (2)$$

$$I\ddot{\theta} + D_w\dot{\theta} + \tau = 0 \quad (3)$$

where (x_G, y_G) is the airship's coordinates, and (F_x, F_y) and τ are the resultant force vector and torque, respectively. I is the rotate inertia moment $l_1^2 M_1 + l_2^2 M_2$. These equations are used to simulate the behavior of the airship. Simulations are carried out with the Runge-Kutta method of 0.001 [sec] step. The control sampling step is 1.0 [sec] step.

3. CMAC BASED Q-LEARNING ALGORITHM

We applied reinforcement learning to acquire a control policy for the airship. The state space of airship control problem is continuous with an infinite number of states. When the problem space is too large to explore completely, the learner must deal with the generalization problem. We applied the *CMAC-based Q-learning* algorithm (Q-CMAC) (Ono and Fukuta, 1996) which is a *Q-learning* algorithm using *CMAC* (Sutton, 1996) as an approach to the generalization problem.

The structure of the state space in Q-CMAC consists of overlapped tilings, as shown in Figure 4. Each tiling represents the state space segmentation by dividing each state variable into some regions. Thus, the Q-values are distributed into these tilings as Q-tables. These tilings are allowed to have different state space segmentations. There-

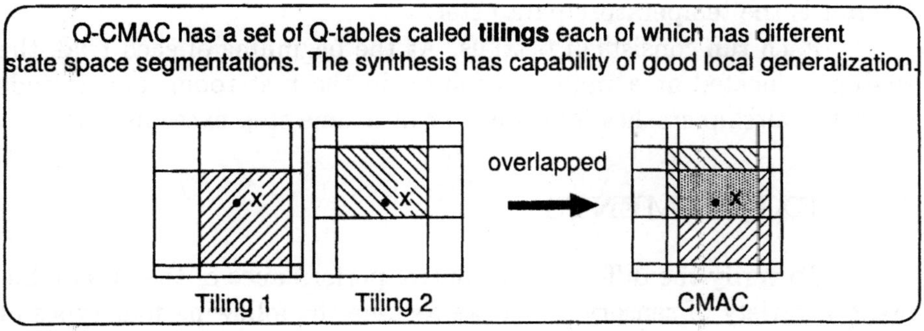

Figure 4. CMAC-based Q-learning [Ono, 1996] (Q-CMAC)

fore, the Q-CMAC can generate a large number of state-action values from a combination of overlapped tilings even if each tiling consists of a relatively small Q-table.

Let $q_i(x,a)$ be a Q-value for taking action a in state x held by the i-th tiling in the Q-CMAC, then the state-action value $Q(x,a)$ can be written as

$$Q(x,a) = \frac{1}{N}\sum_{i=1}^{N} q_i(x,a), \tag{4}$$

where N is the number of tilings. It is assumed that the force produced by each propeller has three states, F_{max}, 0, or $-F_{max}$. Accordingly, the number of actions in this Q-CMAC is $3 \times 3 = 9$.

When the airship takes action a from state x, it receives reward r, and then it moves to the next state y. In this sequence, $q_i(x,a)$ $(i = 1, ..., N)$ is updated as follows:

$$q_i(x,a) \leftarrow q_i(x,a) - \alpha e, \tag{5}$$

$$e = Q(x,a) - (r + \gamma \max_{b \in actions} Q(y,b)) \tag{6}$$

where $\alpha (0 < \alpha < 1)$ is the learning rate and $\gamma (0 \le \gamma < 1)$ is the discount parameter. A current state x consists of d_i $(i = 1, ..., k)$, where d_i is the distance to wall which is detected by i-th sensor.

In this problem, the airship will receive a positive reward, r_s, when it enters the next room. Given a current state, x, and an available action, a_i, the airship selects each action, a, with the probability determined by the Boltzmann distribution:

$$p(a_i|x) = \frac{e^{q(x,a_i)/t}}{\sum_{k \in actions} e^{q(x,a_k)/t}}, \tag{7}$$

where t is the temperature parameter.

Each run consists of t_t trials. At the beginning of each trial, the airship is located at a random position in the first room, and it ends when the airship crashes into walls, or at a time step exceeding t_o.

4. EXPERIMENTS

To verify the difference in control performance of the airship between a variety of sensory setup, we assume the following four types of sensory information that detects distance from walls:

type 1 Four sensors at an angle of $\pm\varphi$ degrees and $(\pm\varphi + 180)$ degrees to a line of the airship as shown in Figure 5. We carried out the simulation with the following three types of φ: (a) $\varphi = 15$, (b) $\varphi = 30$, (c) $\varphi = 45$.

type 2 Excluding two sensors from type 1, two sensors at an angle of $\pm\varphi$ degrees to a line of the airship. Here, φ was set to 30.

type 3 Four sensors as shown in Figure 6. The two lines crossed perpendicularly.

type 4 Adding two sensors to type 1 as shown in Figure 7. Here, $\varphi 1$ and $\varphi 2$ were set to 20.

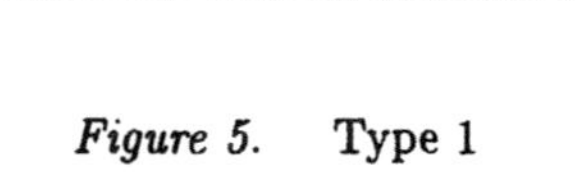

Figure 5. Type 1

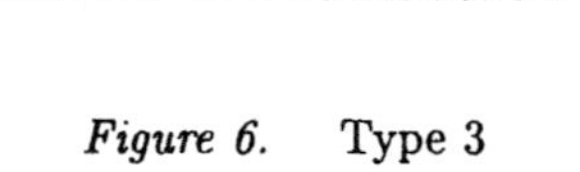

Figure 6. Type 3

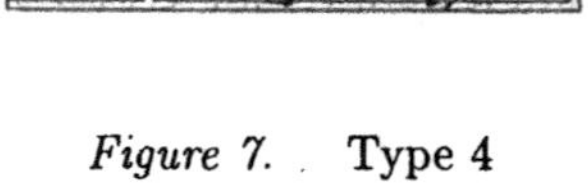

Figure 7. Type 4

The learning started at a random direction and position in the first room in each trial, and ended when the airship entered the third room or crashed into walls, or when the time step exceeded the step limit $t_o = 5000$. The reward was assigned as $r_s = 100.0$ when the airship entered the next room. Each simulation was run for $t_t = 5000$ trials. The state space segmentation of CMAC was set as follows: The number of tilings was 3, and the number of segmentations of each sensor was 9 excepted sensory type 4. In type 4, the number of segmentations concerning $d1 \sim d4$ was 6 and $d5, d6$ was 3, in order to avoid increasing the state space. The length of each segmentation stretch was fixed to 1.0[m] (1.2[m] in type 4) and shifted at regular intervals within the length of a segmentation stretch.

The learning parameters were determined such as $\alpha = 0.2, \gamma = 0.98, T = 0.3$. The simulation began with Q-tables of the airship initialized to 0.

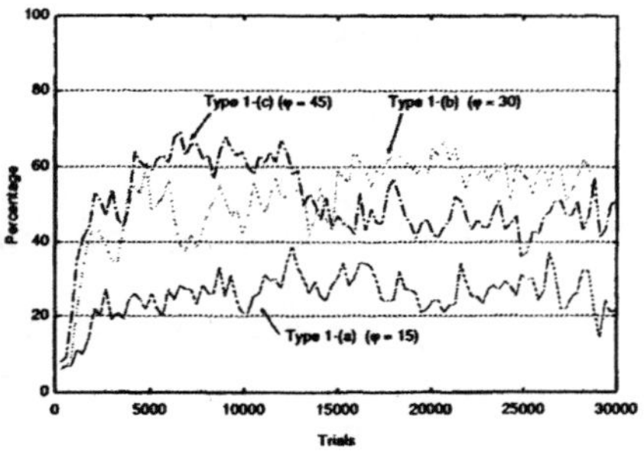

Figure 8. Performance

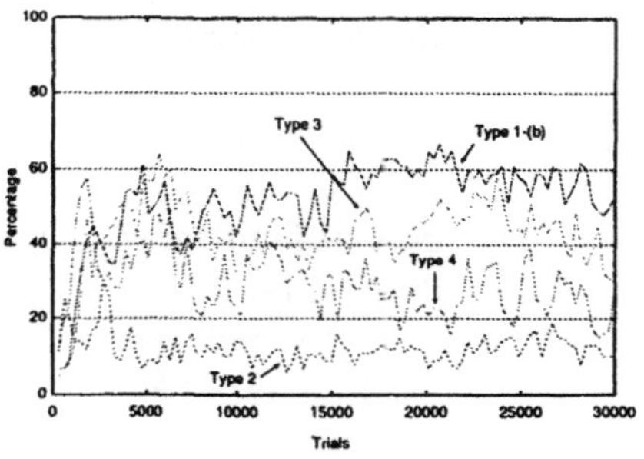

Figure 9. Performance

Figure 8 shows a simulation result for three different value φ in sensory type 1. The vertical axis in Figures 8 and 9 indicates a percentage of the airship that could enter the second room plotted every 300 trials. As shown in Figure 8, the airship with sensor type 1-(a) ($\varphi = 15$) has worse performance than the other two types. This indicates that a small value of φ makes the nearest two sensors result in quite similar sensor information.

Figure 9 shows a simulation result for learning with a different sensory type. In this experiment, the airship with only two sensors (type 2) could hardly learn the given task as expected. However, the airship with six sensors also did not show a very good performance. Since computer memory was restricted, we reduced a number of CMAC's segmentation when the airship had 6 sensors in order to avoid increasing the state space. Increasing a number of sensors or CMAC's segmentations may make possible fine control. However, it is important to consider a balance between fine control and costs for learning. Thus, the learning performance is not always proportional to the amount of sensory information. This problem will be possible when a real learning robot is constructed, and we must be cautious of sensory settings. It was found through these experiments that the airship accomplishes a task even if it

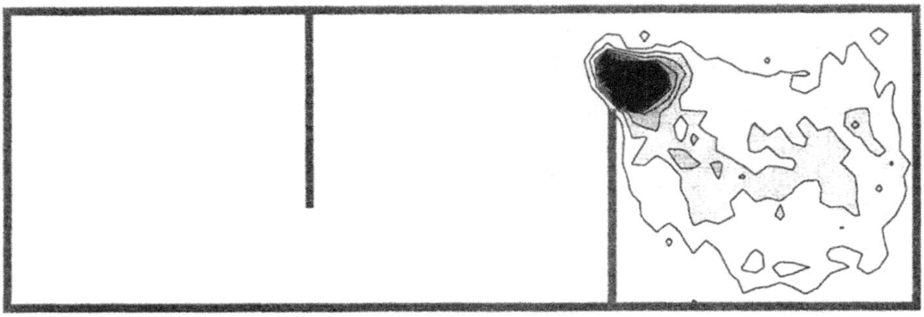

Figure 10. Distribution of airship's positions using sensor type 1-(b)

is placed in a difficult environment such as perceptual aliasing problem which easily occurs.

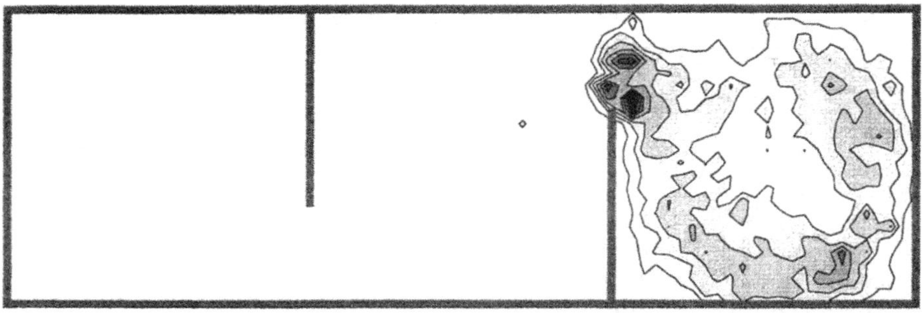

Figure 11. Distribution of airship's positions using sensor type 2

Figures 10 and 11 show the distribution of the airships' positions with sensor type 1 and type 2 respectively, in 100 trials after learning. The darker areas correspond to the higher existence and lighter areas to the lower. Figure 10 shows that the airship with sensor type 1-(b) flew to the second room through center of the first room, while the airship with sensor type 2 flew toward walls and found a loophole as shown in Figure 11. Comparing the two sensor types, the former airship could learn effectively as shown in Figure 9.

Thus, different behavior was acquired according to different sensory setups. However, the airship with these sensors both rambled near the entrance to the second room, and could not go to the third room. The airship seemed to be unable to discriminate states around here, and could not generalize similar sensor inputs. Moreover, since rewards were assigned only when the airship entered the next room, the airship felt sure that it was better being near the entrance. Namely, quite high Q values were assigned the entrance to the second room. In order to settle this problem and make the airship enter the third room, it is necessary to give the way of rewards' assignment more carefully.

5. CONCLUSION

In this paper, the research verified the difference in control performance of the airship between a variety of sensory setups. We simulated a small airship with CMAC as a reinforcement learning method which enabled us to deal with generalization problem, on the assumption that we use ultrasonic transducers afterward.

The experimental results showed that it is not always the case that a learning performance is proportional to the number of sensory information, and different behavior was acquired according to a difference in the sensory setup. This problem will be more severe when a real learning robot is constructed, and we must be cautious of sensory settings.

References

Lin, L. -J. Scaling Up Reinforcement Learning for Robot Control. Proceedings of the 10th International Conference on Machine Learning 1993: 182-189

H. Murao, I. Kitagawa and S. Kitamura. Adaptive State Segmentation for Q-learning (in Japanese). Proceedings of SICE'97 1997: 45-48

N. Ono and Y. Fukuta. Learning to Coordinate in a Continuous Environment. Proceedings of the Second International Conference on Multi-agent Systems, 1996

Singh, S. P. Reinforcement Learning with Replacing Eligibility Traces. Machine Learning 1996;22,1/2/3: 123-158

Sutton, R. S. Learning to predict by the methods of temporal differences. Machine Learning 1988;3: 9-44

Sutton, R. S. Generalization in reinforcement learning: Successful examples using sparce coarse coding. In D.S.Touretzky, M.C.Mozer and M.E.Hasselmo(eds.), Advances in Neural Information Processing Systems 1996: 1038-1044

Y. Takahashi, M. Asada and K. Hosoda. Reasonable Performance in Less Learning Time by Real Robot Based on Incremental State Space Segmentation. Proceedings of the IEEE/RSJ International Conference on Intelligent Robots and Systems 1996: 1038-1044

Ming Tan. Multi-Agent Reinforcement Learning: Independent vs. Cooperative Agents. Proceedings of the Tenth International Conference on Machine Learning 1993: 330-337

Wiering, M. and Schmidhuber, J. Speeding up Q(λ)-learning. Proceedings of the Tenth European Conference on Machine Learning 1998: 352-363

ON THE USE OF A STRUCTURAL MODELLING FOR COLLABORATIVE LEARNING USING THE CONCEPT MAP

Hajime Saito[1], Azuma Ohuchi[2] and Takashi Maeda[1]

[1]*Faculty of Information Media, Hokkaido Information University, 59-2, Nishi-Nopporo, Ebetsu, Hokkaido, 069-8585 Japan [hajime, maeda] @do-johodai.ac.jp*

[2]*Graduate School of Engineering Hokkaido University Nishi 8, Kita 13, Kita-ku, Sapporo, Hokkaido, 060-8628 Japan ohuchi@complex.eng.hokudai.ac.jp*

Abstract In education, it is in general desirable that the educational process is not to cram in knowledge but to build up the knowledge based on learner's understanding of the related concept. Concept maps have been presented in pedagogical literature, and it can more easily and rapidly represent a part of learner's understanding than the written and oral examination. Though the concept map is a powerful and useful tool for teachers to know learner's understanding, supporting method to create and correct it has never been considered. In addition, it is also difficult to objectively evaluate and compare the concept maps created by multiple learners. In this paper, we present a method to support collaborative learning using the concept map based on structural modelling, FISM (Flexible Interpretive Structural Modelling). By using this method, learners are supported to construct the consistent concept map, and to combine some concept maps into a consensus model among learners. Teachers can know a minimum leading point by comparing his/her concept map with a consensus model of learners.

Keywords: Collaborative Learning, Concept Map, FISM.

1. INTRODUCTION

Now, many researchers and educators regard constructive learning as important (Inaba, 1999). As a related idea "concept map" to support building up the knowledge based on learner's understanding of related concepts, has been presented in pedagogical literature. It can more easily and rapidly represent part of learner's understanding than the written or oral examination, so teachers have often used it as a method of examination.

Though the concept map is a powerful and useful tool for teachers to determine learner's understanding, it has never been considered as a supporting method to create and correct it. It is also difficult to objectively evaluate and compare concept maps created by learners.

On the other hand, currently, CSCL (computer supported collaborative learning) has been one of the main topics in educational application of the Internet. In the CSCL system, it is not easy to grasp the state of the learner's understanding because the teacher cannot communicate with the learner directly as in a face-to-face talk. We consider that grasping the learner's understanding is one of the most important subjects and the concept map might become useful and powerful tool in such a field too (Tsai, 2001; Saito, 2000).

In this paper, we present the following method to support teachers and learners in collaborative learning using the concept map on structural modelling.

1. Expression and identification of the concept labels with explanatory note;
2. Construction of the consistent concept map;
3. Calculation of the minimum leading point.

By using this method, learners might get some support to construct the consistent concept map, and to combine some concept maps constructed by them into a consensus model. Teachers can know a minimum leading point by comparing his/her concept map with a consensus model among learners. Related work, FIRS (Fuzzy Item Relational Structure Analysis) (Itoh, 1994) supports creating a consensus model for teaching material among teachers using some graphs based on FISM constructed by items of lecture and their relations, and using feedback of examination after the class. In the FIRS, the graphs are only used by teachers to create a consensus model, but we would like to let learners use the graphs corresponding to our concept map as a kind of teaching aid.

In the next section, we define the concept map referred to in this paper. In the third section, we present a collaborative learning process. Next, we describe the application of a proposed method as an example. Finally, we conclude and describe future work.

2. CONCEPT MAP

The concept map, in general, is a two-dimensional diagram that shows some concept labels, which are keywords to show learning content, and their links. The proposition that might be a simple means of linking is drawn on the map.

There are two types of concept map, hierarchical or nonhierarchical. The hierarchical concept maps, which more higher/lower relation and upper/lower position among the concept labels are given equal meaning, are effective for learnin\g with memory. We especially focus on the hierarchical concept map, which is defined as follows.

Definition 1: set of concept labels

A set of concept label cl^t presented by teacher. m is a number of cl^t.

$$CL^t = \{cl_1^t, cl_2^t, \ldots, cl_m^t\} \tag{1}$$

A set of concept label cl^s expressed by learners. n is a number of cl^s.

$$CL^s = \{cl_1^s, cl_2^s, \ldots, cl_n^s\} \tag{2}$$

Definition 2: Concept map

Concept map CM^t, which is two-dimensional Euclid space arranged and linked the elements of CL^t by teacher. $CM^{t'}$ is drawn only concept labels. CM^s and $CM^{s'}$ are drawn by learners.

Definition 3: Relation among the concept labels

Binary matrix R^t expresses whether concept labels are linked or unlinked on the concept map. R^s is the same as R^t, where R^s (S = 1,2,...,N) is a three dimensional matrix when the number of learners is N. R^s is decided by collaborative learning among the learners.

To be constructed the hierarchical concept map needs to consist of the transitive relationship between the concept labels. So, we propose to introduce the FISM (Flexible Interpretive Structural Modeling) (Ohuchi, 1989) to create the hierarchical Concept map.

FISM is an extended and improved version of ISM (Interpretive Structural Modeling) developed by Warfield (Warfield, 1976). The computer algorithm of FISM is based on the partially filled reachability matrix (PR-matrix) model. The PR-matrix model is an extension of a reachability matrix (R-matrix) model and has great utility in all phases of ISM.

In this paper, we suppose that CM^t is always right; therefore R^t must be a partial ordering matrix that is reflexive, transitive and anti-symmetric matrix. On the other hand, we suppose the learner has ordinal thinking that might include misunderstandings; therefore CM^s needs to be a more flexible representation than CM^t. So, R^s is pseudo-ordering matrix that is, reflexive and transitive matrix and that includes partial order and equivalence relation.

In the following, we describe the collaborative learning with the purpose that the learners acquire CM^t.

3. COLLABORATIVE LEARNING PROCESS

The following are two methods to create the concept map.

1. Teacher indicates his/her own concept labels to learners, and learners decide the relationship between them.

2. Learners express their concept labels, and teacher and learners decide the relationship by collaborative learning.

Compared to Concept maps, method 1 is easier than 2, but we select method 2 to build up the knowledge and change the concepts for learners. In this research, we propose following the collaborative learning process for method 2.

[Learning process of learners]

The learning process of learners consists of five phases.

1. Expression of the concept labels with explanatory notes;
2. Identification of the concept labels among the learners;
3. Decision of relations among the concept labels on a concept map.
4. Making a consensus model among learners; and
5. Rewriting a concept map of learners.

[Learning process of teacher]

The learning process of the teacher also consists of five phases.

i. Indication of the concept labels with explanatory note;
ii. Identification of the concept labels among teacher and learners;
iii. Decision of relations among the concept labels of teacher;
iv. Making an opinion graph to lead;
v. Leading to correct knowledge from the elements of disagreement.

In this paper, we assume the following collaborative learning environment aided by a computer that learners can use.

- Each learner can use a computer.
- Learners can communicate by chat.

In the next section we explain detail of their phases.

3.1 Expression and identification of the concept labels with explanatory notes

Both teacher and learners express the concept labels with explanatory note. They identify the concept labels. If they judge concept labels those are not same notation but similar to expression are the same, then they can identify such labels. Here, we introduce the resemblance based on heuristic that "the documents which include many shared keywords have closely connection" to support identification of the concept labels. In the following, we explain the case that the teacher identifies among the labels of the teacher and learners. This method can also apply to the case to identify the labels among learners.

Here, we define the vector to calculate the resemblance, which is made from the frequency of keywords these are nouns extracted from the explanation of concept labels as a result of using the Japanese morphological analysis system "Chasen". cl^t_i $(i = 1,2,...,m)$ is a vector to represent a concept label of teacher, where m is a number of concept labels of teacher.

$$cl_i^t = (w_{i1}^t, w_{i2}^t, \ldots, w_{iX}^t) \tag{3}$$

Where X is a number of keywords.

cl^s_j $(j = 1,2,...,n)$ is a vector to represent a concept label of learners, where n is a number of concept labels of learners.

$$cl_j^s = (w_{j1}^s, w_{j2}^s, \ldots, w_{jX}^s) \tag{4}$$

Then, S_{ij} represents a resemblance calculated by the inner product of cl^t_i and cl^s_j.

$$S_{ij} = cl_i^t \cdot cl_j^s \tag{5}$$

In the next phase, the teacher identifies the concept labels. Here, we propose two method to support identification at the same time. One is a method to maximize the sum total of S_{ij}, another one is a method using the multiple-dimensional scaling (MDS).

3.1.1 Deciding combinations using the Hungarian method

We introduce the Hungarian method (HM) to decide the combination of concept labels of the teacher and learners. Using the HM, the number of concept labels of the teacher and learners must be equal, so that we adopt a method to make the labels the same using some temporary labels, whose resemblance is 0, and which do not show on the concept map.

3.1.2 Two-dimensional expression using MDS

In this paper, we use the MDS (Bryan, 1986; Hori, 1994) to identify a concept with some concept labels in the phase 2 and ii. Here, we define the standard function "*Stress*" to optimize the location of the concept labels. *Stress* is used to minimise the deformation when multiple-dimension comes down to lower dimension.

$$Stress = \sqrt{\sum_{j=1}^{n} \frac{\sum_{i=1}^{m} \left(d_{ij} - \hat{d}_{ij}\right)^2}{\sum_{i=1}^{m} d^2_{ij}}} \tag{6}$$

$$\hat{d}_{ij} = \frac{1}{S_{ij}}$$

Where d_{ij} is a real distance between cl^t_i and cl^s_j on the concept map. If $S_{ij} = 0$, then calculable max value is substituted for $\hat{d}_{ij}$.

If the teacher focuses on a concept label $cl^s{}_j$, then $stress_j$ shows the relationship between $cl^s{}_j$ and all elements of CL^t.

$$Stress_j = \sqrt{\frac{\sum_{i=1}^{m}\left(d_{ij} - \hat{d}_{ij}\right)^2}{\sum_{i=1}^{m} d^2{}_{ij}}} \qquad (7)$$

In a general multiple optimisation problem, the main purpose is the calculation of a global optimal solution, but we positively use the same local optimal solutions calculated by *stress*. Namely, we think of some solutions as some viewpoints of the teacher.

In this phase, if the teacher could not identify in spite of the two supports HM and MDS, then he/she can add and delete some concept label of the learners to let equal to the total number of concept labels between teacher and learners.

3.2 Decision of relations among the concept labels

Teacher and learners can decide the relationship between the concept labels by drawing the links between them. The teacher knows the right relationships in advance, so that teacher might draw the right partial ordering matrix. Learners are supported by the connotation algorithm in FISM. This algorithm can keep the following points;

1. Known elements are not rewritten;
2. Relations can be partially filled reachability matrix after the application of algorithm.

Learners can finally decide the reachability matrix by taking over this algorithm

4. EXAMINATION

In this paper, we explain the case of collaborative learning between a teacher and two learners using a sample that is the "simple classification problem of the rock". Here, the teacher indicates twelve labels, and learner1 expresses ten labels, and learner 2 expresses fourteen labels.

Phase 1 and i: Expression of the concept labels with explanatory note
In the first place, teacher and learners express the plural labels with an explanatory note concerning the rock shown in table 1.

Table 1. Concept labels expressed by teacher and learners

NO	Teacher	Learner 1	Learner 2
1	rock	rock	rock
2	metamorphic-rock		metamorphic-rock
3	sedimentary-rock	sedimentary-rock	sedimentary-rock
4	slate	slate	slate
5	magma	magma	magma
6	igneous-rock		igneous-rock
7	lime	lime	lime
8	shale	shale	shale
9	marble	marble	marble
10	pumice	pumice	pumice
11	lave	lave	lave
12	granite	granite	granite
13			aqueous-rock
14			plutonic-rock

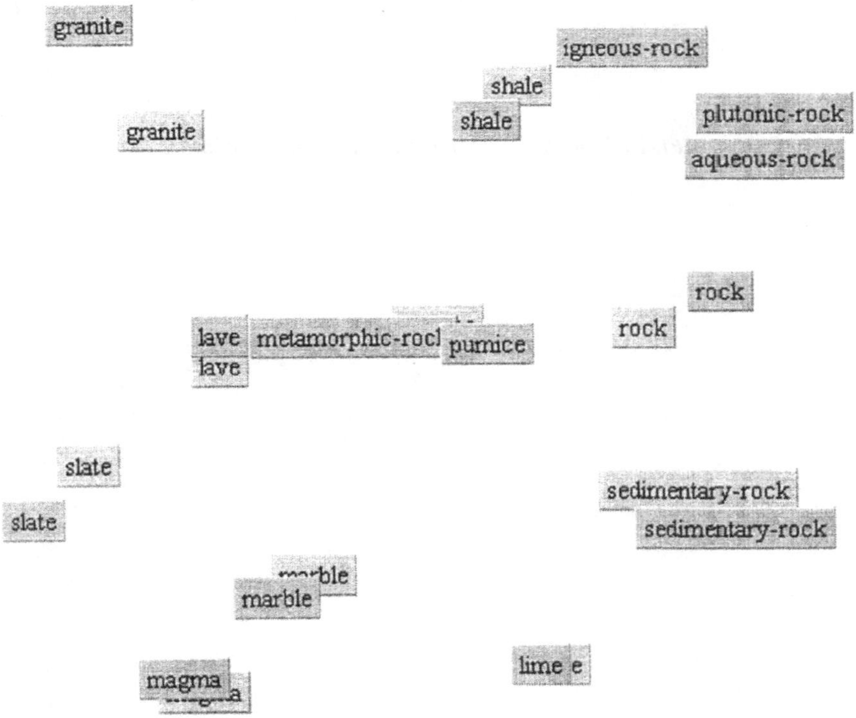

Figure 1. Unlinked concept map to identify concept labels of learners.

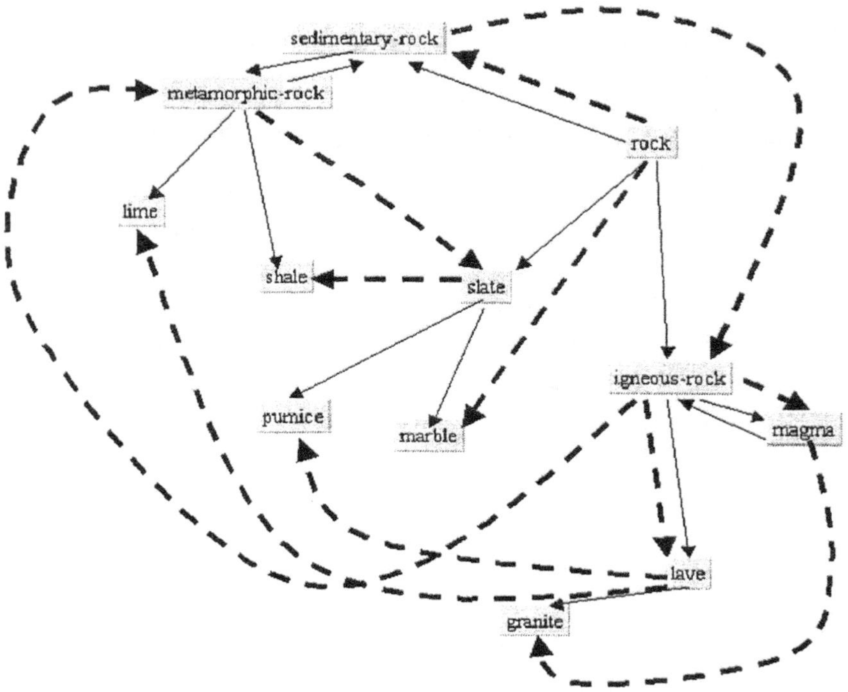

Figure 2. A concept map linked by learner1 and leaner2.

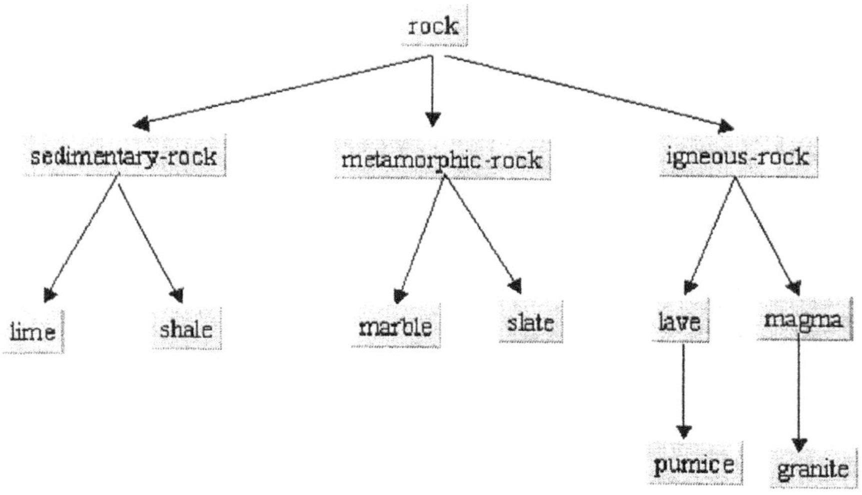

Figure 3. Concept map of Teacher.

Phase 2: Identification of the concept labels among the learners
Learners discuss to make common Concept labels from each concept labels. They can use an unlinked concept map calculated and arranged by the MDS (Figure 1). The concept map displays concept labels of higher resemblance to near place on it.

Phase ii: Indication of the concept labels with explanatory note of teacher and learners
Next, the teacher decides the final concept labels that use the following procedure using unlinked concept map that displays concept labels of teacher and common labels of learners. The teacher returns the final labels to learners.

Phase 3: Decision of relations among the concept labels on a concept map
In this phase, each learner decides each relation among the concept labels on a shared concept map shown Figure 2. Learner 1 draws solid lines, and learner 2 illustrates with broken lines. The positions of labels are decided by discussion among the learners.

Phase iii: Decision of relations among the concept labels of teacher
Teacher also draws the concept map to decide the relations among the concept labels (Figure 3).

Phase 4: Making a consensus model among learners
We use the consensus model (Ohuchi, 1991) in FISM to make a common concept map among learners. The algorithm to make a consensus mode is following.

1. Making a comparison matrix of R^1 and R^2;
2. While disagreement point $\neq \phi$ do begin;
   2.1. Calculating the consensus relation;
   2.2. Drawing opinion graph by substituting R^1 and R^2 for consensus relation
   2.3. Discussion about the disagreement on opinion graph for consensus.
3. End.

Here, we use follow algorithm to create comparison matrix C from plural learners.

$$if\ S^1_{ij} = S^2_{ij}\ (= S^3_{ij}),\ then\ C_{ij} = S^1_{ij} = S^2_{ij}\ (= S^3_{ij});$$
$$else\ if\ S^1_{ij} \neq S^2_{ij}\ \text{or}\ S^2_{ij} \neq S^3_{ij}\ \text{or}\ S^3_{ij} \neq S^1_{ij},\ then\ C_{ij} = y.$$

Table 2. A comparison matrix between learner 1 and 2

Compare Matrix	(1)	(2)	(3)	(4)	(5)	(6)	(7)	(8)	(9)	(10)	(11)	(12)
(1) rock	1	1	1	1	1	1	1	1	1	1	1	1
(2) metamorphic-rock	0	1	1	0	0	0	y	1	0	0	0	0
(3) sedimentary-rock	0	1	1	0	0	0	y	1	0	0	0	0
(4) slate	0	0	0	1	0	0	0	y	y	y	0	0
(5) magma	0	0	0	0	1	y	0	0	0	0	y	1
(6) igneous-rock	0	y	y	y	1	1	y	y	0	y	1	1
(7) lime	0	0	0	0	0	0	1	0	0	0	0	0
(8) shale	0	0	0	0	0	0	0	1	0	0	0	0
(9) marble	0	0	0	0	0	0	0	0	1	0	0	0
(10) pumice	0	0	0	0	0	0	0	0	0	1	0	0
(11) lave	0	0	0	0	0	0	y	0	0	y	1	y
(12) granite	0	0	0	0	0	0	0	0	0	0	0	1

Table 2 describes the comparison matrix in this case. Brackets show the case of three learners.

The consensus relation is the relation that an element becomes 1 or 0 when other element of disagreement is given 1 or 0. The algorithm to calculate consensus relation is the following.

$$
\begin{aligned}
& \textit{if } c_{ij} = 1, \textit{ then set } c_{ij} = 1; \\
& \quad \textit{set } c_{ij} = 1 \textit{ for all } (p,q) \in CW11\,(c_{ij}); \\
& \quad \textit{set } c_{ij} = 0 \textit{ for all } (p,q) \in CW10\,(c_{ij}); \\
& \textit{if } c_{ij} = 0, \textit{ then set } c_{ij} = 0; \\
& \quad \textit{set } c_{ij} = 0 \textit{ for all } (p,q) \in CW00\,(c_{ij}).
\end{aligned}
$$

Where *CW11*, *CW10*, and *CW00* are as follows.

$$
\begin{aligned}
CW11\,(c_{ij}) &= \{(p,q) \mid (p,q) \in Z,\ C_{pi}C_{jq} = 1\}, \\
CW10\,(c_{ij}) &= \{(p,q) \mid (p,q) \in Z,\ C_{iq}C_{jp} + C_{qi}C_{pj} = 1\}, \\
CW00\,(c_{ij}) &= \{(p,q) \mid (p,q) \in Z,\ C_{ip}C_{qj} = 1\}
\end{aligned}
$$

Figure 4 shows final consensus model among learners.

Phase iv: Making an opinion graph to lead

The teacher uses an opinion graph to determine the minimum leading point. The algorithm for the opinion graph is as follows.

1. Making a comparison matrix of R^t and R^s;
2. While disagreement point $\neq \phi$ do begin;
3. Creating the consensus model;
    4. Substituting R^t for consensus model
    5. Drawing opinion graph
    6. Leading to correct knowledge
7. End

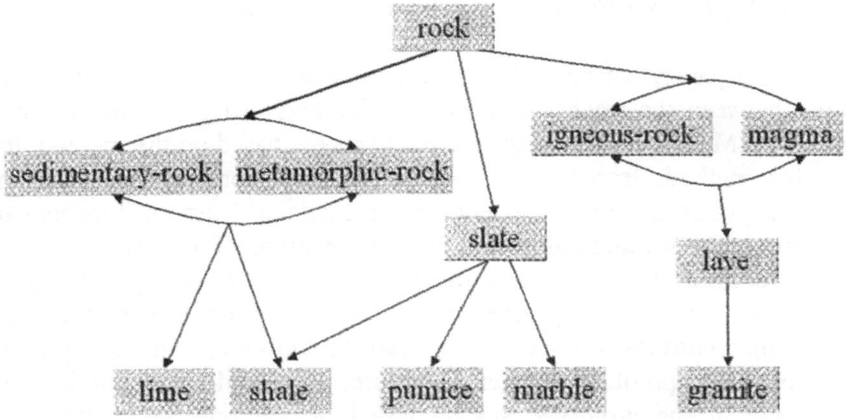

Figure 4. Consensus model among learners.

Figure 5 is an opinion graph that illustrates a minimum leading point. The arrow means that relations had to be linked, and the arrow of reverse type shows consensus relation. In this graph, the minimum leading point is a relation between "metamorphic-rock" and "marble" that is concerned with other three relationships between "sedimentary-rock" and "metamorphic-rock", "metamorphic-rock" and "slate", and "slate" and "pumice".

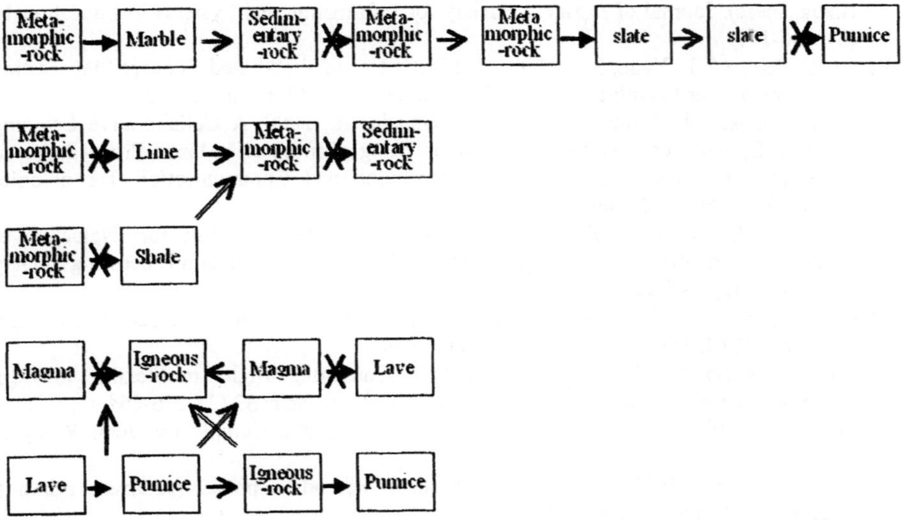

Figure 5. Opinion graph of teacher.

Figure 5 is an opinion graph that illustrates a minimum leading point. The arrow means that relations had to be linked, and the arrow of reverse type shows consensus relation. In this graph, the minimum leading point is a relation between "metamorphic-rock" and "marble" that is concerned with other three relationships between "sedimentary-rock" and "metamorphic-rock", "metamorphic-rock" and "slate", and "slate" and "pumice".

5. CONCLUSION AND FUTURE WORK

In this paper, we have described a method to support collaborative learning between the teacher and multiple learners using the concept map based on FISM, and explained the learning process based on the method. By using this method, learners get some support to construct a consistent concept map of each learner, and a consensus model of learners. Teacher can know a minimum leading point from the opinion graph of teacher.

To confirm the usefulness of the proposed supporting method in collaborative learning, we have to research at least effectiveness of constructing a consensus model among learners and minimum leading point on the opinion graph of the teacher. Therefore, we consider construction and examination of the prototype system based on the proposal method. In addition, to know more detailed learner's situation of understanding, we will extend FISM to express multiple relations and fuzzy relations among concept labels on the concept map.

References

Inaba, A., Toyoda, J., 1999. Underlying Learning Theories and Recent Researches on CSCL (in Japanese). Journal of Japanese Society for Information and Systems in Education 16 (3), 166-175.

Tsai, C.C., Lin, S.S.J., Yuan, S.M., 2001. Student's Use of Web-based Concept Map Testing and Strategies for Learning. Journal of Computer Assisted Learning 17, 72-84.

Saito, H., Tanaka, N., Ohno, T., Maeda T., Ohuchi, A., 2000. A Collaborative Learning Support System for Knowledge Building using Conceptual Maps. Proceedings of Conference on Educational Uses of Information and Communication Technologies (ICEUI 2000) 50-57, Beijing, China.

Itoh, K., Ohuchi, A., 1994. Fuzzy Item Relational Structure Analysis for Evaluation of Students by Performance Scores (in Japanese). Journal of Information Processing Society of Japan 35 (5), 945-954.

Ohuchi, A., Kaji, I., 1989. Correction Procedures for Flexible Interpretive Structural Modeling. IEEE Trans, SMC, SMC-19 (1), 85-94.

Ohuchi, A., Kurihara, M., 1991. Developing a Consensus Model in Flexible ISM (in Japanese). Journal of Information Processing Society of Japan 32 (2), 256-264.

Warfield, J.N., 1976. Societal Systems: Planning, Policy and Complexity, John Wiley & Sons.

Bryan F. J. Manly., 1986. Multivariate Statistical Methods a primer (Japanese Edition). Chapman and Hall Ltd, England.

Hori, K., 1994. A System for Aiding Creative Concept Formation. IEEE Trans, SMC, 24 (6), 882-894.

Shawn M. Glynn, et al., 1991. The psychology of Leaning Science. LEA, PUBLISHERS Hillsdale, New Jersey.

GLOBAL CONVERGENCE PROPERTY OF ERROR BACK-PROPAGATION METHOD FOR RECURRENT NEURAL NETWORKS

Keiji Tatsumi[1], Tetsuzo Tanino[1] and Masao Fukushima[2]

[1] *Graduate School of Engineering, Osaka University,*
Yamada-Oka 2-1, Suita, Osaka 565-0871, Japan.
tatsumi@eie.eng.osaka-u.ac.jp tanino@eie.eng.osaka-u.ac.jp

[2]*Graduate School of Informatics, Kyoto University*
Kyoto 606-8501, Japan
fuku@amp.i.kyoto-u.ac.jp

Abstract Error Back-Propagation (BP) method and its variations are popular methods for the supervised learning of neural networks. BP method can be regarded as an approximate steepest descent method for minimizing the sum of error functions, which uses exact derivatives of each error function. Thus, they have the global convergence property under some natural conditions. On the other hand, Real Time Recurrent Learning method (RTRL) is also one of variations of BP method for the recurrent neural network (RNN) which is suited for handling time sequences. Since, for the real-time learning, this method cannot utilize exact outputs from the network, approximate derivatives of each error function are used to update the weights. Therefore, although RTRL is widely used in practice, its global convergence property is not known yet. In this paper, we show that RTRL has the global convergence property under almost the same conditions as other variations of BP.

Keywords: real time recurrent learning, global convergence property, recurrent neural network.

1. INTRODUCTION

Error Back propagation method (BP) (Rumelhart et al., 1986) is one of the most popular methods for multi-layer neural networks (MNN) and recurrent neural networks (RNN), which realize several or many pairs of input and training signals. Especially, *on-line* back-propagation method (on-line BP) for MNN has been applied to various problems in practice, because of its simplicity of implementation. Therefore, many speed-up algorithms have been proposed for on-line BP algorithm (Hertz et al.,1991; Mak et al., 1999). Moreover, under various assumptions, the convergence analysis has been studied in the stochastic case (Lju, 1977; Gaivoronski, 1994) or in the deterministic case (Luo and Tseng, 1994; Bertsekas, 1996). In particular, Mangasarian and Solodov (1994) or Gaivoronski (1994) showed the global convergence of deterministic on-line BP.

In this paper, we focus on Real Time Recurrent Learning (RTRL), which is a popular variation of on-line BP for RNN. Since RTRL is the real-time learning, it suits for handling time sequences. However, RTRL cannot utilize exact outputs from the network. Thus, approximate derivatives of each error function are used in order to modify the weights. Although it is widely used in practice, for example, adaptive control (Narendra and Parthasarathy, 1990), forecasting (Connor et al., 1994) and modeling complex process dynamics (Parlos et al., 1994), its global convergence property is not known yet.

Therefore, we show that RTRL has the global convergence property under almost the same conditions as other variations of on-line BP for MNN. In addition, we propose a new hybrid method with a safeguard for the learning rate. It can be incorporated in the simple RTRL with any heuristically-determined learning rate, while maintaining the global convergence property.

2. RECURRENT NETWORK

In this paper, we consider the general recurrent network (RNN) shown in Figure 1. In this model, all pairs of units have connections with (synaptic)

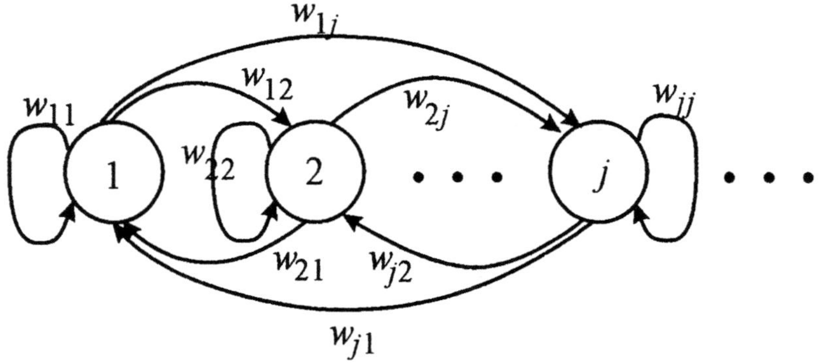

Figure 1: Recurrent neural network(RNN)

weights. In addition, there exists a connection even from a unit to itself. The output of unit j is governed by the rule

$$y_j(t+1) := f\left(\sum_i w_{ij} y_i(t) + x_j(t)\right), \quad t = 0,\dots,T, \tag{1}$$

where w_{ij} denotes the weight from unit i to unit j, $y_j(t)$ is the output from unit j at time t and $x_j(t)$ is the external input into the unit j at time t. We use an appropriate value as an initial output $y_j(0)$ from unit j. The output function $f(\cdot)$ is continuously differentiable. For example, a sigmoid function or a linear function is often used as an output function. If there is no input signal, we let $x_j(t) := 0$.

The learning of this network is to realize the following P pairs of time sequences:

Inputs $x_j^p(t),\ t = 0,\dots,T,\ j = 1,\dots,N,\ p = 1,\dots,P,$

Desirable outputs $d_j^p(t),\ t = 1,\dots,T,\ j \in O^p(t),\ p = 1,\dots,P,$

where $O^p(t)$ is the set of units to which the desirable output $d_j^p(t)$ is specified at time t. Here, an input sequence $x_j^p(t)$, $j = 1,\dots,N,\ t = 0,\dots,T$ is called a *pattern* p and the corresponding desirable sequence $d_j^p(t)$, $j \in O^p(t),\ t = 1,\dots,T$ is called the *training sequence* for pattern p. Moreover, we let $y_j^p(t)$ denote the output from unit j at time t for pattern p.

To measure a gap between the training signals and the outputs of RNN, we define the output error function $E(w)$ by

$$\begin{aligned} E(w) &= \sum_{p=1}^{P}\sum_{t=1}^{T}\sum_{k\in O^p(t)} E_k^{p,t}(w) \\ &= \sum_{p=1}^{P}\sum_{t=1}^{T}\sum_{k\in O^p(t)} \frac{1}{2} e_k^{p,t}(w)^2 \\ &= \sum_{p=1}^{P}\sum_{t=1}^{T}\sum_{k\in O^p(t)} \frac{1}{2}(y_k^p(t) - d_k^p(t))^2 \end{aligned}$$

Thus, the learning of RNN can be formulated as the following minimization problem:

Minimize $E(w)$ (2)

3. RTRL

In order to solve problem (2), many nonlinear optimization methods can be applied. Among them, the steepest descent method, which is called batch Back Propagation (batch BP) method, is the simplest method and modifies the weights as follows:

$$w^{n+1} := w^n - \eta^n \nabla E(w^n), \tag{3}$$

where η^n denotes the learning rate (step-size) at iteration n. As shown in Theorem 1 below, this method has the global convergence property under the following assumptions.

Assumption 1 *The sequence of learning rates* $\{\eta^n\}$ *satisfies*

$$\lim_{n\to\infty} \eta^n = 0, \sum_{n=0}^{\infty} \eta^n = \infty.$$

Assumption 2 *The nonnegative error function E(w) is continuously differentiable and, for each p, t and j, the function* $\nabla E_j^{p,t}(w)$ *is Lipschitz continuous.*

Theorem 1 (Bertsekas, 1999) *Assume that the sequence* $\{w^n\}$ *generated by* (3) *are bounded and Assumptions 1 and 2 are satisfied. Then, either there exists an* $\bar{n}$ *such that* $\|\nabla E(w^{\bar{n}})\| = 0$ *or the sequence* $\{E(w^n)\}$ *converges and* $\|\nabla E(\overline{w})\| = 0$ *for any accumulation point* $\overline{w}$ *of the sequence* $\{w^n\}$.

It is widely known that bach BP methods have also the global convergence property under some other assumptions (Bertsekas, 1999), while convergence properties for real-time BP methods are not known yet. This is mainly because real-time BP methods use approximate derivatives of each error function to modify the weights since exact outputs from the network are unavailable. In this paper, we focus on RTRL (Real Time Recurrent Learning) which is one of popular real-time BP methods and show that it has the global convergence property under the similar assumptions to batch BP.

First, in this section, we describe RTRL in detail. RTRL updates the weights for each input pattern p at each time t during the learning. The output from each unit j is given by

$$y_j^p(t+1) := f\left(\sum_{i=1}^{N} \phi_{ij}^{n,p}(t) y_i^p(t) + x_j^p(t)\right),$$

where the weights at iteration n and time t for pattern p are represented by $\phi_{ij}^{n,p}(t),\ i, j = 1,\dots,N$. Then, the update of weights $\phi_{ij}^{n,p}(t)$ is executed by

$$\phi_{ij}^{n,p}(t) := \phi_{ij}^{n,p}(t-1) + \Delta\phi_{ij}^{n,p}(t),$$

$$\Delta\phi_{ij}^{n,p}(t) := -\eta^{n} \sum_{k \in O^{p}(t)} \bar{e}_{k}^{p,t}(\phi^{n,p}(t-1))\rho_{ij}^{p,k}(t),$$

$$\bar{e}_{k}^{p,t}(\phi^{n,p}(t-1)) := y_{k}^{p}(t) - d_{k}^{p}(t),$$

where $\rho_{ij}^{p,k}$ is an approximation of $\partial y_{k}^{p}(t)/\partial\phi_{ij}$ and is given by

$$\rho_{ij}^{p,k}(1) := 0,$$

$$\rho_{ij}^{p,k}(t+1) := f'(h_{k}^{p}(t))\left\{\delta_{kj} y_{k}^{p}(t) + \sum_{l} \phi_{lk}^{n,p}(t)\rho_{ij}^{p,l}(t)\right\}, t \geq 1$$

where δ_{kj} denotes Kronecker delta and $h_{k}^{p}(t) := \sum_{l=1}^{N} \phi_{lk}^{n,p}(t) y_{l}^{p}(t) + x_{k}^{p}(t)$. Since the weights are updated for each input pattern p and each time t, the output $y_{j}^{p}(t)$ and the output error $\bar{e}_{k}^{p,t}(\phi^{n,p}(t-1))$ are not equivalent to $y_{j}^{p}(t)$ in (1) and $e_{k}^{p,t}$ in (2). Thus, we use $\bar{e}_{k}^{p,t}(\phi^{n,p}(t-1))\rho_{ij}^{p,k}(t)$ as an approximation of $\partial E_{k}^{p,t}(\phi)/\partial\phi_{ij} = e_{k}^{p,t}(\phi)\,\partial y_{k}^{p}(t)/\partial\phi_{ij}$.

Note that $\bar{E}_{k}^{p,t}(\phi^{n,p}(t-1))$ and $\rho_{ij}^{p,k}(t)$ are functions of $\phi^{n,p}(0), \ldots, \phi^{n,p}(t-1)$. Let us define $\bar{\nabla}E^{p,t}$ and $\bar{\nabla}E^{p}$ as follows:

$$\bar{\nabla}E_{k}^{p,t}(\phi^{n,p}(t-1)) := \sum_{j \in O^{p}(t)} \bar{e}_{j}^{p,t}(\phi^{n,p}(t-1)) \times \left(\rho_{11}^{p,j}(t), \rho_{12}^{p,j}(t), \ldots, \rho_{NN}^{p,j}(t)\right)^{\mathrm{T}}, \quad (4)$$

$$\bar{\nabla}E^{p}(\phi^{n,p}) := \sum_{t=1}^{T} \bar{\nabla}E^{p,t}(\phi^{n,p}(t-1)). \quad (5)$$

Although $\bar{\nabla}E^{p,t}(\phi^{n,p}(t-1))$ and $\bar{\nabla}E^{p}(\phi^{n,p})$ should be formally represented as $\bar{\nabla}E^{p,t}(\phi^{n,p}(0),\ldots,\phi^{n,p}(t-1))$ and $\bar{\nabla}E^{p}(\phi^{n,p}(0),\ldots,\phi^{n,p}(T-1))$ respectively, we simplify the notation to avoid possible complication.

After all, RTRL can be summarized in the following algorithm.

RTRL

Step 1 Select initial weights w^{0} and $n := 0$

Step 2 Put $\phi^{n,1}(0) := w^{n}$ and $p := 1$.

Step 3

1. Set $t := 1$.
2. Update $\phi^{n,p}(t)$ by

$$\phi^{n,p}(t+1) := \phi^{n,p}(t) - \eta^{n}\bar{\nabla}E^{p,t}(\phi^{n,p}(t)). \quad (6)$$

3. If $t = T$, then put $\phi^{n,p+1}(0) := \phi^{n,p}(T+1)$ and go to **Step 4**. Else, let $t := t+1$ and go to **2**.

Step 4 If $p = P$, then put $w^{n+1} := \phi^{n,P+1}(0)$ and go to **Step 5**. Otherwise, let $p := p+1$ and go to **Step 3**.

Step 5 If $\left\|\nabla E(w^{n+1})\right\| \le \varepsilon$, stop. Otherwise, let $n := n+1$ and go to **Step 2**.

4. GLOBAL CONVERGENCE

In this section, we show that RTRL has the global convergence property under natural assumptions. First, we state a classical lemma (Ortega, 1972).

Lemma 1 *If $f : \Re^n \mapsto \Re$ is differentiable and $\nabla f : \Re^n \mapsto \Re^n$ is Lipschitz continuous with constant $L > 0$, then*

$$\left| f(y) - f(x) - \nabla f(x)^{\mathrm{T}}(y - x) \right| \le \frac{L}{2}\|y - z\|^2 \quad \forall x, y \in \Re^n.$$

Next, we show a lemma that plays an important role in the following argument.

Lemma 2 *Suppose that nonnegative sequences $\{e_n\}$, $\{a_n\}$, $\{\tau_n\}$, $\{u_n\}$ and a positive constant M satisfy*

$$e_{n+1} - e_n \le -a_n u_n + a_n \tau_n, \tag{7}$$

$$\left| u_{n+1} - u_n \right| \le M a_n, \tag{8}$$

$$\sum_{n=0}^{\infty} a_n = \infty, \quad \lim_{n\to\infty} a_n = \lim_{n\to\infty} \tau_n = 0. \tag{9}$$

Moreover, let

$$\bar{e} := \liminf_{n\to\infty} e_n. \tag{10}$$

We assume that there exist a pair of positive constants ε_0 and δ such that if $\bar{e} + \varepsilon < e_n < \bar{e} + \varepsilon_0$ for any ε satisfying $0 < \varepsilon < \varepsilon_0$, then

$$\delta \le u_n. \tag{11}$$

Then, $\{e_n\}$ converges to $\bar{e}$ and $\{u_n\}$ converges to 0.

Proof : First, we show $\lim_{n\to\infty} e_n = \bar{e}$. Suppose that this is not true. Then there exist a positive constant $\varepsilon_1 < \varepsilon_0$ and a subsequence $\{n_s\}$ such that

$e_{n_s} > \bar{e} + \varepsilon_1$. Now, let us define *j(s)* := $\max\{j \,|\, j < n_s, e_j \le \bar{e} + \varepsilon_1\}$. By (10), we can easily verify the existence of such *j(s)*. Especially, $e_{j(s)} \le \bar{e} + \varepsilon_1$ and $e_{j(s)+1} > \bar{e} + \varepsilon_1$. In addition, (7) and (9) yield:

$$\max\{0, e_{n+1} - e_n\} \to 0.$$

Thus, for any ε_2 satisfying $0 < \varepsilon_2 < \varepsilon_1$, when *s* is sufficiently large, we have $\bar{e} + \varepsilon_2 < e_{j(s)} \le \bar{e} + \varepsilon_1 < \bar{e} + \varepsilon_0$. Therefore, by (11), there exists some positive constant δ such that $u_{j(s)} > \delta$. By (9), for sufficiently large *s*, we have $\tau_{j(s)} < \delta/2$. This along with (7) yield

$$e_{j(s)+1} \le e_{j(s)} - a_{j(s)}(u_{j(s)} - \tau_{j(s)}) < e_{j(s)} - \frac{\delta}{2} a_{j(s)} \le \bar{e} + \varepsilon_1,$$

which contradicts the above assumption $e_{j(s)+1} > \bar{e} + \varepsilon_1$. So, we have $\lim_{n\to\infty} e^n = \bar{e}$.

Next, we show that $u_n \to 0$. If this is not true, there exist some positive constant δ and some subsequence $\{n_t\}$ such that $u_{n_t} \ge 3\delta$. Moreover, for any n_t there exists $m_t > n_t$ such that $u_{m_t} \le 2\delta$. Otherwise, there must exist t_1 satisfying $u_n > 2\delta$ for any $n > n_{t_1}$. Without loss of generality, we can assume that $\tau_n \le \delta$ for $n \ge n_{t_1}$. Then, for any $l > k > n_{t_1}$

$$e_l \le e_k - \sum_{n=k}^{l-1} a_n(u_n - \tau_n) \le e_k - \delta \sum_{n=k}^{l-1} a_n.$$

This fact and $\sum_{n=s}^{\infty} a_n = \infty$ imply $e_l \to -\infty$, which contradicts that $\{e_n\}$ are nonnegative. Thus, for any n_t, there exists $m_t > n_t$ such that $u_{m_t} \le 2\delta$. In addition, without loss of generality, we may suppose $u_n \ge 2\delta$ for all *n* satisfying $n_t < n < m_t$. Moreover, if *n* satisfies $n_t < n < m_t$ for sufficiently large *t*, then $\tau_n < \delta$. Thus, we obtain

$$e_{m_t} \le e_{n_t} - \sum_{n=n_t}^{m_t-1} a_n(u_n - \tau_n) \le e_{n_t} - \delta \sum_{n=n_t}^{m_t-1} a_n. \tag{12}$$

On the other hand, by (8), we have

$$\delta = 3\delta - 2\delta \le u_{n_t} - u_{m_t} = |u_{m_t} - u_{n_t}|$$
$$\le \sum_{n=n_t}^{m_t-1} |u_{n+1} - u_n| \le M \sum_{n=n_t}^{m_t-1} a_n.$$

Therefore, we get

$$\sum_{n=n_l}^{m_l-1} a_n \geq \frac{\delta}{M} \tag{13}$$

By (12) and (13), we obtain

$$e_{m_l} \leq e_{n_l} - \frac{\delta^2}{M}.$$

This contradicts the fact $\lim_{n\to\infty} e_n = \bar{e}$ shown earlier. Therefore, we have $\lim_{n\to\infty} u_n = 0$. **Q.E.D.**

Gaivoronski (1994) claims that the conclusion of the above lemma holds with the condition (10) being replaced by $\bar{e} := \inf_n e_n$. However, this does not seem valid. In fact, if we choose

$$e_n := \sum_{k-1}^{n} \frac{eP}{k}, \quad a_n := \frac{1}{\sqrt{n+1}},$$

$$\tau_n := \frac{2}{\sqrt{n+1}}, \quad u_n := \frac{1}{\sqrt{n+1}},$$

then we have $\inf_n e_n = 1$ and

$$\left| u_n - u_{n+1} \right| = \frac{1}{\sqrt{n+1}} - \frac{1}{\sqrt{n+2}} \leq \frac{1}{\sqrt{n+1}} = a_n.$$

Thus, although (7)-(9) and (11) are satisfied, we obtain $\lim_{n\to\infty} e_n = \infty$. Next, we prove another lemma under the following assumption.

Assumption 3 *The number of values of $E(\overline{w})$ at stationary points $\overline{w}$ of the error function E(w) is finite, that is, the cardinality of I_E is finite, where* $I_E := \{E(w) \in \Re \,|\, \|\nabla E(w)\| = 0\}$.

We may expect that Assumption 3 usually holds in practice.

Lemma 3 *If Assumption 3 is satisfied, then $\overline{\nabla} E^{p,t}(\cdot)$ and $\overline{\nabla} E^{p}(\cdot)$ are Lipschitz continuous. Moreover, the following inequality holds:*

$$\left\| \nabla E(w^n) - \sum_{p=1}^{P} \overline{\nabla} E^p(\phi^{n,p}) \right\| \leq L \sum_{p=1}^{P} \sum_{t=1}^{T} \left\| w^n - \phi^{n,p}(t-1) \right\|,$$

where $L > 0$ is a constant.

Proof : We can easily verify the first assertion of the lemma by Assumption 2 and the definition of $\overline{\nabla} E^{p,t}(\phi^{n,p}(t-1))$ and $\overline{\nabla} E^{p}(\phi^{n,p})$. Next, since we have $\sum_{t=1}^{T}\sum_{k\in O^p(t)}\nabla E_k^{p,t}(w^n)=\overline{\nabla}E^p(w^n,...,w^n)$ and $\overline{\nabla}E^P(\phi^{n,p})$ is Lipschitz continuous, the following inequality holds.

$$\begin{aligned}&\left\|\nabla E(w^n)-\sum_{p=1}^{P}\overline{\nabla}E^p(\phi^{n,p})\right\|\\ &\le\sum_{p=1}^{P}\left\|\overline{\nabla}E^p(w^n,...,w^n)-\overline{\nabla}E^p(\phi^{n,p},...,\phi^{n,p})\right\|\\ &\le L\sum_{p=1}^{P}\sum_{t=1}^{T}\left\|w^n-\phi^{n,p}(t-1)\right\|.\end{aligned}$$

This completes the proof. **Q.E.D.**

We are now ready to prove the following convergence theorem.

Theorem 2 *Assume that the sequences* $\{\phi^{n,p}(t)\}$ *and* $\{w^n\}$ *generated by* **RTRL** *are bounded and Assumptions 1, 2 and 3 are satisfied. Then, either there exists an* $\bar{n}$ *such that* $\|\nabla E(w^{\bar{n}})\|=0$ *or the sequence* $\{E(w^n)\}$ *converges and* $\|\nabla E(\overline{w})\|=0$ *for any accumulation point* $\overline{w}$ *of the sequence* $\{w^n\}$.

Proof : It suffices to show that the assumptions of Lemma 2 are satisfied. Since $\overline{\nabla}E^{p,t}(\phi^{n,p}(t))$ and $\nabla E(w)$ are continuous and $\{w^n\}$ and $\{\phi^{n,p}(t)\}$ are bounded for all *p*, *n*, *t*, there exist some positive constants C_1, C_2 such that

$$\left\|\overline{\nabla}E^{p,t}(\phi^{n,p}(t-1))\right\|\le C_1, \tag{14}$$

$$\left\|\nabla E(w^n)\right\|\le C_2. \tag{15}$$

By (6) and (14), we have

$$\begin{aligned}\left\|\phi^{n,p}(t)-w^n\right\|&=\eta^n\sum_{\tau=1}^{t}\left\|\overline{\nabla}E^{p,\tau}(\phi^{n,p}(\tau-1))\right\|\\ &\le\eta^nTC_1.\end{aligned} \tag{16}$$

Similarly, by (5), (6) and (14), we obtain

$$\begin{aligned}w^{n+1}-w^n&=-\sum_{p=1}^{P}\sum_{t=1}^{T}\eta^n\overline{\nabla}E^{p,t}(\phi^{n,p}(t-1))\\ &=-\eta^n\sum_{p=1}^{P}\overline{\nabla}E^p(\phi^{n,p}),\end{aligned} \tag{17}$$

and

$$\left\| w^{n+1} - w^n \right\| \le \eta^n PTC_1. \tag{18}$$

By (15) and (17), the following holds:

$$\begin{aligned}
&\nabla E(w^n)^{\mathrm{T}} (w^{n+1} - w^n) \\
&\quad = \nabla E(w^n)^{\mathrm{T}} \left(-\eta^n \nabla E(w^n) + \eta^n \nabla E(w^n) + (w^{n+1} - w^n) \right) \\
&\quad \le -\eta^n \left\| \nabla E(w^n) \right\|^2 + \eta^n PC_2 \left\| \nabla E(w^n) - \sum_{p=1}^{P} \overline{\nabla} E^p(\phi^{n,p}) \right\|.
\end{aligned} \tag{19}$$

Moreover, by Lemma 3 and (16), we have

$$\begin{aligned}
\left\| \nabla E(w^n) - \sum_{p=1}^{P} \overline{\nabla} E^p(\phi^{n,p}) \right\| &\le L \sum_{p=1}^{P} \sum_{t=1}^{T} \left\| w^n - \phi^{n,p}(t-1) \right\| \\
&\le \eta^n LPT^2 C_1.
\end{aligned} \tag{20}$$

Thus, (19) and (20) imply

$$\nabla E(w^n)^{\mathrm{T}} (w^{n+1} - w^n) \le -\eta^n \left\| \nabla E(w^n) \right\|^2 + (\eta^n)^2 C_1 C_2 L (PT)^2. \tag{21}$$

By (18), (21) and Lemma 1, we obtain

$$\begin{aligned}
E(w^{n+1}) - E(w^n) &\le \nabla E(w^n)^{\mathrm{T}} (w^{n+1} - w^n) + \frac{L}{2} \left\| w^{n+1} - w^n \right\|^2 \\
&\le -\eta^n \left\| \nabla E(w^n) \right\|^2 + \frac{1}{2} (\eta^n)^2 C_3,
\end{aligned}$$

where $C_3 := C_1 L (PT)^2 [C_1 L/2 + C_2]$. Let $e_n := E(w^n)$, $u^n := \left\| \nabla E(w^n) \right\|^2$, $a_n := \eta^n$ and $\tau_n := \eta^n C_3 / 2$. Then (7) is satisfied. In addition, Assumption 1 means that (9) holds. By (15), we have

$$\begin{aligned}
u_{n+1} - u_n &= \left\| \nabla E(w^{n+1}) \right\|^2 - \left\| \nabla E(w^n) \right\|^2 \\
&\le 2C_2 \left\| \nabla E(w^{n+1}) - \nabla E(w^n) \right\| \\
&\le 2\eta^n C_1 C_2 LPT,
\end{aligned}$$

where the last inequality follows from the Lipschitz continuity of ∇E and (18). Thus, (8) holds.

Since $\{w^n\}$ is bounded, $\{E(w^n)\}$ is also bounded. Thus, $\{E(w^n)\}$ has accumulation points and there exists $\overline{E} := \liminf_{n\to\infty} E(w^n)$. Thus, (10) holds. From Assumption 3, there exists some positive constant δ satisfying (11). Since all conditions of Lemma 2 are satisfied, $\{E(w^n)\}$ converges to

$\bar{E}$ and $\{\|\nabla E(w^n)\|\}$ converges to 0. Moreover, since $\nabla E(w)$ is continuous, at any accumulation point $\bar{w}$ of $\{w^n\}$, $\|\nabla E(\bar{w})\| = 0$ holds. **Q.E.D.**

Moreover, we consider the case that the learning rate η^n in (6) is varied for each input pattern p, time t and unit j, respectively. Specifically, let $\{\eta_j^{n,p,t}\}$ be the sequence of learning rates.

Assumption 4 *The sequence of learning rates* $\{\eta_j^{n,p,t}\}$ *satisfies*

$$\lim_{n\to\infty} \eta_j^{n,p,t} = 0, \quad \sum_{n=0}^{\infty} \eta_j^{n,p,t} = \infty. \tag{22}$$

Moreover, there exists a positive sequence $\{\bar{\eta}^n\}$ *such that*

$$\lim_{n\to\infty} \frac{\eta_j^{n,p,t}}{\bar{\eta}^n} = 1, \quad \lim_{n\to\infty} \bar{\eta}^n = 0, \tag{23}$$

$$\bar{\eta}^n \geq \frac{1}{2PTN} \sum_{p=1}^{P} \sum_{t=1}^{T} \sum_{j=1}^{N} \eta_j^{n,p,t}. \tag{24}$$

In this case, we can prove the following theorem.

Theorem 3 *Assume that the sequences* $\{\phi^{n,p}(t)\}$ *and* $\{w^n\}$ *generated by* **RTRL** □*are bounded and Assumptions 2, 3 and 4 are satisfied. Then, either there exists an* $\bar{n}$ *such that* $\|\nabla E(w^{\bar{n}})\| = 0$ *or* $\{E(w^n)\}$ *converges and* $\|E(\bar{w})\| = 0$ *for any accumulation point* $\bar{w}$ *of the sequence* $\{w^n\}$.

Proof : Let $\alpha_j^{n,p,t} := \eta_j^{n,p,t} / \bar{\eta}^n$. Then, by (24) and the fact that $\eta_j^{n,p,t}$ are positive for all n, p, t, j, we have

$$0 < \alpha_j^{n,p,t} \leq 2PTN.$$

Then, in a way similar to the proof of Theorem 2, we can show that

$$\begin{aligned}
&\nabla E(w^n)^{\mathrm{T}}(w^{n+1} - w^n) \\
&\leq -\bar{\eta}^n \|\nabla E(w^n)\|^2 + \bar{\eta}^n C_4 \sum_{p=1}^{P} \sum_{t=1}^{T} \sum_{j=1}^{N} |1 - \alpha_j^{n,p,t}| \\
&\quad + (\bar{\eta}^n)^2 C_5 \sum_{p=1}^{P} \sum_{t=1}^{T} \sum_{j=1}^{N} \alpha_j^{n,p,t} \\
&\leq -\bar{\eta}^n \|\nabla E(w^n)\|^2 + \bar{\eta}^n \xi^n C_4 + 2(\bar{\eta}^n)^2 C_5 (PTN)^2,
\end{aligned}$$

where C_4 and C_5 are some positive constants and

$$\xi^n := \sum_{p=1}^{P}\sum_{t=1}^{T}\sum_{j=1}^{N}\left|\alpha_j^{n,p,t}-1\right|.$$

Since $\lim_{n\to\infty}\xi^n = 0$, we can show that the assumptions of Lemma 2 are satisfied. **Q.E.D**

5. HYBRID METHOD

In this section, we propose a new simple method which improves upon the general RTRL with an arbitrary heuristic learning rate. First, we consider the general RTRL with positive learning rates $\zeta_j^{n,p,t}$ which are heuristically determined by some procedure (Hertz et al., 1991). By imposing upper and lower bounds for the learning rate, we propose a hybrid method that has the global convergence property.

In the proposed method, the learning rate is determined by

$$\eta_j^{n,p,t} := \mathrm{mid}\{\beta_j^{n,p,t}, \gamma_j^{n,p,t}, \zeta_j^{n,p,t}\}, \tag{25}$$

where $\{\zeta_j^{n,p,t}\}$ is an arbitrary positive sequence, $\mathrm{mid}\{\cdot,\cdot,\cdot\}$ denotes the middle value of three values and $\{\beta_j^{n,p,t}\}$ and $\{\gamma_j^{n,p,t}\}$ are positive sequences satisfying

$$\sum_{n=0}^{\infty}\beta_j^{n,p,t} = \infty,\ \sum_{n=0}^{\infty}\gamma_j^{n,p,t} = \infty, \tag{26}$$

$$\lim_{n\to\infty}\beta_j^{n,p,t} = 0,\ \lim_{n\to\infty}\gamma_j^{n,p,t} = 0, \tag{27}$$

$$\beta_j^{n,p,t} \neq \gamma_j^{n,p,t}. \tag{28}$$

Moreover, we suppose there exist $\bar{\beta}^n$ and $\bar{\gamma}^n$ such that

$$\bar{\beta}^n := \frac{1}{PNT}\sum_{p=1}^{P}\sum_{j=1}^{N}\sum_{t=1}^{T}\beta_j^{n,p,t} = \frac{1}{PNT}\sum_{p=1}^{P}\sum_{j=1}^{N}\sum_{t=1}^{T}\gamma_j^{n,p,t} =: \bar{\gamma}^n, \tag{29}$$

$$\lim_{n\to\infty}\frac{\beta_j^{n,p,t}}{\bar{\beta}^n} = \lim_{n\to\infty}\frac{\gamma_j^{n,p,t}}{\bar{\gamma}^n} = 1. \tag{30}$$

Then, the following theorem holds :

Theorem 4 *Assume that the sequences* $\{\phi^{n,p}(t)\}$ *and* $\{w^n\}$ *generated by* **RTRL** *that uses the modified learning rate given by* (25) *are bounded and Assumptions 2 and 3 are satisfied. In addition, if the positive sequences* $\{\beta_j^{n,p,t}\}$ *and* $\{\gamma_j^{n,p,t}\}$ *satisfy* (26) - (30), *then either there exists an* $\bar{n}$ *such* *that* $\|\nabla E(w^n)\| = 0$ *or* $\{E(w^n)\}$ *converges and* $\|\nabla E(\overline{w})\| = 0$ *for any accumulation point* $\overline{w}$ *of the sequence* $\{w^n\}$.

Proof: First, let us define

$$\overline{\eta}^n := \overline{\beta}^n = \overline{\gamma}^n. \tag{31}$$

We shall show that the assumptions of Theorem 3 are satisfied.

By (25), for all *n*, *p*, *t*, we have

$$\min\{\beta_j^{n,p,t}, \gamma_j^{n,p,t}\} \le \eta_j^{n,p,t} < \beta_j^{n,p,t} + \gamma_j^{n,p,t}. \tag{32}$$

Thus, (27) implies that (22) is satisfied. By (29) and (31), (24) is satisfied.

Next, by (29) and $\beta_j^{n,p,t} > 0$, we obtain

$$\sum_{n=0}^{\overline{n}} \overline{\eta}^n = \sum_{n=0}^{\overline{n}} \frac{1}{PNT} \sum_{p=1}^{P} \sum_{j=1}^{N} \beta_j^{n,p} > \frac{1}{PNT} \sum_{n=0}^{\overline{n}} \beta_{\overline{j}}^{n,\overline{p}}$$

where $\overline{n}$, $\overline{p}$, $\overline{j}$ are some positive integers. As $\overline{n} \to \infty$, we have $\sum_{n=0}^{\overline{n}} \overline{\eta}^n \to \sum_{n=0}^{\infty} \overline{\eta}^n = \infty$ by (26). Moreover, by (25), (30) and (31), we have for each *p*, *j* and *t*,

$$\lim_{n\to\infty} \left| \frac{\eta_j^{n,p,t}}{\overline{\eta}^n} - 1 \right| \le \lim_{n\to\infty} \max\left\{ \left| \frac{\beta_j^{n,p,t}}{\overline{\beta}^n} - 1 \right|, \left| \frac{\gamma_j^{n,p,t}}{\overline{\gamma}^n} - 1 \right| \right\}$$

$$\le \lim_{n\to\infty} \left(\left| \frac{\beta_j^{n,p,t}}{\overline{\beta}^n} - 1 \right| + \left| \frac{\gamma_j^{n,p,t}}{\overline{\gamma}^n} - 1 \right| \right) = 0.$$

Threfore, the assumptions of Theorem 3 are all satisfied. **Q.E.D**

An example of sequences $\{\beta_j^{n,p,t}\}$ and $\{\gamma_j^{n,p,t}\}$ that satisfy conditions (26) - (30) is given by

$$\beta_j^{n,p,t} := \frac{1}{n+1} + \frac{L_j^{p,t}}{(n+1)^2},$$

$$\gamma_j^{n,p,t} := \frac{1}{n+1} + \frac{l_j^{p,t}}{(n+1)^2},$$

where, for all $p = 1,\ldots,P$, $j = 1,\ldots,N$ and $t = 1,\ldots,T$, $L_j^{p,t}$ and $l_j^{p,t}$ satisfy

$$L_j^{p,t} \ne l_j^{p,t},\ 1 > \max\left\{ \left|L_j^{p,t}\right|, \left|l_j^{p,t}\right| \right\},$$

and

$$\sum_{p=1}^{P}\sum_{j=1}^{N}\sum_{t=1}^{T} L_j^{p,t} = 0, \quad \sum_{p=1}^{P}\sum_{j=1}^{N}\sum_{t=1}^{T} l_j^{p,t} = 0.$$

6. APPLICATIONS

RTRL is widely applied to many practical problems. In this section, we show some concrete examples to which RNN with RTRL can be applied.

6.1 Forecasting

Forecasting problems arise in various fields such as finance, economics, industry, and so on (Connor et al.,1994). Forecasting is to construct the (stochastic) models to predict the value of next observation by using current and previous observations. For example, when we have observed time series $x(0),...,x(\tau-1),x(\tau)$, we would like to predict $x(\tau+t), t\geq 1$. For this purpose, we can train RNN with RTRL by using $x(0),...,x(\tau-1)$ and $x(1),...,x(\tau-1),x(\tau)$ as an input sequence and training signals, respectively. Then, RNN is expected to predict future observations $x(\tau+t), t\geq 1$. Since RNN has the nonlinear and feedback structure, it has an advantage over other conventional linear models.

6.2 Identification

RNN with RTRL can be used effectively for the identification of a nonlinear dynamical system (Narendra and Parthasarathy, 1990). Consider the following discrete-time dynamical system

$$y(t+1) = \Phi(y(t), x(t)),\ z(t) = \Psi(y(t)),\ t \geq 0,$$

where $\Phi(\cdot)$ and $\Psi(\cdot)$ are unknown nonlinear functions. The vector $y(t) \in \Re^m$ denotes the state of the system at time t. At every t, given the input $x(t) \in \Re^l$ along with state $y(t)$, the output $z(t) \in \Re^n$ and the next state $y(t+1) \in \Re^m$ are determined completely. The aim is then to construct an appropriate identification model which generates an approximation $\hat{z}(t)$ of $z(t)$ when receiving input $x(t)$. Such identification can be realized by using RNN with RTRL, in which an input sequence and training signals are $x(t)$ and $z(t)$, $t \geq 0$, respectively. RNN with RTRL can also be applied to the adaptive control of a nonlinear dynamical system (Narendra and

Parthasarathy, 1990). Since identification and control of a nonlinear dynamical system are often executed on-line, RTRL is particularly suited.

There are many other applications such as speech recognition and so on (Hertz et al., 1991; Parlos et al., 1994). The above examples suggest that RNN has the potential and capability of dealing with time sequences and RTRL is useful in applications that require an adaptation property. Therefore, the global convergence property of the RTRL established in this paper is significant since it guarantees the robustness of the method when applied to various practical problems.

7. CONCLUSION

In this paper, viewing RTRL as an approximate gradient method for error functions, we have showed that RTRL has the global convergence property under almost the same conditions as other variations of BP. Moreover, we have proposed a new simple method which improves upon the general RTRL with an arbitrary heuristic learning rate. This method uses upper and lower bounds on the learning rate of general RTRL and has the global convergence property.

Back-Propagation Through Time (BPTT) is also a popular on-line BP method for RNN (Hertz et al., 1991). Since BPTT can be regarded as a special case of RTRL, the convergence theorems shown in this paper hold for BPTT as well.

References

Bertsekas, D.P.. Incremental least squares methods and the extended Kalman filter. *SIAM Journal on Optimization* 1996; 6(3): 807-822.

Bertsekas, D.P.. *Nonlinear Programming : 2nd Edition*, Athena Scientific, Belmont, MA, 1999.

Connor, J. T., Martin R. D., Atlas, L.E.. Recurrent neural networks and robust time series prediction. *IEEE Transactions on Neural Networks* 1994; 5(2): 240-254.

Gaivoronski, A. A.. Convergence properties of backpropagation for neural nets via theory of stochastic gradient methods, part 1. *Optimization Methods and Software* 1994; 4(2): 117-134.

Hertz, J., Krogh, A., Palmer, R. G.. *Introduction to the Theory of Neural Computation*, Addison-Wesley, Redwood City, 1991.

Ljung, L.. Analysis of recursive stochastic algorithms. *IEEE Transactions on Automatic Control* 1977; AC-22(4): 551-575.

Luo, Z.Q., Tseng, P.. Analysis of an approximate gradient projection method with applications to the back propagation algorithm. *Optimization Methods and Software* 1994; 4(2): 85-101.

Mak, M. W., Ku, K. W., Lu, Y. L.. On the improvement of the real time recurrent learning algorithm for recurrent neural networks. *Neurocomputing* 1999; 24: 13-36.

Mangasarian, O.L., Solodov, M.V.. Serial and parallel backpropagation convergence via nonmonotone perturbed minimization. *Optimization and Software* 1994; 4(2): 103-116.

Narendra, K. S., Parthasarathy, K.. Identification and Control of dynamical systems using neural networks. *IEEE Transactions on Neural Networks* 1990; 1(1): 4-27.

Ortega, J. M.. *Numerical Analysis: A Second Course*, Academic Press, NewYork, NY, 1972.

Parlos, A. G., Chong K. T., Atiya, A. F.. Application of the recurrent multilayer perceptron in modeling complex process dynamics. *IEEE Transactions on Neural Networks* 1994; 5(2): 255-266.

Rumelhart, D. E., Hinton, G. E., Williams, R. J.. Learning internal representations by error propagation. in D. E. Rumelhart, J. L. McClelland and the PDP Research Group Eds., *Parallel Distributed Processing* MIT Press, Cambridge, MA, 1986; 318-362.

A MATERIAL PLANNING MODEL FOR MIXED MODEL ASSEMBLY LINES

E. Kozan and P. Preston

School of Mathematical Sciences
Queensland University of Technology
GPO Box 2434 Brisbane Qld 4001 Australia
{e.kozan,p.preston}@qut.edu.au

Abstract A stochastic material planning (SMP) model is developed to incorporate uncertainties in timing and amount in demand, and availability of correct parts when needed to satisfy production. SMP combines a master production schedule (MPS), which determines the optimal production schedule based on inventory, backorder, overtime and slack time costs. SMP uses the bill of material (BOM) to generate parts requirements for weekly production plan determined by MPS which is solved by mixed integer programming. The structure of the BOM is quite complex due to the number and type of variants, and timely use of SMP information assists in the ordering of stock to reduce the risk of delays in production due to stock outs. The SMP model is used to reduce this complexity and to improve the accuracy of a multi-product production plant. The SMP is based on and implemented in a truck production plant, is calculated in a MS-Access database.

Key words: Operations scheduling, production planning and inventory control, stochastic modeling.

1. INTRODUCTION

Master production schedule (MPS) models the production of trucks in a mixed-model production plant. It uses a weekly time-bucket across a six month planning horizon. Stochastic Material Planning (SMP) model uses the bill of material (BOM) to generate parts requirements for the weekly production plan. Timely use of this information assists in the ordering of stocks to reduce the risk of delays in production due to stock outs. The models are furthered improved using lot sizing and batch ordering to reduce freight and other associated ordering costs.

Bhal and Ritzman (1984) developed an integrated model for master scheduling, lot sizing and capacity requirements planning. Gessner (1986) discussed various techniques of master production scheduling for make to stock and make to order companies. Sridharan et al. (1987) and (1988) discussed the effect of freezing on the master production schedule and its stability. Tallon (1989) used simulation to compare four techniques of master production scheduling for assemble-to-order products. Vercellis (1991) developed multi-criteria models to integrate multiple conflicting objectives of the manufacturing strategy. Campbell (1992) developed master production scheduling under rolling planning horizons with fixed order intervals. Lin and Krajewski (1992) developed a rolling horizon master production scheduling model for uncertain environments. Das (1993) developed master scheduling with incremental capacity allocation in a rolling horizon for order driven companies having short delivery lead-time with uncertainty in demand. Venkataraman and Smith (1996) disaggregated production plans to a rolling horizon master production schedule for a minimum production batch size requirement. Lin et al. (1994) modelled the effects of environmental factors such as cost structure, bill of material (BOM), cumulative lead-time, forecast error and changes in MPS on the frozen and replanning intervals, in an uncertain rolling horizon. Chu (1995) developed a linear optimisation model for MRP/MPS and practical computational feasibility and the robustness of the decomposition. Kimms (1998) measured in-stability of finite horizon production planning in a rolling horizon. Yeung et al. (1998) carried out an intensive review of literature on the parameters affecting the effectiveness of MRP and MPS.

2. MODELLING FOR TRUCK PRODUCTION PLANTS

The time interval over which no schedule changes are allowed is five weeks, due to a unique problem of the truck industry where dealers confirm orders just before starting production. This is usually equivalent to the customer delivery lead times. Production lead-time, the minimum time period for producing master scheduled trucks (MST) is four weeks. Customer delivery lead-time, the minimum time period required for delivering the MST to the customer, is five weeks. Cumulative lead-time, the minimum time period for production and procuring its raw material (parts) is 13 weeks. Lead times are graphically presented in Figure 1. In an ideal case, the total lead time should be less than the customer lead time. However, in our case the gap between these lead times is too large and increases the complexity of the problem. MPS is modelled on a weekly basis and uses 26 weeks data in a dynamic rollover approach. It determines an optimal production schedule based on expected demand, inventory costs, labour costs, and backorder costs. MPS is constrained by work centre machine times, availability of labour and calender days per week as variable parameters. SMP uses the production plan developed by MPS in conjunction with the bill of materials (BOM) and variant split percentages to determine parts requirements and to order requirements. The notation outlined below is consistent for all stages of the model.

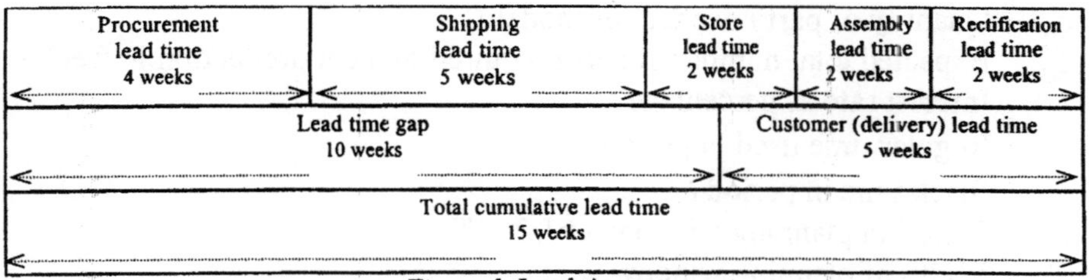

Figure 1. Lead time gap.

Nomenclature:

α Service level parameter ($\alpha = 0$ for no truck supply; $\alpha = 1$ for 100% delivery of trucks ie. no backorders).

a Average absenteeism.

b_{it} Back order cost for truck-model i in period t, $(Mt_i r/52)w + Dl$.

BOM_{pj} Number of part j required in $PMLp$ ($p = 1...5275; j = 1,2,...,J$).

B_t Available capital for inventory at the period t.

c_{it} Carrying cost for truck-model i in period t is a percentage of the total truck cost, $c_{it}=(Tr_i \times r)/52 \times w)$;

c_o Overtime labour cost per hour.

c_s Penalty cost for slackness.

Dl Penalty cost for backorders per period.

D_t Expected demand for period t aggregating over all truck models.

d_{it} Expected demand for truck-model i in period t, as forecast by the marketing department.

h Regular working hours per day.

h_i Labour hours to build truck model i.

i Truck-model group ($i=1, 2...,I$).

I^+_{i0} Beginning inventories of finished truck of model i in period 1.

I^-_{i0} Beginning back orders of finished truck of model i in period 1.

I^+_{it} Expected ending inventories of finished truck of model i in period t.

I^-_{it} Expected ending back order of finished trucks of model i in period t.

m_i Machine hours for truck model i.

M^c_t Maximum available machine hours in period t at work centre c.

Mt_i Material cost for model i.

N Number of truck models.

N_t Number of working days in period t.

o_{mt} Available time as a percentage of regular time.
O_t Overtime used in period t.
p_j The probability that the part j will be in the store by required date.
PML_{pt} Expected requirement of $PML\ p$ in period t.
q_{ji} Quantity of part j needed for model I.
Q_{jt} Expected total number of part j required to meet production in week t.
r Interest rate as percentage.
R_t Regular time used in period t.
S_t Slack time in period t.
t Period in planning window, $t=1,2,...T$.
T Width of planning window.
Tr_i Truck production cost for model i.
w Number of weeks in the period.
W_t Number of workers in period t.
$V_{v,i}$ Probability of variant v on model i (v=1,2,…,V, i = 1,2,...,I).
x_{it} Production of truck-model i finished in period t.
y Maximum inventory capacity.

3. MPS MODEL FOR A TRUCK ASSEMBLY PLANT

The developed model is used to minimise total cost of carrying finished products, back-orders, overtime and slack time costs at the aggregate level. A multi-period rolling plan with a provision for adjustment of forecast errors in the subsequent planning cycles is proposed.
The following assumptions are considered in developing the model:

- Demand forecast is accurate.
- The size of the workforce is fixed for the planning horizon and hence hiring and firing of labour are left to what if analysis;
- Skill required to work in various workstations is uniform. Hence resources are assumed to be available from a pool of skilled labours able to do jobs in any work-station;
- Set up time is negligible because most of the workstations are equipped with general-purpose assembly facilities for all product models and therefore, not included;
- Assembly line is well-balanced, so only individual centres capacities are used in the model; See Burdett & Kozan (2000).
- Subassemblies have work in process inventories; and
- Machines and labour are balanced.

Objective function

Minimise the sum of inventory carrying costs, back order costs, overtime costs and a slack time penalty cost.

$$\textbf{Min}\sum_{t=1}^{T}\sum_{i=1}^{N}(c_{it}I_{it}^{+}+b_{it}I_{it}^{-})+\sum_{t=1}^{T}c_{o}O_{t}+c_{s}S_{t} \tag{1}$$

Constraints

Inventory conservation ensures that beginning stock of finished product plus trucks produced during the period less net demand is equal to the ending inventory (if positive) or ending backorder (if negative).

$$I_{it}^{+}-I_{it}^{-}=I_{it-1}^{+}-I_{i(t-1)}^{-}+x_{it}-d_{it} \quad \forall i,t \tag{2}$$

Regular time constraint ensures that regular time used in production must be less than or equal to total available regular labour-hours after adjustments for absenteeism for any period

$$R_{t}=hN_{t}W_{t}(1-a) \quad \forall t \tag{3}$$

Overtime constraint ensures that overtime used must be less than or equal to overtime limit for any period and is equal to the hours over regular time needed to satisfy planned production.

$$O_{t}\leq R_{t}O_{mt} \quad \forall \mathrm{t} \tag{4}$$

Equation 5 is use3d to make the feasible solution realistic by ensuring that overtime can only be used when there is no slacktime.

$$\mathit{If}\ O_{t}>0\ \mathit{then}\ S_{t}=0 \quad \forall \mathrm{t} \tag{5}$$

Labour hours constraint ensures that regular time and overtime used in any period equals the labour-hours required to build trucks in that period

$$R_{t}+O_{t}-S_{t}=\sum_{i=1}^{N}h_{i}x_{it} \quad t=1,\ 2,\ldots w \tag{6}$$

Service level constraint ensures that backlog in any period must be less than or equal to acceptable limit (where, $0\leq\alpha\leq 1$, say 0. 95 i. e, service level is 95%)

$$\sum_{i=1}^{N}I_{it}^{-}\leq(1-\alpha)\sum_{i=1}^{N}d_{it} \quad \forall\ \mathrm{i}, \tag{7}$$

Finished goods inventory constraints ensures that total number of finished trucks at the any period should be less than the rest of the planning period's of demand. This is expressed as a fixed amount based on storage space for finished trucks and as capital tied up as set in the budget

$$\sum_{i=1}^{N}I_{it}^{+}Tr_{i}\leq B_{t} \quad \forall_{t} \tag{8}$$

$$\sum_{i=1}^{12}I_{it}^{+}\leq y \quad \forall_{I} \tag{9}$$

The completion constraints ensure that the total demand is met for each truck model over the planning horizon.

$$\sum_{t=1}^{T} x_{it} = \sum_{t=1}^{T} d_{it} - I_{i0}^{+} + I_{i0}^{-} \quad \forall_t \tag{10}$$

Work Centre capacity constraint ensures that the time required at a work centre for planned production is matched by available hours at that centre. Where M_t^c is equal to regular time + overtime – slack time at work centre *c*.

$$\sum_{i=1}^{N} x_{it} m_i \leq M_t^c \quad \forall\ t, c \tag{11}$$

Equation (12) ensures that production, inventory, back order, regular time production and overtime production can not be negative.

$$S_t, R_t, O_t \geq 0 \quad \forall_t \tag{12}$$

Equation (13) ensures that demand, production, inventories and backorders are integer.

$$d_{it},\ x_{it}, I_{it}^{+}, I_{it}^{-} \text{ are non-negative integers} \tag{13}$$

4. SMP MODEL FOR TRUCK ASSEMBLY PLANT

Any truck model is composed of standard parts, variants and optional items. 'Must' variants, such as automatic or manual drive, are parts or groups of parts required for assembling any vehicle. One of the variants must be selected to make a valid product. Optional variants, such as bug screen or air conditioner may or may not be selected by the customer. The customer specifies optional items and variants (configurations) so detailed planning should be undertaken after the customer gives a firm order. Unfortunately this isn't possible as the industry has a 5 week delivery lead-time while the factory needs a 15 week lead-time. To overcome this the production planner uses expected percentages for each of the variants on each of the truck models. If none of the options are selected then the decision is to make a standard product and use a product from stock items.

$$PML_{pt} = \sum_{i}^{I} [d_{it} * \prod_{v} v_{vi}] \quad \forall p, t \text{ and } v \mid v_{vi} \in PML_p \tag{14}$$

A PML is a planning bill of materials – an artificial grouping of parts under a non-engineering part number used to help streamline the material planning process. It is a method of grouping variants but does not exist as an entity itself. To determine the demand for each PML, we multiply through variant mix and demand for each PML; and sum over different models. We assume parts arrive on time or are satisfied from the safety stock.

$$Q_{jt} = PML_{pt} * BOM_{pj} \quad \forall\ t, j. \tag{15}$$

This gives the expected requirements of each part for each week of the planning interval. Due to the stochastic nature of the variant mix these values are only expected requirements and unlikely to be integer. So a number of rounding strategies are trialed as outlined in the results section.

5. APPLICATION TO THE TRUCK PLANT

The model is applied for a 6 month cycle using next years demand forecast. The model is designed to be run once a week using a roll-over dynamic approach. In this approach, week 1's data is removed and the other weeks data are moved forward, week 2 becomes 1 and week 27 is added as week 26. Starting inventory and backorder values are changed as necessary. The net demand for any part is calculated by subtracting available inventory from the gross demand and orders are released based on lead-time for supply of parts. Uncertainties in availability of parts and associated decision of airfreight or rectification make the problem further complicated.

A complete list of the model's input is detailed below:
Labour hours(h_i), material cost (Mt_i) and production costs (Tr_i) for 12 truck models are calculated using the data of a truck assembly plant in Australia. Machine hours (m_i) are assumed to be balanced with man-hours. Grouping of trucks is done based on following factors: Engine capacity (7, 10, 14 and 16 litres); Configuration (Rigid body /Tractor); Control (Forward /Normal); Cab hight (Low/Medium/High); Wheel x Drive (4 x 2, 6 x 4, 8 x 4). Some parameter values of the model is as follows: a=2.5%; r=9 %; Dl=$7,000; h=7.6 hours; y=100; $\alpha = 95\%$; $b_{it} = ((Mt_i \times r)/52 \times w + Dl)$; $B_t = \$5000\,000$; $c_{it} = (Tr_i \times r)/52 \times w)$; c_o=$28.5; c_s=$19.83; O_{mt}=25%; W_t=80 workers; $I^+_{i0}=0$ trucks; $I^-_{i0}=0$ trucks. T=26 weeks; V=559; P=527; I=12; J=4740. Material Cost, Production Cost for different truck type and labour hours per truck for different truck types is given in Table 1.

5.1 Application of Master Production Scheduling Model to a Truck Plant

A Master Production Schedule for twenty-six weeks is developed. It is set up for a 26-week cycle using the first six months of this years weekly demand forecast. MPS is designed to be run weekly using a roll-over dynamic approach. However, since 6 weeks production is locked in, these are input as fixed parameters and the remaining 20 weeks production is optimally determined based on this. The work centre hours are confidential and as such are not included with this paper.

Table 1. Model parameters

Truck models(i)	Labour hours (h_i)	Material costs (Mt_i)	Production costs (Tr_i)
1	141	64813	92496
2	148	77528	103559
3	133	75431	96316
4	141	86669	109783
5	148	96633	120902
6	141	92215	115398
7	141	97259	121416
8	139	92215	115398
9	146	111206	135015
10	158	93431	118447
11	158	100351	126173
12	161	127012	152833

The demand forecast is provided in Table 2.

Table 2. Expected demand (d_{it})

	Period (months)											
Models(i)	1	2	3	4	5	6	7	8	9	10	11	12
1	2	2	2	0	1	2	1	0	1	1	1	0
2	4	2	3	3	2	8	5	1	2	2	4	1
3	0	1	0	1	1	1	2	1	1	1	1	0
4	13	10	16	11	16	15	12	10	4	5	10	3
5	3	10	12	12	6	6	7	4	1	2	5	2
6	7	7	8	6	14	16	17	22	18	14	17	10
7	0	1	2	2	2	1	2	4	1	3	1	1
8	0	10	13	10	10	14	10	14	2	5	5	0
9	0	3	2	0	2	1	0	2	2	1	1	0
10	4	2	3	1	3	5	5	7	2	3	5	1
11	3	3	4	2	9	14	7	18	10	15	16	4
12	3	6	7	4	6	9	8	9	1	5	6	2

The MPS model described above was run in GAMS with a typical execution taking around 1 minute to run (although in some cases would need several hours to reach a suitable integer solution). Optimal production levels for each truck model for the first 26 weeks are graphed in Figure 2.

To demonstrate the application of the rolling horizon in Master Production Scheduling the model has been applied rolling over 4 week blocks (nb this should be done weekly but four weeks has been used for simplicity) to the 2000 forecast demand. This was done firstly by running the model for weeks 1-26, then taking the inventory levels after the forth week as initial levels was run for week 5-30. Similarly this process was repeated for weeks 9-34, 13-38, 18-43, 22-47 and 26-51 (no production in week 52) to complete the year. The results of these are shown in Figure 3.

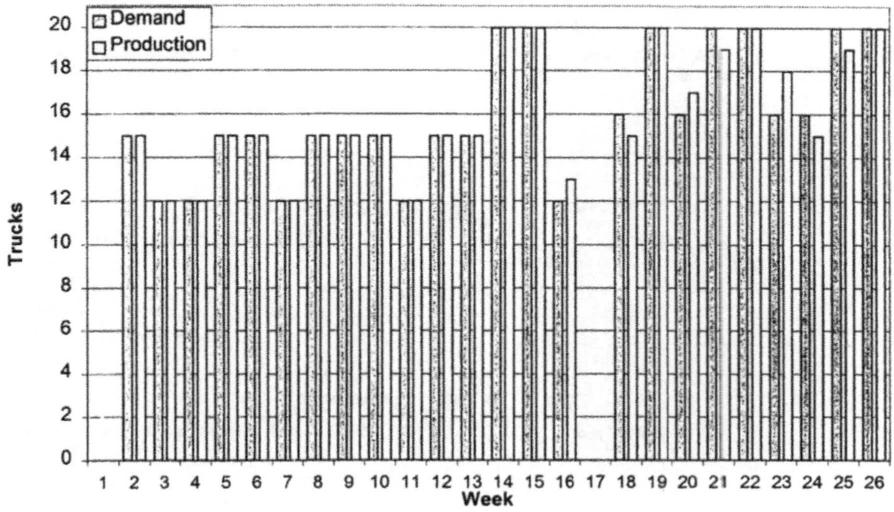

Figure 2. Comparison of optimum weekly productions and demands for the first 26 weeks.

5.1.1 Sensitivity analysis

Sensitivity analysis is performed to determine how changes to the model parameters affect the solution. This can be a valuable tool in determining, for example, the number of workers that should be employed or how the company will be able to cope with increased demand.

Three main areas of what if analysis have been performed on the model, number of workers, absenteeism and labour hours. These results are summarised below.

Analysis of number of workers

To find the optimal number of workers, demand and production need to be analysed. Results are summarised in Figure 4. Five different sized samples of workers namely 72, 76, 80, 84 and 88 are chosen.

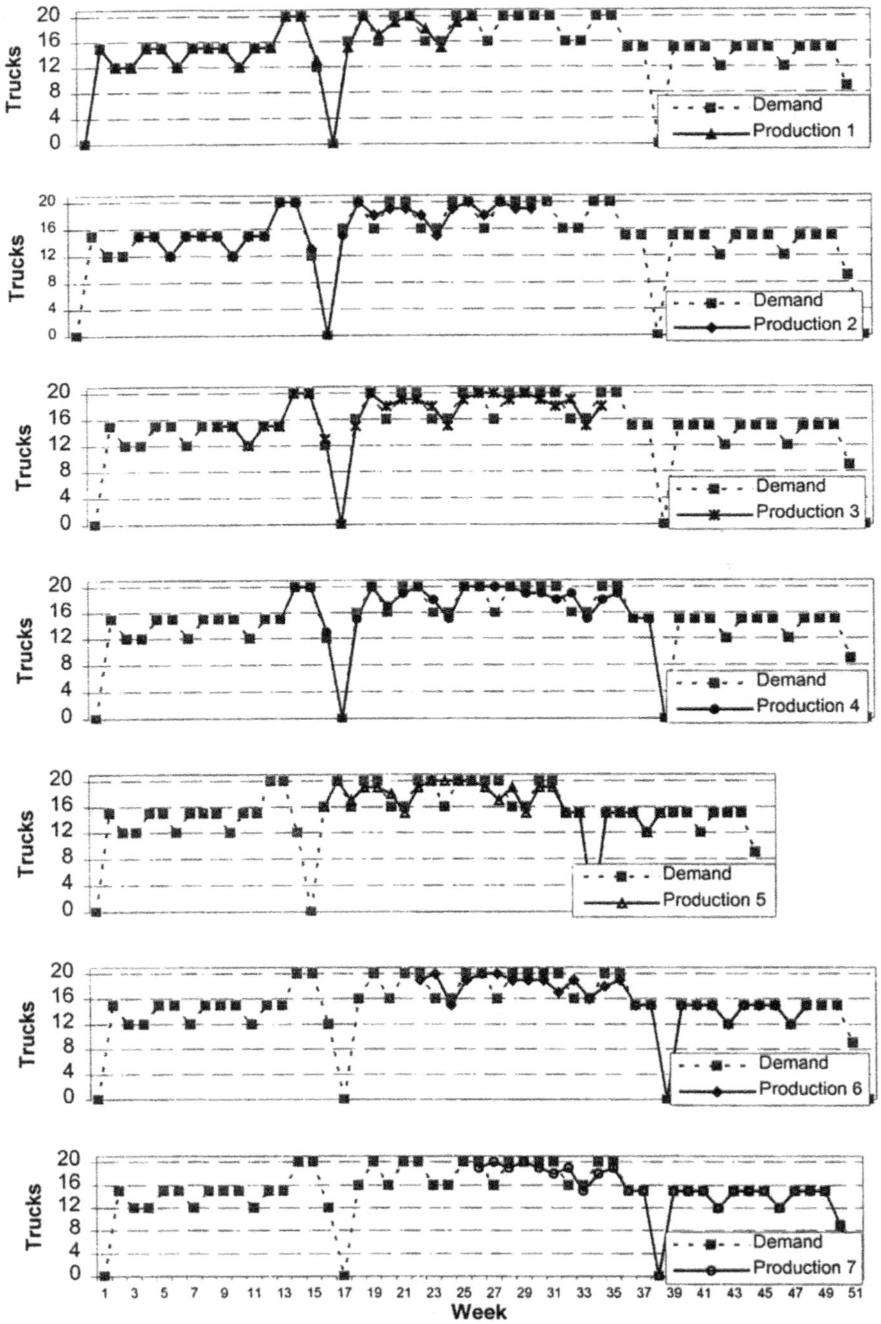

Figure 3. Comparision of production and demand using 4 week rolling horizon.

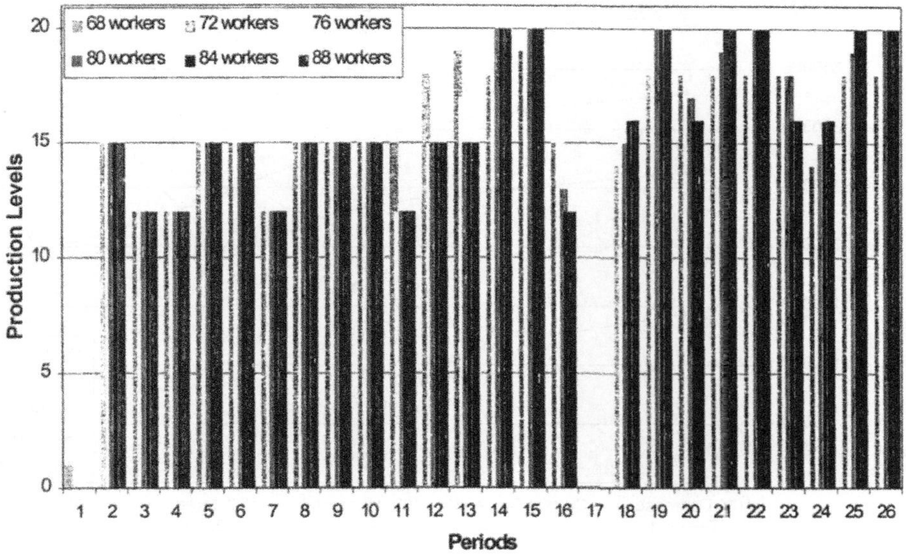

Figure 4. Production level vs number of workers.

The production levels for each week can vary greatly with the changing number of workers. This is particularly noticeable in the second half of the graph. This would suggest that demand in that time period is close to or exceeds production capacity carrying inventory or backordering trucks is necessary. To explore the variation in demand, we need to consider the optimal number of workers for the minimum total cost. Total cost includes all costs involved in the production such as slack time costs, backorder costs, inventory costs and so on. With the current demand, it appears that the optimal number of workers, to minimise the total cost would be 72 workers. This can be clearly noted in Table 3. Slacktime cost is the main cost factor in these results and as such, increasing the number of workers increases the costs due as this just creates more slacktime.

Table 3. Comparison of total costs.

Workers	72	76	80	84	88
Cost ($)	127545	184717	249125	315542	383582

The relationship between demand and the number of workers is examined by increasing the demand by 10%, 15% and 20% and decreasing the demand by 10%. It can be noted that as the demand increases so does the optimal number of workers, in relation to the minimum total cost. This can be observed in Figure 5.

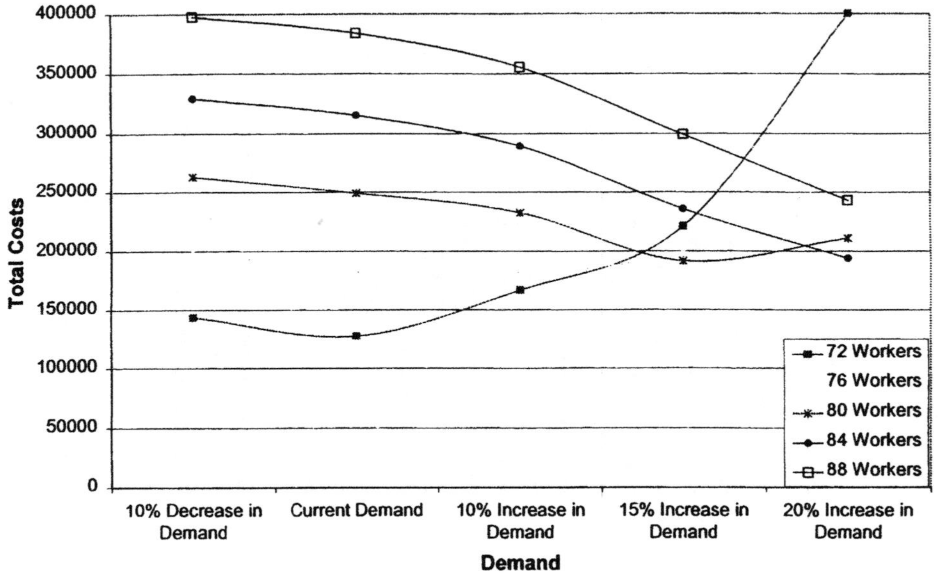

Figure 5. Inventory and related costs for variations in demand and number of worker.

Analysis of absenteeism

The main concept of this analysis is to observe the effects of absenteeism on production levels and costs. The analysis was conducted on 80 workers. It can be noted that the costs vary slightly with the different levels of absenteeism. Overall, the total cost of overtime, inventory, backorder and slacktime, seem to increase with the decrease of absenteeism. However, the effects on the production levels by absenteeism were minimal. In a few periods the production level differed by only one truck. Hence the change in cost can be attributed to a reduction in slack time and it is noted that the factory is overstaffed for large periods during the year with 80 workers and the current demand level.

Table 4. Total cost versus absenteeism variation.

Absent (%)	0.0216	0.0243	0.027	0.0297	0.0324
Cost ($)	256 113	252 504	249 125	245 988	242 608

Efficiency of factory labour

Labour hours per truck model, describes the number of hours taken to construct that particular model of truck. To find the optimal number of labour hours, two aspects need to be considered: - costs and the production level. This analysis was conducted on 80 workers. There were seven different sized samples of labour hours chosen for analysis. These were a 20%, and 10% decrease and 0%, 10%, and 20%, increase on the original labour and work centre hours. A 20% increase is found to minimise the total cost. This can be seen in Figure 9.

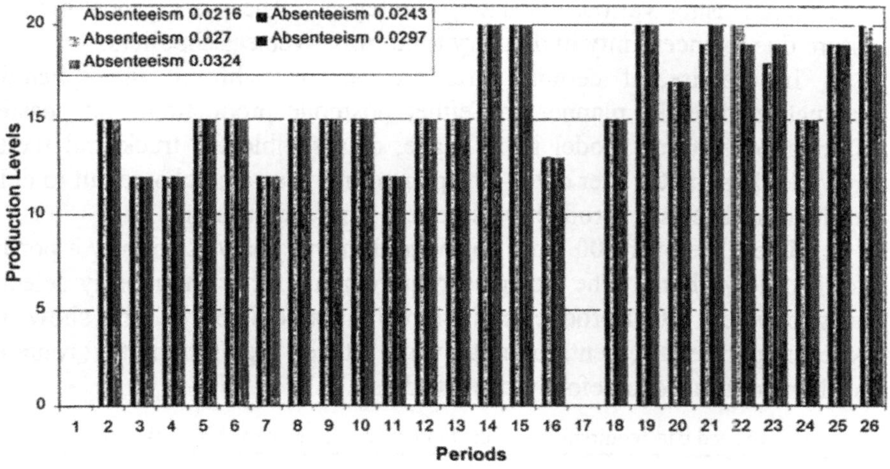

Figure 6. Production levels against absenteeism

Table 5. Costs for varying labour hours per truck

Change in workcentre hours (%)	-20	-10	0	10	20
Cost ($) for 80 workers	470 163	358 832	249 125	148 836	816 12

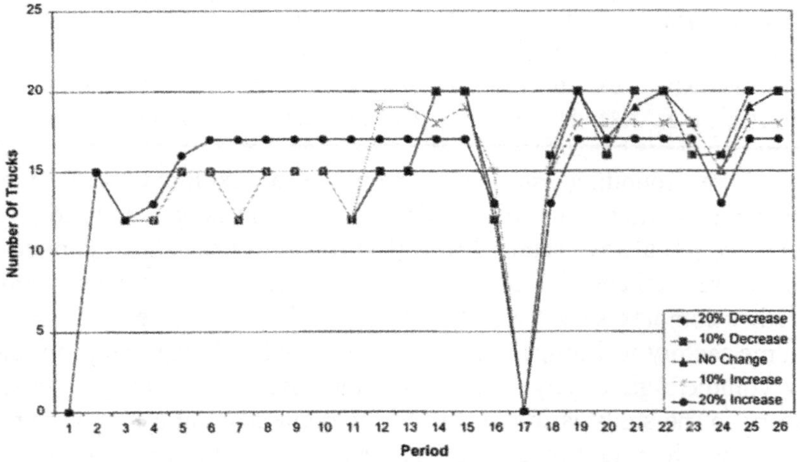

Figure 7. Production levels for varying labour hours per truck

5.2 Materials Plan (MP) Model for Truck Assembly Plant

The MP is solved in MS-Access by importing the bill of materials (BOM) database, inputting the production plan from MPS and running in a visual basic module. MP solves for expected Q_{jt} (part requirements) values to meet planned production. Order quantities for individual parts are then determined using various lot sizing strategies. These are probabilistic in nature due to uncertainty in delivery time from overseas suppliers.

In practise, if certain parts are not available at the scheduled production date the planner can either postpone production, and possibly insert another truck model in its place, or assemble the truck and fix the unavailable part at a later date. The planner may have no choice but to delay production if certain "production stopper" parts are missing.

There are over 500 parts considered in this analysis so only a portion can be shown here. The expected requirements for ten randomly selected parts for weeks 39-51 production is shown in Table 6 below. This shows the expected part requirements are not always integer, highlighting that rounding mechanism is needed before placing orders.

Table 6. Expected part requirements of the randomly selected parts for weeks 39-51

Part	Total	39	40	41	42	43	44	45	46	47	48	49	50	51
1062012	0.39	0.04	0.03	0.04	0	0.04	0.04	0	0	0	0.04	0.1	0	0
1062066	28.05	3	3.15	0	2.2	2	5	2	3.15	1.3	1.15	3.2	2	0
1062115	40.38	5.37	5.02	3.16	2.2	2.98	4.58	2.4	3.01	1.6	3.77	3.8	2.4	0
1062138	34	3	4	0	3	2	5	2	4	3	2	4	2	0
1062206	161.5	21.5	20.1	12.6	8.8	11.9	18.3	9.8	12	6.5	15.1	15	9.7	0
1062365	96.3	8.55	11.6	0	8.5	5.5	14.2	5.5	11.2	8.7	5.65	11	5.7	0
1062567	116	10	11	12	7	11	8	10	7	10	12	11	7	0
1062632	5.95	0	0.85	0	0.9	0	0	0	0.85	1.7	0.85	0.9	0	0
1062756	0.19	0.03	0.03	0.01	0	0.01	0.03	0	0.03	0	0.01	0	0	0
1062827	44.37	5.82	2.88	5.82	1	3.84	3.84	2.9	0	2.9	4.83	6.7	3.9	0

Five rounding strategies have been explored with the total requirements shown in Table 7. The strategies investigated are to (I) always round up, (II) always round down, (III) round to nearest, (IV) round up the cumulative total and (V) the current system. The system currently employed is rather complex so we won't go into details but was designed for a large volume factory in Europe and is acknowledged as being inappropriate for a low volume/high variety factory as is the case for this study. As such the factory in the study is in the process of implementing a new system. This shows that the round total up is the approach that most closely matches the raw totals.

The four new rounding schemes provide different levels of accuracy over a full years production. Table 8 shows the average percentage error for the full list of parts over the 52 weeks of 2000. The parts have been split into 3 different categories based on the amount required.

From Table 8 it is seen that the percentage error increases as the number of parts required becomes smaller. These small orders should be calculated by hand as it is most likely that they will only be ordered in when a customer requires them. There are only a small number of these parts so the extra work should be minimal.

Table 7. Comparison of rounding strategies for 10 random parts

Part	Raw Totals	I	II	III	IV	V
1062012	0.39	10	0	0	1	10
1062066	28.05	33	27	27	29	39
1062115	40.38	47	35	40	41	48
1062138	34.00	34	34	34	34	34
1062206	161.52	168	156	161	162	192
1062365	96.30	102	91	98	97	97
1062567	116	116	116	116	116	116
1062632	5.95	7	1	7	56	6
1062756	0.19	10	0	0	61	17
1062827	44.37	46	35	46	45	36

Table 8. The percentage errors for the 4 different rounding schemes

Number of Parts	I	II	III	IV
< 1 per month	4500.375	100.000	100.00	79.137
> 1 per month & < 1 per week	65.536	41.840	12.568	0.971
> 1 per week	2.676	2.956	0.403	0.068
All parts	160.455	6.822	4.030	2.839

Order quantities for individual parts would then be determined using various economic lot sizing strategies. These are probabilistic in nature due to uncertainty in delivery time from overseas suppliers.

In practise if certain parts are not available at the scheduled production date the planner can either postpone production, and possibly insert another truck model in its place, or assemble the truck and fix the unavailable part at a later date. The planner may have no choice but to delay production if certain "production stopper" parts are missing.

6. CONCLUSION

This paper is focused on the master production scheduling with a stochastic materials plan model. The stochastic model for mixed model assembly line make-to-order truck company is developed and implemented at its Australian plant. The problems due to uncertainties in demand and availability of parts are included in the model. The MPS model is formulated as a mixed integer programming problem in a time window and solved using GAMS. SMP is solved by MS Access using weekly production schedule from MPS.

A rolling horizon technique is used to update the next plan on the basis of improved information. Sensitivity analyses of various changes to inputs of the models are applied. This approach improves the minimisation of the sum of inventory carrying costs, back order costs, overtime costs and slack time penalty costs.

There is enough scope for future research in the following areas: cost-benefit analysis for airfreight such as build with shortages or deferment; development of strategy for inventory reduction; common parts analysis for standard orders based on budgeted production; and detail programming for each truck model to determine the daily programs.

When implemented at the Brisbane plant of a multi-product, make-to-order truck company, the approach improves the accuracy and produces a good production schedule based on the established constraints.

Further research continues into investigating the problem of detail programming. Authors are currently working on this area (see Burdett and Kozan, 2000) and the detail programming part will be merged with the system in the near future.

7. ACKNOWLEDGEMENTS

This work was partly supported by the Australian Research Council under a SPIRT grant and by Volvo Truck Australia. This support is greatly acknowledged.

References

Bhal, H.C. and Ritzman, L.P. An Integrated Model for Master Scheduling, Lot Sizing and Capacity Requirements Planning. Journal of the Operational Research Society 1984; 35: 389-399.

Burdett, R. L. and Kozan E. Evolutionary algorithms for Flowshop Sequencing with Non-Unique Jobs, International Transactions in Operations Research 2000; 7.5: 401-418.

Campbell G.M. Master production scheduling under rolling planning horizons with fixed order intervals. Decision-Sciences 1992; 23(2): 312-331.

Chu, S.C. (1995). A Mathematical Programming Approach Towards Optimised Master Production Scheduling. International Journal of Production Economics 1995; 38(2) :269-279.

Das S.K.. Master Scheduling with Incremental Capacity Allocation and a Rolling Horizon. Management Science 1993; 21(3): 353-361.

Gessner, R. A. Master Production Schedule Planning. John Willy and Sons, New York, 1986.

Kimms A. Stability Measures for Rolling Schedules with Applications to Capacity Expansion Planning-Master Production Scheduling and Lot Sizing. Omega 1998; 26(3): 355-366.

Lin N.P. and Krajewski, L.J. A Model for Master Production Scheduling in Uncertain Environments. Decision-Sciences 1992; 23(4): 839-861.

Lin N.P., Krajewski, L.; Leong, G.K. and Benton, W.C. The effects of environmental factors on the design of master production scheduling systems. Journal of Operations Management 1994; 11(4): 367-384.

Sridharan V; Berry, W.L. and Udayabhanu V. Freezing the master Production schedule under rolling planning horizons. Management Science 1987; 33(9): 1137-1149.

Sridharan, V. Berry, W.L. and Udayabhanu V. Measuring master Production schedule stability under rolling planning horizons. Decision Sciences 1988; 19(2): 147-166.

Tallon, W. J. A Comparative Analysis of Master Production Scheduling Techniques for Assemble-to-Order Products. Decision-Sciences 1989; 20(3): 492-506.

Venkataraman, R. and Smith, S.B. Disaggregation to a Rolling Horizon Master Production Schedule with Medium Batch-Size Production Restriction. International Journal of Production Research 1996; 34(6): 1517-1537.

Vercellis, C. (1991). Multi-Criteria Models for Capacity Analysis and Aggregate Planning in Manufacturing Systems. International Journal of Production Economics 1991; 23(1): 261-272.

Yeung, J.H.Y; Wong, W.C.K and Ma, L. Parameters Affecting the Effectiveness of MRP Systems: a Review. International Journal of Production Research 1998; 36(2), 313-331.

A FUZZY PROGRAMMING MODEL FOR A CROSS-BORDER LOGISTICS PROBLEM UNDER AN UNCERTAIN ENVIRONMENT IN HONG KONG

Stephen C. H. Leung and K. K. Lai

Department of Management Sciences, City University of Hong Kong,
83, Tat Chee Aveune, Kowloon Tong, Hong Kong.
mssleung@cityu.edu.hk, mskklai@cityu.edu.hk

Abstract In this study, we consider a logistics problem involving the transportation of products between Hong Kong and China. The logistics here differ from logistics as traditionally defined because of the existence of a cross-border variable, inevitable in any transaction between Hong Kong and China. Logistics management often involves fuzzy or vague data; we therefore develop two fuzzy models for different logistic problems in an uncertain environment. In one problem the demand is fuzzy while the cost components are crisp. In the other problem all the costs and demand are fuzzy. The parameters in both problems are characterized as triangular fuzzy numbers and the optimal solution is achieved via the fuzzy ranking function of the fuzzy numbers with respect to their total integral value. A set of data from a Hong Kong-based manufacturing company is used to test the robustness and effectiveness of the proposed models.

Keywords: Logistics, triangular fuzzy number, ranking of fuzzy numbers, optimization.

1. INTRODUCTION

Transportation was one of the earliest problems to which linear programming was applied in the area of logistics. Developed by Hitchcock (1941) it concerns the allocation of homogeneous products from m sources to n destinations. The sources are production facilities with an available supply of goods, a_i, $i = 1,2,...,m$, while the destinations are sales outlets, warehouses, etc., with a certain demand, b_j, $j = 1,2,...,n$. The classical transportation problem is stated as follows: each a_i from source i is transported to n destinations to satisfy demand b_j. A cost function, c_{ij}, is associated with the transportation of one product unit from source i to destination j. The cost function comprises operating costs, the number of products transported, delivery time, unfulfilled demand, etc. The decision-maker (DM) determines the quantity of products x_{ij} to be transported from source i to destination j such that the total transportation cost is minimized. An extension of the transportation problem is the solid (multi-index) transportation problem, in which the product can be carried by l different modes of transport or conveyance (Haley, 1962;1963). Instead of considering x_{ij}, DM determines the quantity of product x_{ijk} to be transported from source i to destination j using conveyance k, $k = 1,2,...,l$, such that the total transportation cost is minimized.

However, in the real world, many logistics problems are not solved as easily as transportation problems. Logistics management is unlikely to be satisfied with the results derived by the above methods because many real-world planning problems involve imprecise, fuzzy or vague data. A model to reconcile the contradictions between real-world data and the realm of formulation is needed. Stochastic linear programming (Dantzig, 1955), fuzzy programming (Zimmermann, 1978) and chance-constrained programming (Charnes and Cooper, 1959) were developed to incorporate uncertainty data into an optimization model. A number of models have been developed to solve transportation problems according to practical problems encountered and solved in logistics strategies. Chang (2000) estimated an optimal revenue using a linear demand function in which the demand quantity is a triangular fuzzy number. Teodorovic (1994) stated that uncertainty may be due to factors such as traffic conditions, the type of driving, weather conditions.

Cheung and Powell (1996) proposed a two-stage model using stochastic programming to formulate the distribution and inventory problems with uncertain demand. The model can be solved using a framework of dynamic networks with random arc capacities.

In recent years, fuzzy set theory has been given a place in the realm of logistics. Teodorovic (1999) reviewed the use of state-of-the-art fuzzy set theory techniques to solve vehicle and crew routing, and scheduling and dispatching problems. Chanas et al (1993) studied a classic transportation problem with fuzzy numbers characterized as interval numbers. Different models such as the interval transportation problem with equality constraints, inequality constraints and mixed constraints were considered. Chanas and Kuchta (1996) proposed an algorithm to optimally solve the transportation problem with interval cost coefficients, in which supply and demand are precise. Later, Chanas and Kuchta (1998) proposed an algorithm to solve the transportation problem with fuzzy supply and demand as well as fuzzy goals, which transforms an interval problem into a classic problem.

Das et al (1999) studied a solution to the multi-objective transportation problem with interval cost, source and destination parameters, transforming the constraints with interval parameters into deterministic ones. The problem was then solved by fuzzy programming techniques. Jimenez and Verdegay (1998) considered an interval solid transportation problem and a fuzzy solid transportation problem. Yager's index was used as a ranking function of a fuzzy number. Genetic algorithms have also been proposed to solve a multi-objective solid transportation problem with fuzzy numbers (Gen et al, 1995; Li et al, 1997).

After China began its economic reforms in 1979, many manufacturers moved their production facilities from Hong Kong over the border to cities in southern China (Shenzhen, Dongguan, Zhuhai, Huaizhou, Xiamen) to benefit from lower wage and rental costs. After the products are manufactured in China, the finished and packed products are exported to Hong Kong in the manufacturing companies' own lorries, or hired lorries or by carrier, and are then shipped overseas or sold locally. A unique characteristic of the logistics problems experienced by Hong Kong-based companies operating in Hong Kong and China is the border separation between Hong Kong and China. All vehicles traveling between Hong Kong and China must cross the border at one of three control points: in the north-east New Territories at Sha Tau Kok, in the northern New Territories at Man Kam To, or in the north-west New Territories at Lok Ma Chau, paying a cross-border toll.

This study looks at a transportation problem faced by a Hong Kong manufacturer. This is a typical logistics problem for manufacturers operating both in Hong Kong and China. The specific problem for the company in this study is as follows. The company has its headquarters in Hong Kong and its warehouse is located at a container terminal in Hong Kong. The production facility of the company is situated in Dongguan in China. All vehicle trips between Dongguan and Hong Kong must pass the Lok Ma Chau control point (Figure 1). Every day finished products are transported from Dongguan to Hong Kong to fulfill daily sales orders and future demand. The company has three of its own lorries (hereafter, private lorries) available for this transport route, which originates at the depot in Hong Kong. The lorries are registered in Hong Kong and can operate on both sides of the border. The company hires more lorries to handle excess demand. Two types of lorries are hired. One is registered in China (hereafter, China lorries) and can only operate in China. The other is hired in Hong Kong (hereafter, HK lorries), which has a larger capacity and can operate on both sides of the border. With the information presently available there are currently three routing alternatives:

1) use private lorries to transport the products directly from Dongguan to Hong Kong;

2) hire HK lorries to transport the products directly from Dongguan to

Hong Kong; or

3) hire China lorries to transport the products from Dongguan to Shenzhen first, switching to the private lorries between Shenzhen and Hong Kong in order to enable more round-trips to be made so that more products can be delivered in a single day under a travel time constraint.

Moreover, apart from operating costs based on the routes chosen, and hiring costs associated with lorries hired, other costs will be considered in this study. The first of these, inventory costs, are incurred for storing any excess products in Hong Kong. Also, owing to the fact that each trip across the border also involves a complicated inspection procedure, an additional payment in addition to a basic salary is given to drivers who make more than one round-trip border crossing a day.

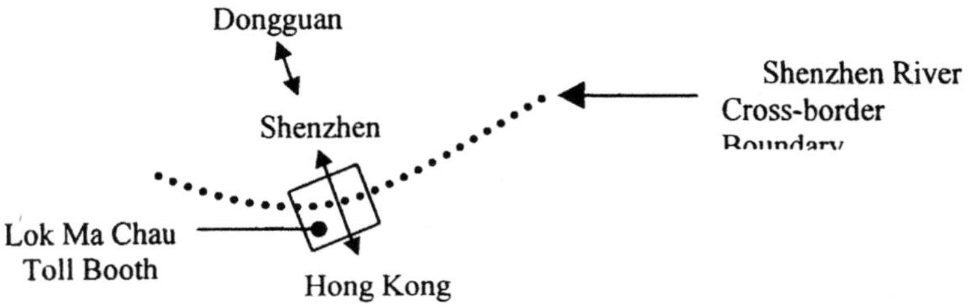

Figure 1. Location of Boundary between Hong Kong and China.

The company currently adopts transportation strategy 3. However, from experience the company knows that this strategy involves higher transportation costs and requires more storage space, but lower additional costs. In order to improve transportation efficiency and increase the company's competitive edge, the aim is to find an optimal plan with vehicle composition and route assignment to handle imprecise demand and fuzzy cost components. To this end we aim to develop two fuzzy models each applicable to a different logistics problem. In the first model, the cost components are crispy and demand is fuzzy or vague. Thus, demand is characterized by a fuzzy number. In the second model, we assume that all the cost components and demands are fuzzy or vague. These are characterized by corresponding fuzzy numbers. The purpose of this study is to find an optimal long-term transportation strategy, including the optimal delivery routes and the optimal vehicle fleet composition for a weekly planning horizon to minimize total transportation costs under an uncertain environment based on fuzzy ordering of fuzzy numbers with respect to their total integral values. The models may serve as a basis for the study of transportation problems in cross-border logistics systems. The simplicity of the implementation and operation of the model should enable decision-makers to manage the flow of products in terms of allocation and routing without having to learn complex operations and programming procedures.

This study is organized as follows. After this introductory section, the concept of fuzzy sets theory is presented. Two fuzzy models to solve the cross-border logistic problems mentioned above are formulated in Section 3. A set of data from a Hong Kong factory is used to test the effectiveness and efficiency of the models proposed in Section 4. Conclusions are given in the final section.

2. PRELIMINARY CONCEPTS

The problem considered in this study arises when a decision-maker has vague or imprecise data at hand. At this point the coefficients in the problem can be defined by fuzzy numbers (Lai and Hwang, 1992; Zimmermann, 1996). In this study, the demand and cost components are L-R triangular fuzzy numbers (Dubois and Prade, 1980). Before developing an optimization model to treat such a cross-border logistics problem, some basic concepts of fuzzy sets theory and ranking fuzzy number functions are described to provide an understanding of the fuzzy models developed below.

2.1 Fuzzy Sets and Fuzzy Numbers

Definition 2.1 *Let X be a universe of discourse. A fuzzy subset $\tilde{A}$ of X is defined by a membership function $\mu_{\tilde{A}} : X \to [0,1]$. The function value $\mu_{\tilde{A}}(x)$ represents the grade of membership of x in $\tilde{A}$. The closer the value of $\mu_{\tilde{A}}(x)$ is to 1, the more x belongs to A.*

Definition 2.2 *Let $F(\Re)$ be a set of all L-R triangular fuzzy numbers in real line $\Re$. A triangular fuzzy number $\tilde{A} \in F(\Re)$ is a fuzzy number with the membership function $\mu_{\tilde{A}} : \Re \to [0,1]$ parameterized by a triplet (a_L, a, a_R) where a_L and a_R denote the lower and upper limits of the support of a fuzzy number $\tilde{A}$ with mode a:*

$$\mu_{\tilde{A}}(x) = \begin{cases} \mu_{\tilde{A}}^{L}(x) = \dfrac{x - a_L}{a - a_L}, & \text{if } a_L \le x \le a \\ 1, & x = a \\ \mu_{\tilde{A}}^{R}(x) = \dfrac{x - a_R}{a - a_R}, & \text{if } a \le x \le a_R \\ 0, & \text{otherwise.} \end{cases} \tag{1}$$

where $a_L \le a \le a_R$ are real numbers, and $\mu_{\tilde{A}}^{L}(x)$ and $\mu_{\tilde{A}}^{R}(x)$ are the left membership function and right membership function of the fuzzy number $\tilde{A}$.

$\mu_{\tilde{A}}^{L}:[a_L,a]\rightarrow[0,1]$ *is continuous and strictly increasing and* $\mu_{\tilde{A}}^{R}:[a,a_R]\rightarrow[0,1]$ *is continuous and strictly decreasing (Figure 2).*

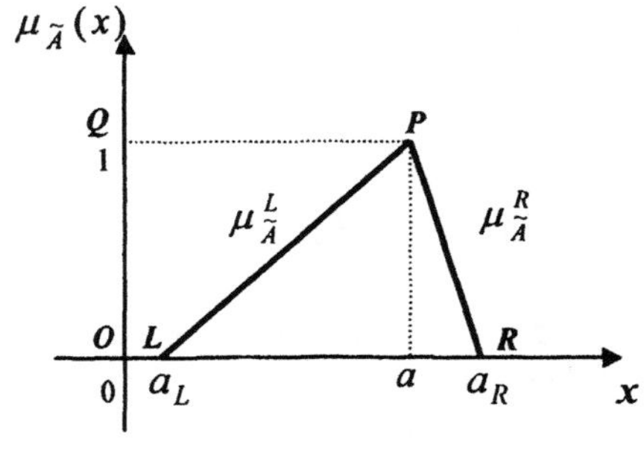

Figure 2. Triangular Fuzzy Number.

In triangular fuzzy number $\tilde{A}$ denoted by $\tilde{A}=(a_L,a,a_R)$, the parameter a gives the maximal grade of $\mu_{\tilde{A}}(a)$, i.e. $\mu_{\tilde{A}}(a)=1$; it is the most possible value of the evaluation data. a_L and a_R are the lower and upper bounds of the available areas of the evaluation data. For a non-fuzzy number A, this can be expressed as $\tilde{A}=(a,a,a)$. A triangular fuzzy number is therefore a generalization of a non-fuzzy number.

By the extension principle (Zadeh, 1965), an arithmetic operation $\tilde{W}=\tilde{U}*\tilde{V}$ of two triangular fuzzy numbers $\tilde{U}$ and $\tilde{V}$, is a triangular fuzzy number whose membership function $\mu_{\tilde{W}}(x)$ is defined by Dubois and Prade (1980) and Kaufmann and Gupta (1991) as:

$$\mu_{\tilde{W}}(z)=\mu_{\tilde{U}*\tilde{V}}(z)=\sup_{z=x*y} min\{\mu_{\tilde{U}}(x),\mu_{\tilde{V}}(y)\}$$

where $*$ denotes one of the arithmetic operations +, – or $\times$. Then, the result of fuzzy addition $\tilde{u}\oplus\tilde{v}=(u_1+v_1,u_2+v_2,u_3+v_3)$ is also a triangular fuzzy number. The operation of fuzzy subtraction $\tilde{u}-\tilde{v}=(u_1-v_1,u_2-v_2,u_3-v_3)$ and fuzzy scale multiplication $k\times\tilde{v}=(kv_1,kv_2,kv_3)$ also holds.

2.2 Ranking Fuzzy Numbers with Respect to their Total Integral Value

Before making a decision, decision-makers have to assess the alternatives with fuzzy numbers and rank these fuzzy numbers correspondingly (Liou and Wang, 1992). It can be seen that ranking fuzzy numbers is a very important procedure in solving the fuzzy programming problem. In reality, decision-makers having different viewpoints will give different ranking outcomes under the same situation. Therefore, a number of methods have been proposed for ranking fuzzy numbers (Bortolan and Degani, 1985; Fortemps and Roubens, 1996; Liou and Wang, 1992; Saade and Schwarzlander, 1992). A relatively simple computation and easily understood method proposed by Liou and Wang (1992) is considered in this study. Compared with other approaches, this method has the advantage of being relatively simple to compute and allows the decision-maker to decide his/her own level of optimism. Other ranking order methods to manipulate the fuzzy programming model such as Yager indexes, the α-preference index, the Baas-Kwakernaak method, the Jain method, Adamo index and so forth (Bortolan and Degani, 1985; Cadenas and Verdegay, 2000) are also applicable to the problem.

Definition 2.3 *Let $\tilde{A}$ be a triangular fuzzy number with membership function (1). Suppose that $g_{\tilde{A}}^{L}(y)$ and $g_{\tilde{A}}^{R}(y)$, $y \in [0,1]$, are the inverse function of $\mu_{\tilde{A}}^{L}(x)$ and $\mu_{\tilde{A}}^{R}(x)$, $x \in \Re$, respectively, then the left integral value of $\tilde{A}$ is defined as*

$$I_L(\tilde{A}) = \int_0^1 g_{\tilde{A}}^{L}(y)dy$$

and the right integral value of $\tilde{A}$ is defined as

$$I_R(\tilde{A}) = \int_0^1 g_{\tilde{A}}^{R}(y)dy .$$

Recall that $g_{\tilde{A}}^{L}(y)$ and $g_{\tilde{A}}^{R}(y)$ are the inverse functions of $\mu_{\tilde{A}}^{L}(x)$ and $\mu_{\tilde{A}}^{R}(x)$, respectively, and are formulated as

$$g_{\tilde{A}}^{L}(y) = a_L + (a - a_L)y \text{ and } g_{\tilde{A}}^{R}(y) = a_R + (a - a_R)y \text{ where } y \in [0,1].$$

Thus, we have

$$\begin{aligned} I_L(\tilde{A}) &= \int_0^1 g_{\tilde{A}}^{L}(y)dy \\ &= \int_0^1 [a_L + (a - a_L)y]\,dy = \frac{1}{2}(a_L + a), \end{aligned}$$

and

$$\begin{aligned} I_R(\tilde{A}) &= \int_0^1 g_{\tilde{A}}^{R}(y)dy \\ &= \int_0^1 [a_R + (a - a_R)y]\,dy = \frac{1}{2}(a + a_R) \end{aligned}$$

The left integral value $I_L(\tilde{A})$ and the right integral value $I_R(\tilde{A})$ of a triangular fuzzy number $\tilde{A}$ are geometrically significant, i.e. they are the areas of trapezoids *OLPQ* and *ORPQ* respectively (Figure 2).

Definition 2.4 *If $\widetilde{A}$ is a triangular fuzzy number with membership function (1), then the total λ-integral value of $\widetilde{A}$ with index of optimism $\lambda \in [0,1]$ is defined as*

$$I_T^\lambda(\widetilde{A}) = \lambda I_R(\widetilde{A}) + (1-\lambda) I_L(\widetilde{A}) = \frac{1}{2}[\lambda a_R + a + (1-\lambda) a_L] \qquad (2)$$

where $I_R(\widetilde{A})$ and $I_L(\widetilde{A})$ are the right and left integral values of $\widetilde{A}$, respectively.

The left integral value is used to reflect the pessimistic viewpoint and the right integral value is used to reflect the optimistic viewpoint of the decision-maker. The total λ-integral value is a convex combination of right and left integral values through an index of optimism.

Remark *When $\widetilde{A}$ is a crisp number, denoted by (a, a, a), then*

$$I_T^\lambda(\widetilde{A}) = \lambda I_R(\widetilde{A}) + (1-\lambda) I_L(\widetilde{A}) = \frac{1}{2}[\lambda a + a + (1-\lambda) a] = a$$

It is observed that the total integral value of a non-fuzzy number A is a for all values of λ.

Definition 2.5 *Let $S = \{\widetilde{A}_1, \widetilde{A}_2, \cdots, \widetilde{A}_n\}$ be a set of convex fuzzy numbers, and that ranking function R $R: S \to \Re$ is a mapping from S to the real line $\Re$. For any distinct $\widetilde{A}_i, \widetilde{A}_j \in S$, the ranking function has the following properties (Liou and Wang, 1992; Cadenas and Verdegay, 2000)*

1) $R(\widetilde{A}_i) < R(\widetilde{A}_j)$ *implies* $\widetilde{A}_i < \widetilde{A}_j$,
2) $R(\widetilde{A}_i) = R(\widetilde{A}_j)$ *implies* $\widetilde{A}_i = \widetilde{A}_j$,
3) $R(\widetilde{A}_i) > R(\widetilde{A}_j)$ *implies* $\widetilde{A}_i > \widetilde{A}_j$.

In this study, the total λ-integral value of fuzzy numbers is used as the ranking function. Under a given level of optimism $\lambda \in [0,1]$, fuzzy numbers can be ordered by comparing their total integral value with their λ- value. That is, for any distinct $\widetilde{A}_i, \widetilde{A}_j \in F(\Re) \subset S$, we use the following criteria to rank a fuzzy number based on Definition 2.5:

1) $I_T^\lambda(\widetilde{A}_i) < I_T^\lambda(\widetilde{A}_j)$ implies $\widetilde{A}_i < \widetilde{A}_j$ ($\widetilde{A}_i$ is smaller than $\widetilde{A}_j$),
2) $I_T^\lambda(\widetilde{A}_i) = I_T^\lambda(\widetilde{A}_j)$ implies $\widetilde{A}_i = \widetilde{A}_j$ ($\widetilde{A}_i$ is equal to $\widetilde{A}_j$),
3) $I_T^\lambda(\widetilde{A}_i) > I_T^\lambda(\widetilde{A}_j)$ implies $\widetilde{A}_i > \widetilde{A}_j$ ($\widetilde{A}_i$ is greater than $\widetilde{A}_j$).

Obviously, the ranking function for fuzzy numbers is affected by the decision-maker's degree of optimism, which is represented by the parameter

$\lambda \in [0,1]$. A larger λ specifies a higher degree of optimism. For instance, when $\lambda = 1$, the total integral value $I_T^1(\widetilde{A}) = I_R(\widetilde{A})$ represents an optimistic decision-maker's point of view. On the other hand, when $\lambda = 0$, the total integral value $I_T^0(\widetilde{A}) = I_L(\widetilde{A})$ indicates a pessimistic decision-maker's point of view. When $\lambda = 0.5$, $I_T^{0.5}(\widetilde{A}) = \frac{1}{2}(I_R(\widetilde{A}) + I_L(\widetilde{A}))$ indicates a moderately optimistic decision-maker's point of view and is well qualified to be a defuzzification of the fuzzy number $\widetilde{A}$.

3. FUZZY MODEL FOR A CROSS-BORDER LOGISTICS PROBLEM WITH UNCERTAINTY

Here we develop the two fuzzy models for a cross-border problem with imprecise demand and imprecise cost components. Under the company's current logistics strategy, the road network consists of three routes: route 1, connecting Dongguan and Shenzhen; route 2, connecting Shenzhen and Hong Kong; and route 3, connecting Hong Kong and Dongguan (Figure 3).

In this study, similar to other local manufacturers who have moved their production base to China, the company may operate private lorries *i*, $i \in I^1$, China lorries *i*, $i \in I^2$ that can operate only in China, and/or HK lorries *i*, $i \in I^3$ that can operate on both sides of the border. This gives $I = I^1 \cup I^2 \cup I^3$. According to an optimal schedule plan, these lorries may travel route *r*, $r \in R = \{1,2,3\}$ with *n*-th round-trips, $n \in N$ restricted by a working time limitation *T*. For cost effectiveness, a routing and schedule plan involves six working days, *p*, $p \in P$, in which the decision-maker determines the daily volume of products to be delivered by various lorries to fulfill sales demand, D^p, in Hong Kong. Let x_{irn}^p be the volume of products loaded on lorry *i* which chooses route *r* on *n*-th round at period *p*. Let y_{irn}^p be 0-1 variable (binary variable) and equal to 1 if lorry *i* is operated to deliver goods via route *r* on *n*-th round at period *p* ($y_{irn}^p = 1$ whenever $x_{irn}^p > 0$) and 0 if otherwise ($y_{irn}^p = 0$ whenever $x_{irn}^p = 0$).

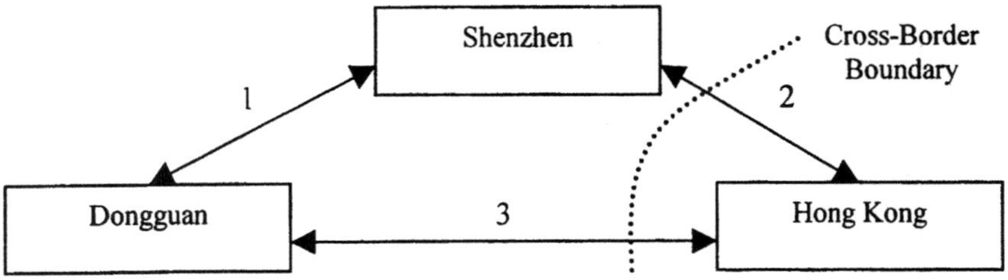

Figure 3. Road Network Structure.

3.1 Fuzzy Model for a Cross-Border Logistics Problem with Imprecise Demand and Crisp Cost Components

3.1.1. Notations

Parameters:

S^p = the volume of products supplied from the factory in Dongguan at period p

$\widetilde{D}^p$ = the fuzzy volume of products required by the sales outlet in HK at period p

q_i = the capacity of lorry i

u^0_{DG} = the initial volume of products stored in the Dongguan warehouse

v^0_{HK} = the initial volume of products stored in the HK warehouse

oc_{ir} = the operating cost associated with lorry i choosing route r

hc_i = the hiring cost associated with lorry i

ic_{HK} = unit inventory cost at the HK warehouse

ac = additional payment to drivers making more than one cross-border round-trip a day

t_r = the time spent on one round trip via route r

T = maximum number of driver hours

M = maximum daily volume of products produced in the factory in Dongguan.

Variables:

x^p_{irn} = the volume of products loaded by lorry i which travels via route r on n-th round at period p

$$y^p_{irn} = \begin{cases} 1, & \text{if lorry } i \text{ is operated to deliver the goods choosing route } r \text{ on } n-\text{th round at period } p, \\ 0, & \text{otherwise.} \end{cases}$$

u^p_{DG} = the volume of products stored in the Dongguan warehouse at period p

v_{HK}^{p} = the volume of products stored in the HK warehouse at period p

ε_{i}^{p} = the number of round trips over 1.

It is noted that $i \in I^2 \cup I^3$ are lorries hired on a per-order basis. In this model, only one round trip is considered for all hire lorries (HK lorries and China lorries), which gives y_{ir1}^{p} for all $i \in I^2 \cup I^3$.

3.1.2 Objectives

In the literature, a considerable number of papers consider transportation costs that consist of different cost components such as travel time, trip cost and so on, and are formulated into a single linear function that has to be minimized. Alternatively, however, the decision-maker might be interested in multiple-objective decision-making, in which the cost components are optimized at weighting and preemptive levels using goal programming, interactive programming and so on. In this study, the transportation costs take the linear form:

Transportation costs = Operating costs + Hiring costs + Inventory costs + Additional costs

The corresponding cost components are described below.

Operating costs:

These costs are associated with fuel, maintenance, ticket fine subsidies, loading costs, etc. on each route. The determination that has the greatest relation with other transportation problems is based on the volume of products to be transported using route *r* by lorry *i*, $i \in I^1$. $i \in I^2 \cup I^3$ are lorries hired on a per-order basis, hence no operating costs are incurred. The mathematical expression for these operating costs is:

$$\sum_{p \in P} \sum_{i \in I^1} \sum_{r \in R} \sum_{n \in N} oc_{ir} x_{irn}^{p} \tag{3}$$

where oc_{ir} is the operating cost associated with lorry *i* traveling via route *r*.

Hiring costs:

These costs are associated with hiring a HK lorry to deliver the products directly from Dongguan to Hong Kong or hiring a China lorry to deliver products from Dongguan to Shenzhen. It is supposed that each hired lorry makes one round trip for each order, which gives n=1. The cost component is formulated as:

$$\sum_{p \in P} \sum_{i \in I^2 \cup I^3} \sum_{r \in R} hc_{i} y_{ir1}^{p} \tag{4}$$

where hc_i is the hiring cost associated with lorry *i*, $i \in I^2 \cup I^3$.

Inventory costs:

Inventory costs are associated with the costs incurred for the storing of excess products in Hong Kong warehouses. The formula is:

$$ic_{HK}\sum_{p\in P}v^{p} \tag{5}$$

where ic_{HK} is the unit inventory cost at the Hong Kong warehouse.

Additional costs:

The formula for the allowance provided to drivers who make more than one cross-border round trip:

$$ac\sum_{i\in I^1}\sum_{p\in P}\varepsilon_i^p \tag{6}$$

where ac is the additional allowance given to drivers who make more than one cross-border round trip.

3.1.3 Constraints

The first model for a cross-border logistics problem assumes that demand is imprecise and characterized as a triangular fuzzy number, and that the cost components are free from uncertainty as defined above. Constraints are formulated in the model as follows:

$$S^p \le M, \qquad \forall p\in P, \tag{7}$$

$$x_{irn}^p \le q_i y_{irn}^p, \qquad \forall i\in I, r\in R, n\in N, p\in P, \tag{8}$$

$$\sum_{r\in R}\sum_{n\in N} t_r y_{irn}^p \le T, \qquad \forall i\in I^1, p\in P, \tag{9}$$

$$\sum_{r\in R}\sum_{n\in N} y_{irn}^p - \varepsilon_i^p \le 1, \qquad \forall i\in I^1, p\in P, \tag{10}$$

$$S^p + u^{p-1} - \sum_{i\in I}\left(x_{i1}^p + x_{i3}^p\right) - u^p = 0, \qquad \forall p\in P, \tag{11}$$

$$v^{p-1} + \sum_{i\in I}\left(x_{i2}^p + x_{i3}^p\right) - v^p = \widetilde{D}^p, \qquad \forall p\in P, \tag{12}$$

$$x_{irn}^p, y_{irn}^p, \varepsilon^p, u^p, v^p \ge 0, \qquad \forall i\in I, r\in R, n\in N, p\in P. \tag{13}$$

Constraint (7) ensures that the amount of products delivered at period p cannot exceed the maximum amount of manufactured products produced in the factory in Dongguan. Constraint (8) ensures that $y_{irn}^p = 1$ rather than zero whenever $x_{irn}^p > 0$ and $y_{irn}^p = 0$ when $x_{irn}^p = 0$ because of the nature of the objective function. Hence, Constraint (8) ensures that the volume of products carried by lorry i cannot exceed the lorry's capacity. Constraint (9) ensures that the operating hours of drivers cannot exceed each driver's maximum number of working hours. Constraint (10) determines the number of round trips minus 1 for each private lorry. It gives that $\varepsilon_i^p = 0$ if $\sum_{r\in R}\sum_{n\in N} y_{irn}^p \le 1$, and $\varepsilon_i^p = \sum_{r\in R}\sum_{n\in N} y_{irn}^p - 1$ if $\sum_{r\in R}\sum_{n\in N} y_{irn}^p > 1$.

Constraint (11) determines the volume of products stored in the Dongguan warehouse given that u^0 is the initial volume of products stored. Constraint (12) determines the volume of products stored in the Hong Kong warehouse given that v^0 is the initial volume of products stored. $\widetilde{D}_p$ is the fuzzy demand required by the sales outlet in Hong Kong at period p and is characterized as a triangular fuzzy number. For instance, the outlet demand requires about 1,600 units and is characterized as (1,400, 1,600, 2,000). The equality relation in Constraint (12) is the fuzzy relation ranking fuzzy numbers (Campos and Verdegay, 1989). It is noted that if the fuzzy numbers in Constraint (12) become crisp numbers, the relation becomes the conventional equality between real numbers.

The next procedure to solve the proposed model for the cross-border logistics problem is to transform the fuzzy programming model to a linear programming model. Using ranking function, Constraint (12) is reformulated to a crisp linear constraint with respect to a given decision-maker's degree of optimism, λ, as:

$$I_T^{\lambda}\left(v^{p-1} + \sum_{i\in I}\left(x_{i2}^{p} + x_{i3}^{p}\right) - v^{p}\right) = I_T^{\lambda}(\widetilde{D}^{p}),$$
$$\forall p \in P, \tag{14}$$

It is reminded that $I_T^{\lambda}(\bullet)$ is a real number. As long as the cost coefficients and parameters are linear, the optimal solution can be easily obtained using the simplex method. Many other packages such as LINDO (Linear Interactive and Discrete Optimization) can also solve the problem efficiently (Evans, 1993).

3.2 Fuzzy Models for Imprecise Demand and Vague Cost Components

This section considers the cross-border logistics problem with imprecise demand and cost components. By the extension principle of fuzzy sets, the operating costs in (3) will induce the following fuzzy quantities (convex fuzzy sets):

$$\sum_{p\in P}\sum_{i\in I^1}\sum_{r\in R}\sum_{n\in N}\widetilde{oc}_{ir}\, x_{irn}^{p} \tag{15}$$

where $\widetilde{oc}_{ir}$ are the fuzzy operating costs associated with lorry i traveling via route r.

Hiring costs in (4) will become:

$$\sum_{p\in P}\sum_{i\in I^2\cup I^3}\sum_{r\in R}\widetilde{hc}_{i}\, y_{ir1}^{p} \tag{16}$$

where $\widetilde{hc}_i$ is the fuzzy hiring cost associated with lorry i, $i \in I^2 \cup I^3$.

Inventory costs in (5) will be formulated as:

$$\tilde{ic}_{HK} \sum_{p \in P} v^p \tag{17}$$

where $\tilde{ic}_{HK}$ is the fuzzy unit inventory cost at the Hong Kong warehouse.

The additional costs as in (6) become:

$$\tilde{ac} \sum_{i \in I^1} \sum_{p \in P} \varepsilon_i^p \tag{18}$$

where $\tilde{ac}$ is the fuzzy additional allowance given to drivers who make more than one cross-border round trip in one day.

Using ranking function, the proposed fuzzy programming with fuzzy demand and cost coefficients is reformulated to linear programming as:

Minimize

$$\sum_{p \in P} \sum_{i \in I^1} \sum_{r \in R} \sum_{n \in N} I_T^\lambda (\tilde{oc}_{ir}) x_{irn}^p +$$

$$\sum_{p \in P} \sum_{i \in I^2 \cup I^3} \sum_{r \in R} I_T^\lambda (\tilde{hc}_i) y_{ir1}^p$$

$$+ I_T^\lambda (\tilde{ic}_{HK}) \sum_{p \in P} v^p + I_T^\lambda (\tilde{ac}) \sum_{i \in I^1} \sum_{p \in P} \varepsilon_i^p \tag{19}$$

subject to (7) – (11), (13) and (14).

4. APPLICATION TO HONG KONG

In this section, the two models presented in the last section will be applied to a manufacturer in Hong Kong. Under the company's current logistics strategy, three private lorries (lorries 1–3) are operated to deliver the products using routes 2 and 3; six China lorries (lorries 4–9) use route 1; and two HK lorries use route 3 (lorries 10–11).

The operating costs incurred for the use of different routes by private lorries and the hiring costs incurred for the use of two types of hired lorry are shown in Table 1. It is noted that the capacities of a private lorry and a China lorry are same, and the capacity of a HK lorry is larger. The capacities of each of the lorries are also shown in Table 1. The inventory cost for the Hong Kong warehouse is \$0.67 per unit. By experience, it is observed that it takes five hours and eight hours to make a round trip on route 2 and route 3 by private lorry respectively, and it is known that private lorry drivers work 10 hours each day. Roughly, this gives two round trips on route 2 and one round trip on route 3 each day.

Table 1. Characteristics of Three Types of Lorry (US$1 = HK$7.8)

Lorry	Operating Costs			Hiring Costs			Capacity (Units)	Travel Time		
	Route 1	Route 2	Route 3	Route 1	Route 2	Route 3		Route 1	Route 2	Route 3
1	-	$2.18	$3.18	-	-	-	250	-	5 hrs	8 hrs
2	-	$2.18	$3.18	-	-	-	250	-	5 hrs	8 hrs
3	-	$2.18	$3.18	-	-	-	250	-	5 hrs	8 hrs
4	-	-	-	$350	-	-	250	-	-	-
5	-	-	-	$350	-	-	250	-	-	-
6	-	-	-	$350	-	-	250	-	-	-
7	-	-	-	$350	-	-	250	-	-	-
8	-	-	-	$350	-	-	250	-	-	-
9	-	-	-	$350	-	-	250	-	-	-
10	-	-	-	-	-	$1,800	450	-	-	-
11	-	-	-	-	-	$1,800	450	-	-	-

Note: '-' represents no possible trip made by the lorry with respect to the route

4.1 Results of a Single Period with Crisp Data

An example to illustrate the proposed model in a fixed period with 1,600 units of demand, $\tilde{D} = (1600,1600,1600)$ and crisp cost components is given below.

Objective:

1) Operating costs

$$2.18y_{121} + 2.18y_{122} + 3.18y_{131} + 2.18y_{221} + 2.18y_{222} + 3.18y_{231}$$
$$+ 2.18y_{321} + 2.18y_{322} + 3.18y_{331}$$

2) Hiring costs

$$350y_{411} + 350y_{511} + 350y_{611} + 350y_{711} + 350y_{811}$$
$$+ 350y_{911} + 1800y_{1031} + 1800y_{1131}$$

3) Inventory costs

$$0.67v$$

4) Additional costs

$$70\varepsilon_1 + 70\varepsilon_2 + 70\varepsilon_3$$

The routing and the flow of products with fleet management are illustrated in Figure 4. A breakdown of the costs is given in Table 2.

Table 2. The Optimal Solution in a Single Period

Test	Operating Costs	Hiring Costs	Inventory Costs	Additional Costs	Total Costs
One Period	$2,226	$3,600	$0	$0	$5,826
	Route 2[1]	Route 3[2]	Hired[3]	Average Cost[4]	
One Period	0	3	2	$3.64	

1: Number of private lorries using route 2, 2: Number of private lorries using route 3
3: Number of HK lorries hired 4: Average cost = Total transportation cost/Total products transported

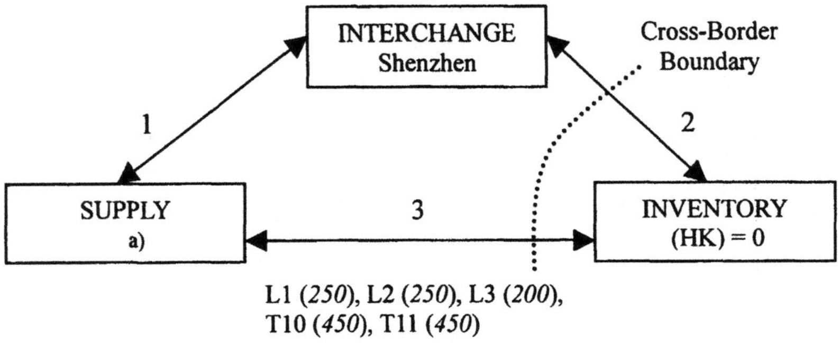

Figure 4. Geographical Presentation of One-period Result.

Note: L1 (*250*) represents lorry 1 transporting 250 units
Supply (Dongguan) represents the amount of products produced in the factory in Dongguan
Inventory (HK) represents the volume of products stored at the warehouse in Hong Kong

In this example, the average transportation cost per unit is \$5,826/1600 = \$3.64, and the inventory level in HK = 0. Private lorries 1, 2 and 3 are assigned to carry 250, 250 and 200 units of products respectively and two HK lorries are hired to carry the remaining 900 units of products using route 3.

4.2 Results of Six Periods with Imprecise Demand and Crisp Cost Components

For cost effectiveness, a weekly route and fleet composition are normally planned according to the daily market demand. Below we consider an optimal transportation planning for six working days. Table 3 shows a fuzzified range of market demand with the lower and higher values corresponding to a period of six working days. The notation of three values in the parentheses is defined in Section 2.

Table 3. Fuzzified Demand for Six Periods

Period	1	2	3
	(1,100,1,200,1,300)	(1,200,1,400,1,600)	(1,900,2,200,2,500)
Period	4	5	6
	(1,600,1,900,2,200)	(1,400,1,600,1,800)	(2,400,2,500,2,600)

It is assumed that periods 3 and 4 experience a greater fluctuation for each week, so the lower and higher limits are 300 units. The width of both limits are reduced by 100 units to 200 units in periods 2 and 5. Lastly, the widths of the start and the last periods are 100 units.

We input the fuzzified demands to the model (3)–(13) with various degrees of decision-maker optimism as shown in Table 3, the transportation costs versus the degree of optimism is plotted in Figure 5, and the actual values are presented in Table 4. It is noted that a non-increasing curve is obtained, which indicates that the objective value becomes larger as the value of λ increases because demand for the product in the market increases.

Table 4. Objective Value versus Degree of Optimism

Level of DM's Optimistic	1.0	0.8	0.6	0.4	0.2	0.0
Total Transportation Cost	37710.5	36804.6	35853.8	34860.8	33779.0	32953.5

Figures 6–9 show the various cost components versus the decision-maker's degree of optimism. It is noted that, although market demand for the products increases as λ increases, the trip costs are not strictly increasing (Figure 6). Hiring costs on the other hand increase dramatically (Figure 7). One explanation may be that the extra products are to be delivered by HK lorry. In Figure 8, the inventory costs tend upward as λ increases because market demand for the products is increasing, particularly in periods 3–6. However, only a maximum 2,300 units can be delivered from existing resources daily (500 units if the private lorries use route 2, and 450 units if the HK lorries use route 3). More products will be therefore transported in the earlier periods and stored in the warehouse to meet future demand. Approximately 8–10 trips are made by private lorry via route 2. Thus there is little variation in the additional costs (Figure 9).

A breakdown of the costs and the related output corresponding to λ = 1, 0.5 and 0 representing optimistic, moderate and pessimistic decision-maker outlooks respectively are listed in Table 5. The average cost slightly decreases as λ decreases.

Table 5. The Optimal Solutions in Six Period for Imprecise Demand

Test	Operating Costs	Hiring Costs	Inventory Costs	Additional Costs	Total Costs
$\lambda = 1$	$15,626	$21,500	$234.5	$350	$37,710.5
$\lambda = 0.5$	$15,149	$19,700	$134.0	$350	$35,333.0
$\lambda = 0$	$15,172	$17,200	$301.5	$280	$32,953.5
	Route 2	Route 3	Hire	Average Cost	
$\lambda = 1$	10	13	10	$3.70	
$\lambda = 0.5$	10	13	9	$3.68	
$\lambda = 0$	8	14	8	$3.66	

4.3 Results of Six Periods with Imprecise Demand and Imprecise Cost Components

Below, we consider the cross-border logistics problem with imprecise demand and imprecise cost components. Table 6 shows the fuzzified range of different cost components characterized as triangular fuzzy numbers. It is noted that the additional costs are assumed to be non-fuzzy number.

Table 6. Fuzzified Cost Components

Operating Costs (Route 2)	Operating Costs (Route 3)	Hiring Cost (China)
(1.68, 2.18, 3.18)	(2.68, 3.18, 4.18)	(300, 350, 450)
Hiring Costs (HK)	**Inventory Costs (HK)**	**Additional Costs**
(1,700, 1,800, 2,300)	(0.57, 0.67, 0.97)	(70, 70, 70)

In Figure 10, consistent with the findings in the previous section, it is noted that a non-increasing curve is plotted, which indicates that the objective value becomes larger as the value of λ increases because market demand for the product increases.

Table 7. Multi-period Optimal Solutions for Imprecise Demand and Cost Components

Test	Operating Costs	Hiring Costs	Inventory Costs	Additional Costs	Total Costs
$\lambda = 1$	$18,476.0	$24,500	$287	$350	$43,613.0
$\lambda = 0.5$	$15,870.5	$20,730	$144	$350	$37,094.5
$\lambda = 0$	$15,596.0	$14,150	$248	$770	$30,764.0
	Route 2	Route 3	Hire	Average Cost	
$\lambda = 1$	10	13	10	$4.28	
$\lambda = 0.5$	10	13	9	$3.86	
$\lambda = 0$	22	7	4	$3.42	

A breakdown of the output is presented in Table 7. It can be seen that as λ decreases, hiring costs drop dramatically, but trip costs do not. This is explained by the increase in the number of trips using route 2 made by private lorry, which consequently raises the hiring costs for China lorries. Moreover, the average transportation costs per unit drops significantly as λ decreases, in particular in comparison with the results obtained that are shown in Table 5.

5. CONCLUSION

In this study, two new models are presented for application to multi-period cross-border logistics problems. In the first model, the cost components are crisp but demand is fuzzy. In the other, the costs and demand are considered fuzzy. In both, the fuzzy uncertainties are characterized by triangular fuzzy numbers. Consequently, the resulting objective is fuzzy and optimization is achieved through a fuzzy ranking function with respect to the total integral values of these numbers. These models were applied to a manufacturer in Hong Kong. Decision-makers can arrive at an appropriate transportation schedule with routing and vehicle composition through a λ range varying from 0 to 1 representing the level of optimism of the decision-maker concerned.

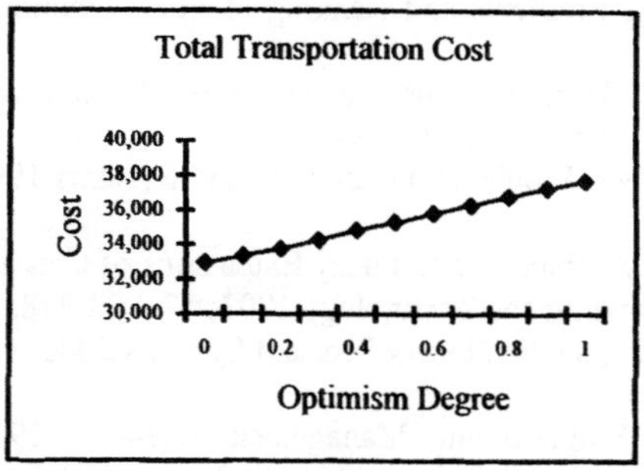

Figure 5. Total Transportation Costs versus Degree of Optimism.

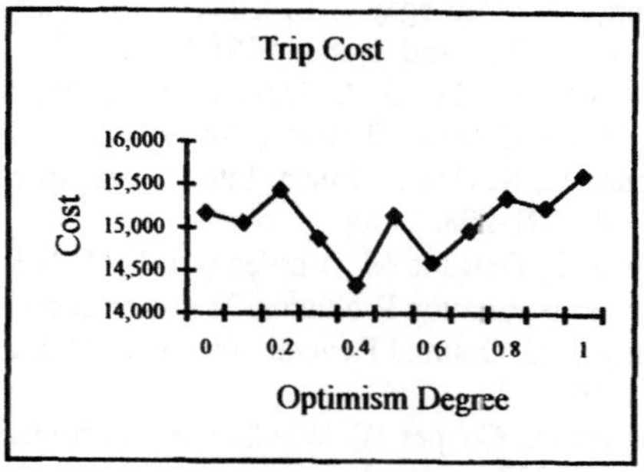

Figure 6. Trip Costs.

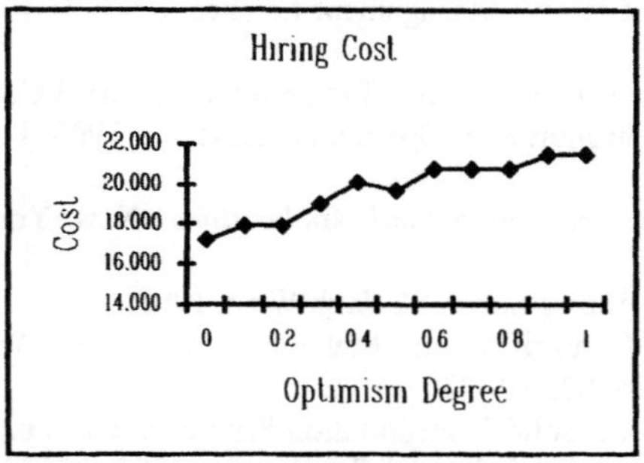

Figure 7. Hiring Costs.

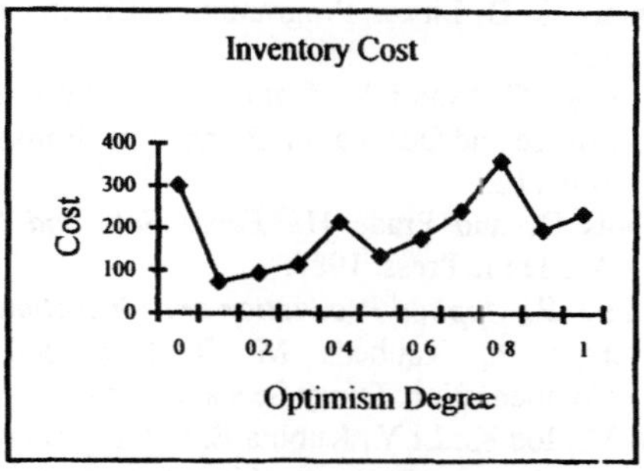

Figure 8. Inventory Costs.

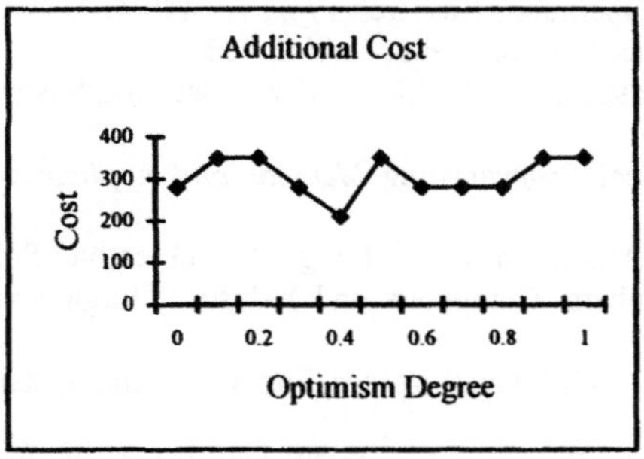

Figure 9. Additional Costs.

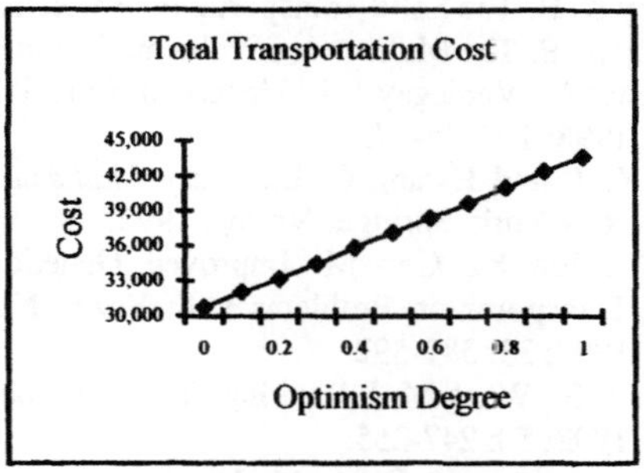

Figure 10. Total Transportation Costs versus Degree of Optimism.

References

Bortolan G., Degani R. A Review of Some Methods for Ranking Fuzzy Subsets. Fuzzy Sets and Systems 1985; 15: 1-19.

Cadenas J. M., Verdegay J. L. Using Ranking Functions in Multiobjective Fuzzy Linear Programming. Fuzzy Sets and Systems 2000; 111: 47-53.

Campos L., Verdegay J. L. Linear Programming Problems and Ranking of Fuzzy Numbers. Fuzzy Sets and Systems 1989; 32: 1-11.

Chanas S., Kuchta D. A Concept of the Optimal Solution of the Transportation Problem with Fuzzy Cost Coefficients. Fuzzy Sets and Systems 1996; 82: 299-305.

Chanas S., Kuchta D. Fuzzy Integer Transportation Problems. Fuzzy Sets and Systems 1998; 98: 291-298.

Chanas S., Delgado M., Verdegay J. L., Vila M. A. Interval and Fuzzy Extensions of Classical Transportation Problems. Transportation Planning and Technology 1993; 17: 203-218.

Chang S. C. Optimal Revenue for Fuzzy Demand Quantity. Fuzzy Sets and Systems 2000; 111: 399-417.

Charnes D., Cooper W. W. Chance-Constrained Programming. Management Science 1959; 5: 197-207.

Cheung R. K. M., Powell W. B. Models and Algorithms for Distribution Problems with Uncertain Demand. Transportation Science 1996; 30: 43-59.

Dantzig G. B. Linear Programming Under Uncertainty. Management Science 1995; 1: 197-206.

Das S. K., Goswami A., Alam S. S. Multiobjective Transportation Problem with Interval Cost, Source and Destination Parameters. European Journal of Operational Research 1999; 117: 100-112.

Dubois D. and Prade H., *Fuzzy Sets and System Theory and Applications*. New York: Academic Press, 1980.

Evans J. R., *Applied Production and Operations Management*. St. Paul: West, 1993.

Fortemps P., Roubens M. Ranking and Defuzzification Methods Based on Area Compensation. Fuzzy Sets and Systems 1996; 82: 319-330.

Gen M., Ida K., Li Y., Kubota E. Solving Bicriteria Solid Transportation Problems with Fuzzy Numbers by a Genetic Algorithm. Computers and Industrial Engineering 1995; 29: 537-541.

Haley K. B. The Solid Transportation Problem. Operations Research 1962; 10: 448-463.

Haley K. B. The Multi-index Problem. Operations Research 1963; 11: 448-463.

Jimenez F., Verdegay J. L. Uncertain Solid Transportation Problems. Fuzzy Sets and Systems 1998; 100: 45-57.

Lai Y. J. and Hwang C. L., *Fuzzy Mathematical Programming Methods and Applications*. New York: Springer-Verlag, 1992.

Li Y., Ida K., Gen M. Improved Genetic Algorithm for Solving Multiobjective Solid Transportation Problems with Fuzzy Numbers. Computers and Industrial Engineering 1997; 33: 589-592.

Liou T. S., Wang M. J. Ranking Fuzzy Numbers with Integral Values. Fuzzy Sets and Systems 1992; 50: 247-255.

Kaufmann A. and Gupta M. M., *Introduction to Fuzzy Arithmetic Theory and Application*. New York: Van Nostrand Reindold, 1991.

Saade J. J., Schwarzlander H. Ordering Fuzzy Sets Over the Real Line: An Approach Based On Decision Making Under Uncertainty. Fuzzy Sets and Systems 1992; 50: 237-246.

Teodorovic D. Fuzzy Sets Theory Applications in Traffic and Transportation. European Journal of Operational Research 1994; 74: 379-390.

Teodorovic D. Fuzzy Logic Systems for Transportation Engineering: The State of the Art. Transportation Research 1999; 33A: 337-364.

Zadeh L. A. Fuzzy Sets. Information Control 1965; 8: 338-252.

Zimmermann H.–J. Fuzzy Programming and Linear Programming with Several Objective Functions. Fuzzy Sets and Systems 1978; 1: 45-55.

Zimmermann H.–J., *Fuzzy Set Theory – And Its Applications*. London: Kluwer Academic Publishers, 1996.

THE DSS ARCHITECTURE BASED ON NON-MATHEMATICAL PROBLEMS SPECIFICATION AND MODEL/SOLVER INDEPENDENCE

Tsuyoshi Mukohara and Yasuki Sekiguchi

Dept. of Business Administration, Hokkaido University
Kita 9 Nishi 7, Kita-ku, Sapporo 060-0809, Japan
muko@pop.econ.hokudai.ac.jp, seki@econ.hokudai.ac.jp

Abstract The proposed DSS architecture includes (1) specifying decision problems with an extended entity-relationship model 'GERM', (2) distinguishing a user-defined model from a standard model, (3) bridging a problem specification and its user-defined model through 'unification condition', and (4) bridging a user-defined model and a standard model through 'connection information'. A prototypal DSS with the proposed architecture helps a decision-maker having insufficient mathematical knowledge to communicate with OR/MS experts and to select and invoke an adequate solver for a decision problem.

Keywords: Decision Support System, Model Management, Modeling, Optimization, Mathematical Programming

1. INTRODUCTION

For decision-makers or OR/MS practitioners ('users', hereafter) who do not understand mathematical models ('models') sufficiently, solving confronted decision problems in the real world ('problem') by using mathematical programming applications ('solvers') is not easy. Even if OR/MS experts support users, it is desirable that users can invoke a suitable solver and solve a problem instance only after specifying the problem in the real world (without explicitly developing a mathematical model). Sekiguchi et al. (2000) and Bao (1998) propose a method to describe a problem in an extended Entity-Relationship Model (GERM problem specification, hereafter), and a method to combine GERM problem specifications and corresponding mathematical models. This paper proposes the architecture of a decision support system (DSS), which applies these proposals and transforms mathematical models linked with GERM problem specifications into numerical data in the format that a solver can accept. With a system based on the architecture, users can invoke a suitable solver from a GERM problem specification, which is a non-mathematical description of a real-world problem. The system also improves reusability of problem specifications, mathematical models and solvers.

This paper is structured as follows: Section 2 gives a summary of preceding research on OR/MS-based DSS focusing on both problems specification and solver invocation. The proposed architecture is outlined and a prototypal system based on the architecture is illustrated in Section 3. Some terminology is also introduced there. The succeeding two chapters are devoted to explaining key ideas of the proposed architecture. Section 4 introduces the extended Entity-Relationship Model called GERM. Section 5 introduces the concepts of user-defined models and standard models, proposes distinguishing them, and explains their relations. Section 6 summarizes the contributions of this paper.

2. PRECEDING RESEARCH AND MOTIVATION

Supporting the whole process of the modeling life cycle is an important issue in the decision support system (DSS) field. However, Geoffrion (1987) pointed out that doing OR/MS tends to be a low productivity activity because the life cycle scope of the computer-based model management system is narrow. From this point of view, modeling languages are the most successful products in this field.

A solver can be invoked from a model described by an algebraic modeling language such as AMPL (Fourer, 1993), GAMS (Brooke et al., 1996), and MPL (Maximal Software, 1994). Model description by these modeling languages does not depend on a specific built-in solver. Although solvers available with these modeling languages are not very many in number, several solvers may be used to solve a model described with these modeling languages. We call this property model/solver independence. These modeling languages realize some degree of model/solver independence. Because of this property, users do not need to change or modify a model description when they exchange a solver into another one, if the new solver is applicable to the modeling language used for describing the model. However, solvers available with a modeling language are restricted to only a few that solve general mathematical programming models such as Linear Programming (LP), Integer Linear Programming (ILP) and Quadratic programming (QP). The range of problems possible to be described with a modeling language is restricted to some part of these MP models depending on the solvers that are planned to use. Thus, these modeling languages cannot deal with any other type of models and solvers.

Model/data independence is another property of modeling languages, and implies that a model description is clearly separated from detailed data used to instantiate it. Model/data independence is important because it allows detailed data instantiating a model to be modified in format, values, dimensions etc., without causing a change in the model description. AMPL alone among these modeling languages provides model/data independence. Structured Modeling (SM) is a framework for modeling and is based on the Entity-Relationship model. The Structured Modeling Language (SML, Geoffrion,1992a, b) was proposed based on SM. SML is fully implemented and is available in an environment called FW/SM and promotes a disciplined approach to modeling. LSM (Chari et al., 1993) is a prototypal modeling language partially implemented in Prolog. Both languages support model/data independence and model/solver independence. LSM also supports meta-level model representation and reasoning. SM enables model-based work to be done with greater productivity and acceptance by non-specialists (Geoffrion, 1987). However, information about qualitative characteristics of a real problem, which are reflected on the structure of its mathematical model, is treated as additional one in SM.

SML and LSM are applicable to plural modeling paradigms (Geoffrion, 1986). Some different architecture enabling a modeling language to treat plural modeling paradigms has been proposed. Abbass et al. (2001) proposes the architecture to integrate different paradigms (i.e., Operation Research and Artificial Intelligence) using Constraint Logic Programming-like meta language. Amatürk et al. (2000) discusses an ILP modeling system which uses relational algebra for activities related to modeling. The system combines model management with database management. ASCEND (1997), although it does not have a meta language, can deal with equation models as well as optimization models. However, it seems difficult to realize model/data independence because the object-orientation tends to unite model description with data.

Bhargava et al. (1993) proposes a method to embed a modeling language into a more general language by using the first order predicate logic. The embedding language makes it possible to represent various non-mathematical information about models such as dimensions and units of variables, the source of the model, and so on. However, the non-mathematical information is simply added onto a mathematical model. This information does not support users and OR/MS experts in constructing a model. Available solvers for a model vary depending on the embedded language.

Yet these modeling languages are far from supporting the entire modeling life cycle. To realize support for the entire modeling life cycle and for non-expert users, it is important to start with support for problems specification. However, such DSSs like this must cope with the fact that users and OR/MS experts do not know a resulting mathematical model while specifying a decision problem to be solved.

Besides, a mathematical model can be equivalently represented into various forms. Thus a mathematical model built for a problem specification and represented in a form does not necessarily correspond evidently with the target mathematical model for the available solver. In order to invoke the available solver from a problem specification, two challenging issues are remaining: One is establishing correspondence between the mathematical model for a problem specification and the target mathematical model for its solver, and the other is transforming numerical data given at the phase of problem specification into solver input in its specified form.

Wu (2000) proposes a model management system that includes the phase of problem analysis. This system is devoted for a specific problem of test construction, and its architecture is not easily applied forDSSs that support the solution process of a wider range of decision problems.

The present paper develops a new architecture where a target solver can be invoked starting from a problem specification. The architecture makes it possible to realize model/data independence and a higher degree of model/solver independence.

3. THE PROPOSED ARCHITECTURE

3.1 An Outline of the Proposed Architecture

The term 'model' is used variously in the OR/MS field. The present paper uses a 'model type' or a 'model' to indicate a mathematical model with some parameters and a 'model instance' or an 'instance' to indicate a numerical example of a model type. Similarly, a 'problem type' and a 'problem' imply a type of problems that includes some unknown parameters in it, and a 'problem instance' and an 'instance' imply a numerical example of a problem type.

Each step in the modeling life cycle has a suitable expression of models or problems (Jones, 1996). From a view point of establishing a support over the modeling life-cycle, the authors of the present paper propose to distinguish the following four types of descriptions of a decision problem: (1) a non-mathematical problem description (GERM which is explained in Section 4), (2) a mathematical model which is developed to mathematically represent the problem in a GERM (user-defined model; UDM), (3) a standard mathematical model that is the target problem for a solver (standard model; SDM), and (4) input data to a solver in the format in which an available solver accepts it (solver model). Notice that a UDM is developed from a GERM and represents a user's view of a decision problem, while a SDM is defined from a solver and represents a view of developers of a solver.

From the viewpoint of model/data independence in model management, the proposed architecture also distinguishes the model type level and the instance level. The architecture is illustrated in Figure 1. It shows how to relate four descriptions stated above and their instances in a DSS system. Through the relations, a suitable solver for a confronted decision problem described in a GERM is found, then instance data of a numerical example is prompted, and the data is accepted and transformed into input data and is fed to the solver for finding a solution.

The architecture is based on databases that store and manage the four types of descriptions and their relationships. The relationship between a GERM and its UDM is described in 'unification condition'(Bao, 1998), that is actually a list of parameters and variables in a GERM combined to respective symbols in its UDM. The relation between a UDM and the SDM of an applicable solver is described in 'connection information' that is a collection of transformation rules of symbols in the UDM into symbols in the SDM. The relation between a SDM and a solver model is rather straightforward and has already established in such software as matrix generators and model translators for modeling languages such as MPL, AMPL and GAMS (Fourer et al., 1990). A modeling language usually provides model translators for a few solvers, and a SDM and its instance described in a modeling language can be solved without any change even if a different solver is used.

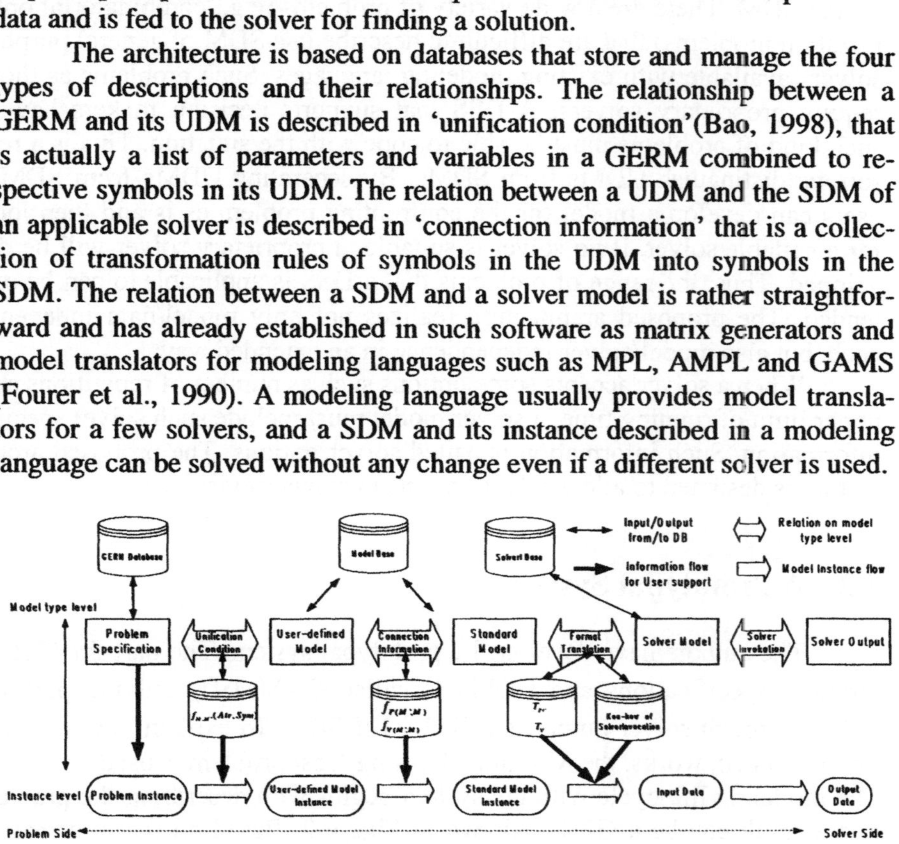

Figure 1: The Proposed System Architecture.

This feature is called 'model/solver independence', which usually means that a SDM is clearly separated from its solver model. The importance of this feature comes form, at least, two reasons. First, the representation used by a solver is often optimized for efficient solution. They are therefore inconvenient for verifying and validating a model formulation, and for communicating it to users and OR/MS experts. Second, if solver models and SDMs are clearly separated, a solver may be changed (or improved) without heavy burden on modeling side, or a description of a decision problem in a SDM may be changed to improve it or to adapt it to the change of the real world without heavy burden on the solver side.

However, the present stage of model/solver independence is not sufficient for establishing the relationship between UDMs and SDMs. In traditional modeling systems based on modeling languages, describing a model in a computer 'executable' form is a fundamental discipline. To realize the discipline, a modeling language is designed for a SDM and is able to describe only a restricted range of UDMs that are transformable into the SDM. However, users do not know in advance whether their problems can be described in the SDM. There are a wide variety of problems (e.g., combinatorial optimization problems) that are difficult to describe in a SDM of general purpose solvers available with existing modeling languages. Such problems as those require proprietary solvers. A DSS that supports decision makers having such kind of problems must be able to cope with the situation. This is a reason we distinguish UDMs from SDMs. By separating UDMs from SDMs, users can develop a model for the confronting problem as is and then look for a suitable solver. If no solver is suitable, a proprietary solver will be developed. Thus, the range of problems that a DSS is applicable to can be extended. The proposed architecture realizes not only model/data independence, but also model/solver independence in an extended way.

When a solver accepts some options such as number of repetitions and upper limit of running time, a solver model must include such solver specific information. Such information is called solver options. The proposed architecture is designed to allow solver options in solver model.

3.2 A Prototypal System

The authors have developed a prototypal system. The part for GERM problem specifications and UDMs is called□CAMMS□, and the part for UDMs through solver output is called □GERSIS□. To explain how the prototypal system works, the Minimize-Drilling-Cost problem is used.

The modeling life cycle with the system begins with describing the problem and a result can be a GERM shown in Figure 2. Based on the information put in the GERM, a UDM is developed. The GERM may be improved through collaboration between users and OR/MS experts. After development of the UDM, the unification condition is input to the system (Figure 3). Comparing Figure 3 to Figure 2, it is clear that locations of holes are coded as x[i] and y[i], and a distance between a pair of holes is calculated by the last equation in the UDM.

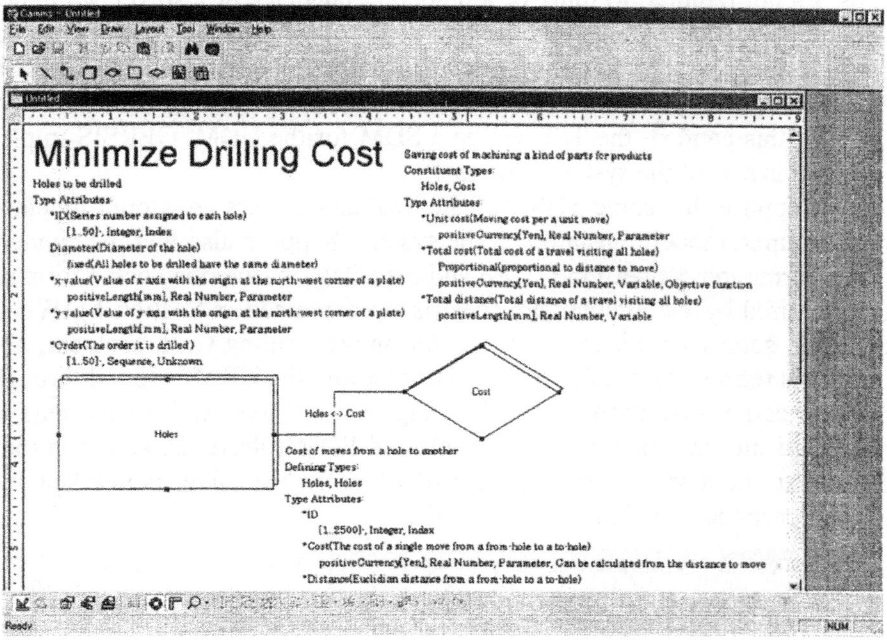

Figure 2: The GERM of the Minimize-Drilling-Cost Problem

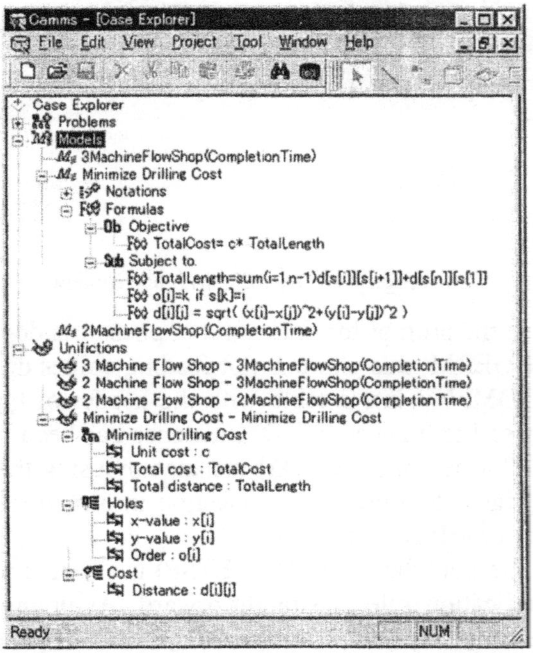

Figure 3: A UDM and Unification Condition of the Minimize-Drilling-Cost Problem.

UDMs do not have to be executable. Therefore, users and OR/MS experts are not required to think of a modeling language or a solver while developing a UDM. Fortunately, the UDM (Formulas in Figure 3) is nothing but a variation of the travelling salesman problem (TSP). Hence, if a dedicated TSP solver or a solver applicable to it is available, the UDM is solvable. In this context, the TSP can be a SDM for the UDM. GERSIS implements this part of the system (Figure 4).

Suppose that some SDMs of the TSP and solvers are stored together with required model translators in the system. Suppose also that the connection information between the UDM and the SDMs is stored in such a form as one required by the architecture (details are explained in Section 5). When the user solves an instance of the Minimize-Drilling-Cost problem, the GERM is retrieved from the GERM database and the UDM is found through the unification condition (phase I in Figure 4). Then, its instance data is translated into instance data of the selected SDM (phase II), and it is then translated into a solver model (phase III). Finally, the solver is invoked and the solver model is fed into it (phase IV).

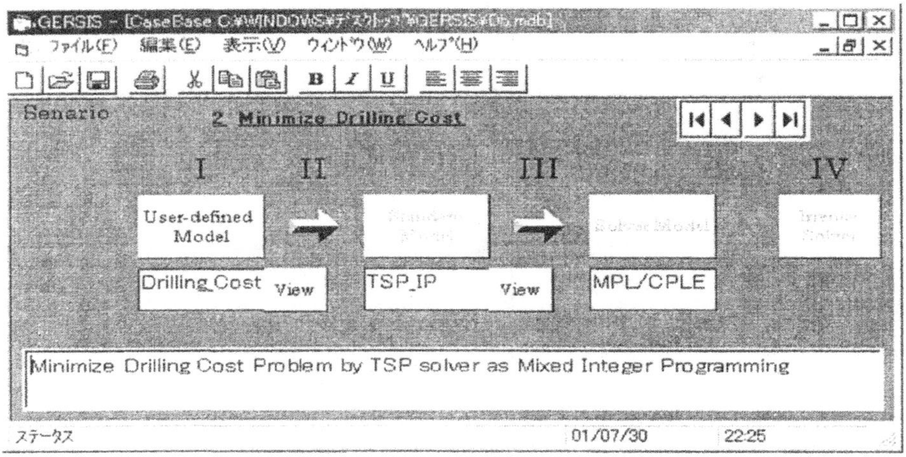

Figure 4. The GERSIS Main Window.

Because the prompt for data input at phase I is done in a style closely related to the GERM, this phase is not feeding values directly into the symbols in the UDM, but into a table with the header of type attribute names. When the phase I is finished, a UDM instance has been built. The UDM instance is transformed into the SDM instance by using the parameter specification rule (Figure 5) stored in 'connection information'. Then, CPLEX (a built-in solver of MPL) is invoked.

When the Euclidian type TSP (ETSP) is selected as the SDM, the parameter specification rule is straightforward. Assuming that GERSIS has already stored the connection information, GERSIS generates ETSP solver model only by selecting a ETSP solver. By following the similar process as one in the case of CPLEX, a ETSP is invoked and solves the problem instance.

The example above shows that the UDM is transformed into two types of SDMs. It is noteworthy that CPLEX can also solve other problems (for example, the assignment problem and the transportation problem), and that the ETSP solver is also applicable to any ETSP equivalent problems. The SDMs of these solvers may be connected to the UDMs of these problems. Suppose, for example, that the GERM of another real problem is modeled into the same UDM as the Minimize-Drilling-Cost problem. It may be unified with the UDM. It seems easy to handle these cases in the proposed architecture, and add required unification conditions and connection information to the databases in the system.

```
SQL2 = □SELECT * FROM T_Distance□
Form1.Data2.RecordSource = SQL2
Form1.Data2.Refresh
Set Rs2 = Form1.Data2.Recordset
Dim disx As Long
Dim disy As Long
Dim dis As Long
    Do Until Rs.EOF
  Rs2.AddNew
  Rs2![From_ID] = Rs![T_Position.ID]
  Rs2![To_ID] = Rs![T_Position_1.ID]
  disx = CLng(Rs![T_Position.Pos_X]) - CLng(Rs![T_Position_1.Pos_X])
  disy = CLng(Rs![T_Position.Pos_Y]) - CLng(Rs![T_Position_1.Pos_Y])
  dis = CLng(Sqr(CDbl(disx * disx + disy * disy)))
  If dis = 0 Then dis = 9999999
      Rs2![Distance] = dis
      Rs2.Update
  Rs.MoveNext
 Loop
 Rs.Close
 Rs2.Close
```

Figure 5. A Parameter Specification Rule.

4. GERM PROBLEMS SPECIFICATION

Entities and relationships are called objects or object instances, and entity types and relationship types are called object types. The basic assumption underlying our proposal is the following: A decision problem can be defined by describing object types, instances and their inter-relationships in the world of discourse. The Generic Entity-Relationship Model (GERM) as shown in Figure 2 is the proposed style to describe decision problems in the real world. This section explains extensions and specializations necessary for the Entity-Relationship Model (ERM) to be expanded to the GERM.

4.1 Type and instance attributes

Object types have a set of names and values (TAttrib and TAValue, cf. Table 1) of type attributes. An identifier (ID) is a type attribute and its value indicates a rule for assigning an ID value to each instance. A value of a type attribute includes three kinds of fields: InstAttrib, QuantitativeChar and QualitativeChar. A value of InstAttrib may be 'Yes' or 'No'. 'Yes' implies that the type attribute becomes an attribute for instances of the object type (i.e., instance attribute) as well as describes a common or a set-theoretic property of instances included in the object type. A value of Quantitative-Char expresses a numerical characteristic of the object type, and it consists of domain (the value set allowed for instance attributes), type (integer, real, date, set, sequence, etc.), status (identifier, constant, parameter, variable, decision variable, etc), dimension [unit] and supplement. When an instance attribute takes a numerical value, the corresponding type attribute has a value to QuantitativeChar. A qualitative characteristic of an object type is described as an explanation to a name given to the characteristic and makes a value of QualitativeChar. The value of QualitativeChar may include plural names and explanations. When an instance attribute expresses a qualitative property, the corresponding type attribute has values to QualitativeChar. Every name of object types or type attributes is attached a quiddity (Explanation) that explains the real-world meaning of an object type.

Table 1. List of objects, object types and generic types in GERMs

Classification	Description
Entity type	(Name, Explanation, TAttrib, TAValue, GName)
Relationship type	(Name, Explanation, DefiningTypes, Relations, Corresp, MinMaxCardi, TAttrib, TAValue, GName)
Entity	(Name, ID, Attrib, AValue)
Relationship	(Name, ID, defID, Attrib, AValue)
System type	(Name, Explanation, ConstituentTypes, Fixed, TAttrib, TAValue, GName, InstanceSets)
System	(Name, Case#, Attrib, AValue)
Generic entity type	(Name, Explanation, TAttrib, TADomain)
Generic relationship type	(Name, Explanation, GDefiningTypes, GRelations, Gcor-resp, GMinMaxCardi, TAttrib, TADomain)
Generic system type	(Name, Explanation, GConstituentTypes, TAttrib, TA-Domain)

(Example 1) Lists of names and values of type attributes for two object types are shown in Figure 2. Type attributes of the Holes-object type is illustrated above the doubled rectangle of holes. Asterisks attached to some type attributes indicate that InstAttrib's of the object types have a value 'Yes'. Diameter has a value named 'fixed' to QualitativeChar. X-value has a value to QuantitativeChar consisting of positive (domain), Length [mm] (dimension [unit]), real number (type) and parameter (status).

The context of description of a type attribute can be classified into two categories: characteristics of a single instance and inter-relationships of instances. The ordinary ERA in information systems analysis (see Leonard, 1992, for example) focuses its analysis to type attributes that become instance attributes, and characteristics of each instance are described as its domain while inter-relationships between instances are described by some integrity conditions. The domain is a part of QuantitativeChar (or explanation in QualitativeChar). An integrity condition describes a logical consistency of values as a data set in an information system. Inter-relationships described by type attributes in the above proposal describe a property of the world of discourse where a decision problem exists, and includes a wider range of inter-relationships than the logical consistency among data. For example, the diameter-type attribute is not described in ERM because it does not have values in the data set of holes. It is noteworthy that type (and instance) attributes having values to QuantitativeChar may become parameters or variables in a mathematical model while those with values to Qualitative-Char may affect the structure of a mathematical models and a solution algorithm.

4.2 Entity types and relationship types

Entity types and relationship types in our proposed approach do not differ conceptually from traditional entity and relationship types, description of their type attributes however has several extensions and specializations to them as introduced in the above. An entity type is a named set of type attributes. The name of the set is the name of the entity type. A relationship type is an object type defined by a set of other object types, that is, entity types and relationship types. Similarly to the ERM, some object types may be defined either as an entity type or as a relationship type, and which to take depends on the judgement of the analysts. The object types that define a relationship type are called its defining types. Correspondence relations such as 1:1, 1:M and M:N are usually described to each defining type on E-R diagrams in the ER Approach. Problems specification requires far more precise description of defining relation, and requires us to additionally designate the minimum cardinality, the maximum cardinality and a keyword to each defining type. The minimum (maximum) cardinality is the minimum (maximum) number of instances of the relationship type that are defined by an instance of the defining type (Lazimy (1988)). A keyword is a label that shows the role of the defining type in definition of the relationship type. The set of these values to all defining types are expressed as Corresp, MinMax-Cardi, and Relations, respectively, in Table 1.

(Example 2) The Minimize-Drilling-Cost problem in Figure 2 consists of two object types: holes and cost. The holes-entity type is defined by an explanation and five type attributes. The cost-relationship type is defined by (a pair of) the holes-entity type (although it is not shown explicitly in the figure). The holes-entity type has the value of the minimum (maximum) cardinality equal to the number of holes and a keyword 'from', and another one has the similar value of the minimum (maximum) cardinality and a keyword 'to' (, but they are not shown in the figure). This may be understood that every 'from' hole has each hole in the holes-entity type as its 'to' hole and vice versa. Note that the rectangle and the diamond except their names are only for illustration and are not essential for the problem specification.

4.3 System types and problem types

Two kinds of object types can be observed in a world of discourse: one is object types whose instance sets change from instance to instance of the decision problem and another one is object types whose instance sets do not change over all instances of the decision problem. Both object types in Figure 2 are of the former kind. An object type of this kind is said to be 'open'. However, the machine-entity types, which are observed in the world of discourse of the sequencing problems, are examples of the latter kind. An object type of this kind is said to be 'fixed'. Generally speaking, type attributes of object types and instance attributes of fixed object types determine the structure of a mathematical model, and instance attributes of open object types become its parameters. When an instance set of every open object type is fixed, parameter values in a mathematical model are determined and a numerical instance of the problem is determined. An entity is a named set of its ID, instance attributes (Attrib) as indicated by InstAttrib and their values (AValues). A relationship is a similar named set with an additional set (defID) of IDs of instances of its defining types (cf. Table1).

A GERM describing a decision problem consists of entity types, relationship types and instance sets of fixed entity and relationship types. Note that the only way for an entity type to have a relationship with other object types is to define a relationship type. If an entity type in a GERM does not define any relationship type, the entity type is isolated from other object (types) and cannot affect a decision problem related to other object (types).

[Proposition 1 (Unity)] Every entity type in a GERM describing a decision problem defines one or more relationship types.

Suppose that a relationship type in a GERM is defined by some object types which are not included in it. This means that the GERM does not hold the whole information necessary to describing a decision problem.

[Proposition 2 (Self-consistency)] All defining types of every relationship type in a GERM describing a decision problem are included in the GERM.

A GERM satisfying the above two propositions and having additionally its name and a set of type attributes like entity types is called a 'system type'. One of type attributes of a GERM is a set of constituent types (ConstituentTypes), that is, the list of names of object types included in the GERM. A system type whose constituent types have type attribute(s) with status 'decision variable' is called a 'problem type'. When we observe the world of discourse of an instance of a decision problem, every object type has their instance set. Therefore, each object type may be either fixed or open. It depends on recognition of decision problems in the world of discourse. Whether a constituent type is fixed or open must be described in the □Fixed□ field (cf. Table 1). A problem type corresponds to a set of problem instances that are generated by assigning an instance set satisfying their type attributes to every open object type. InstanceSets represent the sets of instances of all constituent types.

4.4 Generic object types

In this section, the term 'object type' includes system types as well. Decision problems are extremely diverse. Compared to it, the diversity of managerial resources and activities described as object types in decision problems is far less. Variations of some object types appear repetitively in many decision problems. The diversity of decision problems represents the diversity of characteristics and combinations of some object types. If examples of descriptions of characteristics and combinations of object types that appear in GERMs are classified and stored in a form convenient for reuse, they will become effective information for supporting problem specification process of a new decision problem in a similar world of discourse. We propose to do this by constructing 'generic' object types while building a database of GERMs. Please refer to Sekiguchi et al. (2000) for detailed definitions of generic types.

Generic types of entity, relationship and system types store the know-how of description of respective object types in the real world. A new object type enriches information in generic types. There may be cases where new object types in a problem specification are described only by selecting generic types and a combination of values to type attributes. These are the reasons why we call our model description a GERM. For example, after information of the Minimize-Drilling-Cost problem has been added to generic types, specifying similar problems where diameter of holes vary or cost of drilling is not proportional to tour length would be very easy. (However, developing their mathematical models may not be easy.) The prototypal system explained in the preceding section makes it possible. For decision makers and/or practitioners who have a decision problem but are not experts in the

OR/MS field, such non-mathematical specifications as those illustrated in Figure 2 seem useful in communicating with OR/MS professionals and in understanding mathematical models of their problems (cf. Figure 3 of the UDM and the unification condition).

5. USER-DEFINED MODELS AND STANDARD MODELS

Even if a user has identified the suitable solver for a UDM and such a solver has been prepared, users should arrange or transform instance data of the UDM into one of the SDM for the target solver because of differences between UDMs and SDMs. Although the transformation concerns instance data, it is realized by formulating relations between symbols in a UDM and those in a SDM before instance data is prepared. At least, relations of symbols representing parameters and decision variables between these models must be formalized. The relation of parameters is called 'parameter specification rule', that of decision variables is called 'solution transformation rule', and their combination is called 'connection information'. The essential part of connection information is to arrange values for symbols in a SDM by scaling and/or indexing.

5.1 Mathematical feature of connection information

Parameter specification rules should satisfy the following proposition.
[Proposition 3] All the parameters of the SDM must be specified prior to invocation of an available solver.

The set of values for parameters in a SDM is a part of its solver model. It is evident that a solver cannot solve an instance of a UDM if values of some parameters in the corresponding SDM are unknown. Now let $P = \{p_1, p_2, \ldots, p_m\}$ be the set of symbols for parameters in a SDM, and let $P' = \{p'_1, p'_2, \ldots, p'_n\}$ be the set of symbols for parameters in the corresponding UDM. Then, the projection $f_P : P' \to P$ exists according to proposition 3. If the projection is a bijection, then this type of parameter specification rule is called a 'standard type'. If the projection is an injection and not a surjection, the set $P - f(P')$ of parameters whose values are not determined by the projection must be included in the set of constants of the UDM. This type of parameter specification rule is called a 'special type'. Lastly, if the projection is neither standard nor special, it is called a 'derivation type'. These projections define parameter specification rules.

Similarly to this, solution transformation rules should satisfy the following proposition.

[Proposition 4] Values of all decision variables of a UDM must be calculated from parameters and variables in the respective SDM if a solver to the SDM is applicable to the UDM.

Although the solver output is the set of values of decision variables in the target SDM, users require knowing the values of decision variables in the UDM. If there are some decision variables that are not determined, the UDM has not been solved. Therefore, the necessity of the proposition for a solver to be applied for the UDM is evident. Now, let $V=\{v_1,v_2,...,v_m\}$ be the set of decision variables in the SDM and $V'=\{v'_1,v'_2,...,v'_n\}$ be the set of decision and induced variables in the UDM. Then, the projection $f_V:V\rightarrow V'$ must be a surjection, if it is a solution transformation rule.

[Proposition 5] If the set of symbols for parameters and/or decision variables in a UDM is indexed, the index sets must be properly related to index sets of the corresponding SDM.

Let $P'=\{p'_i;i\in I\}$ be the set of symbols for parameters in a UDM, $P=\{p_j;j\in J\}$ be that of its SDM, we assume that a bijection function $f_I:I\rightarrow J$ (an indexing rule) exits. Similarly, let $V=\{v_i;i\in I\}$ be the set of decision variables in the SDM and $V'=\{v'_j;j\in J\}$ be the set of decision or induced variables in the UDM. We assume that an indexing rule for V and V' is a bijection also.

To help understanding the mathematical features of connection information, the following examples are illustrated.

$$Minimize\quad Z=\sum_{i\in J}\sum_{j\in J} dis_{(i,j)}Rou_{(i,j)}$$

$$Subject\ to\ \sum_{i\in J} Rou_{(i,j)}=1,\ \text{for each } j\in J$$

$$\sum_{j\in J} Rou_{(i,j)}=1,\ \text{for each } i\in J$$

$$\sum_{i\in S}\sum_{j\notin S} Rou_{(i,j)}\geq 1,\ S\subseteq J, S\neq\varnothing, S\neq J$$

$$Rou_{(i,j)}\in\{0,1\},\ \text{for each } i,j\in J$$

Where Z: Objective

$dis_{(i,j)}$: Parameter

$Rou_{(i,j)}$: Decision variable

J: Index set,

S: Subset of J

Figure 6: The SDM of the TSP

(Example 3) The UDM is one for the Minimizing-Drilling-Cost problem (Figure 3), and the SDM is the TSP (Figure 6). Then,

$$\text{UDM: } P'=\{(x[i],y[i],c);i\in I\}$$

SDM: $P=\{dis_{(i,j)};i,j\in J,i\neq j\}$, $V=\{Rou_{(i,j)};i,j\in J,i\neq j\}$

The following parameter specification rule may be used.

$$f_I : I \to J \text{ (bijection)}$$

$$f_P : \{(x[i],y[i],x[j],y[j],c);i,j\in I,i\neq j\}\to\{dis_{(f_I(i),f_I(j))}\},$$

where,

$$f_P(x[i],y[i],x[j],y[j],c)=\sqrt{(x[i]-x[j])^2+(y[i]-y[j])^2}*c$$

This is a 'derivation type'.

Similarly,

$$f_V : \{i_k;i_k\in J,k=1,\ldots,n\}\to\{s[k];s[k]\in I\}$$

$$\text{where } i_1=1,\ Rou(i_{k-1},i_k)=1 \text{ for } k=2,\ldots,n,$$

$$\text{and } s[k]=f_V(i_k)=f_I^{-1}(i_k)$$

Thus, it is confirmed that the above propositions are satisfied by this case. The above rule is implemented as shown in Figure 5.

Minimize $Z=\sum_{i\in J}\sum_{j\in J}dis_{(i,j)}Rou_{(i,j)}$

Subject to $\sum_{i\in J}Rou_{(i,j)}=1$, for each $j\in J$

$\sum_{j\in J}Rou_{(i,j)}=1$, for each $i\in J$

$\sum_{i\in S}\sum_{j\notin S}Rou_{(i,j)}\geq 1$, $S\subseteq J,S\neq\varnothing,S\neq J$

$Rou_{(i,j)}\in\{0,1\}$, for each $i,j\in J$

$dis_{(i,j)}=\sqrt{(Pos\_x(i)-Pos\_x(j))^2+(Pos\_y(i)-Pos\_y(j))^2}$, for each $i,j\in J$

Where

Z: Objective

$dis_{(i,j)}$: Induced variable

$Rou_{(i,j)}$: Decision variable

$Pos\_x(i)$: Parameter

$Pos\_y(i)$: Parameter

J: Index set

S: Subset of J

Figure 7: The Euclidian Type TSP

(Example 4) The UDM is one for the Minimizing-Drilling-Cost problem (Figure 3), and the SDM is the Euclidian type TSP (Figure 7). Then,

SDM: $P=\{(Pos\_x[i],\ Pos\_y[i]);i\in J\}$, $V=\{Rou_{(i,j)};i,j\in J,i\neq j\}$

The following parameter specification rule may be used.

$$f_I : I \to J \text{ (bijection)}$$

$$f_P : \{(x[i],y[i],c);i\in I\}\to\{(Pos\_x[f_I(i)],Pos\_y[f_I(i)])\},$$

where

$$f_P(x[i],y[i],c)=(x[i]*c,\ y[i]*c).$$

This is a 'standard type' because this function can be inversed. Solution transformation rule f_V is similar to that of Example 3. Thus, this case also satisfies the propositions 3 to 5.

The three propositions do not guarantee equivalency of two models, but they are minimum conditions for a SDM and its solver to be applicable to solution of a UDM.

6. CONCLUSION

This paper proposes a new DSS architecture that enables invocation of solvers from non-mathematical problem description. The architecture realizes a DSS that transforms the GERM problem specification and its instances into the solver model and then invokes a suitable solver. By specifying correspondence relations between four descriptions (i.e., GERM problem specifications, UDMs, SDMs and solver models) of a decision problem by unification condition, connection information, and model translators, a suitable solver can be invoked automatically from a GERM problem specification. The advantages of this architecture are as follows. First, because a problem specification is non-mathematical, it is hospitable for users who do not have deep mathematical knowledge. Second, by distinguishing UDMs from SDMs and by using connection information, the variety of available solvers and combinations of UDMs and solvers has in-creased. Third, because unification condition allows many to many relations between GERMs and UDMs, a decision problem may be solved by using more variety of models and solvers in the DSSs based on the proposed architecture than in those based on modeling languages. The effectiveness of the proposed architecture was confirmed by developing a prototypal system (CAMMS and GERSIS) and by applying it to a problem. An extension of this prototypal system is now underway.

There is still room for further investigations. The burden for OR/MS experts and/or the solver developers is heavy in the proposed architecture, because the model transformation procedures are not built automatically. The architecture should become more intelligent by introducing, for example, inference mechanisms for transformations.

7. ACKNOWLEDGEMENTS

This research was partially supported by the Ministry of Education, Science, Sports and Culture, Grant-in-Aid for Scientific Research No.(C)(2), 09680412.

References

Abbass H. A., Towsey M., Finn G., Kozan E. Meta-Representation for Integrating OR and AI in an Intelligent Decision Support Paradigm. International Transactions of Operational Research 2001; 8: 107-119

Atamtürk A., Johnson E.L., Linderoth J.T. Savelsbergh M.W.P. A Relational Modeling System for Linear and Integer Programming. Operations Research 2000; 48: 846-857

Bao J. A Study of Case-base System for Problem Specification and Model Building. <in Japanese>, Ph.D. Dissertation, Faculty of Economics, Hokkaido University, 1998

Bhargava H.K., Kimbrough S.O. Model Management: An Embedded Language Approach. Decision Support Systems 1993; 10: 277-299

Brooke A., Kendrick D., Meeraus A. *GAMS Release 2.25 A USER'S GUIDE*. GAMS Development Corporation, 1996

Chari S., Krishnan R. Toward a Logical Reconstruction of Structured Modeling. Decision Support Systems 1993; 10: 301-317

Fourer R., Gay D.M., Kernighan B.W. A Modeling Language for Mathematical Programming. Management Science 1990; 36: 519-554

Fourer R. *AMPL: Modeling Language for Mathematical Programming*. Belmont, CA: Duxbury Press, 1993

Geoffrion A.M. Integrated Modeling Systems. Working paper No. 343, Western Management Science Institute, University of California 1986; 1-16.

Geoffrion A.M. An Introduction to Structured Modeling. Management Science 1987; 33:547-588

Geoffrion A. M. The SML Language for Structured Modeling: Level 1 and 2. Operations Research 1992a; 40: 38-57

Geoffrion A.M. The SML Language for Structured Modeling: Level 3 and 4. Operations Research 1992b; 40: 58-75

Jones C.V. *Visualization and Optimization*. Dordrecht, The Netherlands: Kluwer Academic Publishers, 1996

Lazimy R. "Knowledge Representation and Modeling Support in Knowledge-Based Systems." In *Entity-Relationship Approach*, S.T. March ed. Elsevier Science Publishers B.V.; 1988: 133-161

Leonard M. *Database Design Theory*. London: Macmmillan Education, 1992

Maximal Software Inc. *MPL User Manual*. Maximal Software Inc., 1994

Sekiguchi Y., Bao J., Mukohara T. Formal Theory of the Generic Entity-Relationship Model (GERM) for Problems Specification. Proceedings of the 2nd Joint International Workshop - ORSJ Hokkaido Chapter and ASOR Queensland Branch -; 2000: 183-190

The Research Group of A. Westerberg ASCEND IV Advanced System for Computations in Engineering Design - A Portable Mathematical Modeling Environment Release 0.8. Department of Chemical Engineering, Carnegie Mellon University, 1997

Wu I.L. Model Management System for IRT-Based Test Construction Decision Support System. Decision Support Systems 2000; 27: 443-458

HYPER SENSITIVITY ANALYSIS OF PRODUCTIVITY MEASUREMENT PROBLEMS

L. Churilov[1] and M. Sniedovich[2]

[1] *School of Business Systems, Monash University, Vic, 3800, Australia*

[2] *Department of Mathematics and Statistics, The University of Melbourne, Vic, 3010, Australia*

Abstract In this paper we introduce a method for conducting a Hyper Sensitivity Analysis (HSA) of productivity and efficiency measurement problems. HSA is an intuitive generalization of conventional sensitivity analysis where the term "hyper" indicates that the sensitivity analysis is conducted with respect to functions rather than numeric values. The concept of HSA is suited for situations where several candidates for the function quantifying the utility of (input, output) pairs are available. Both methodological and technical issues arising in the area of multiple criteria productivity measurement in the context of such an analysis are examined.

Keywords: hyper sensitivity analysis, productivity, efficiency, DEA, multiple objective programming, composite concave programming.

1. INTRODUCTION

Over the years various aspects of *productivity* and *efficiency* management have been the focus of extensive research efforts (Davis (1978), Fried et al (1993), Fuentes et al (1998), Coelli et al (1998), and Avkiran (1999). Traditionally, the area of *productivity management* concerns itself with a variety of management issues such as *productivity planning*, *measurement and evaluation*, *control*, and *improvement*. The *productivity* of a production unit is historically defined as the ratio of the output(s) it produces to the input(s) it uses (see Fried et al (1993), Coelli et al (1998)).

As demonstrated in Fried et al (1993), productivity varies due to differences in production technology, differences in the efficiency of production process, and differences in the environment in which production occurs. Typically, certain efforts are taken to isolate the *efficiency* component and to measure its contribution to productivity. Fried et al (1993) determine the *efficiency* of a production unit as a comparison between observed and optimal values of its output and input. This comparison can take the form of the *ratio* of observed output to maximum potential output obtainable from the given input, or the ratio of minimum potential input to observed input required to produce given output, or some combination of the two.

In these two comparisons the optimum is defined in terms of production possibilities, and efficiency is *technical.* If, in addition, information on prices is available, and a behavioral assumption, such as cost minimization, or profit maximization, is appropriate, it is possible to consider *allocative* efficiency, in addition to *technical* efficiency.

As discussed by Coelli et al (1998), allocative efficiency in input selection involves selecting that mix of inputs (e.g., labour and capital) which produce a given quantity of output at a minimum cost, given the input prices which prevail. *Allocative* and *technical* efficiency combine to provide an overall economic efficiency measure.

According to Fried et al (1998, p.4), the answers to the following questions lay the foundation for successful productivity and efficiency measurement:

Q#1: How many and which outputs and inputs should be included in the analysis?

Q#2: How should the output and inputs be weighted in the aggregation process?

Q#3: How should the potential of the production unit be determined?

It should be noted that from a decision making perspective, whatever the answers to these questions could be, the very way the questions are posed unequivocally suggests the *multiattribute* and *multiobjective* nature of this situation. Various attempts to answer some or all of these questions were made by Caves et al (1982), Sink (1989), Fried et al (1998), Coelli et al (1998), Fuentes et al (1998), and Avkiran (1999). Fried et al (1998) suggest that the majority of research in the area of productivity and efficiency measurement is rooted in either *econometric* or *mathematical programming* approaches. The two approaches are different in many ways, but the essential differences, and the sources of the advantages of one approach or the other, boil down to two characteristics: the *stochastic* and *parametric* nature of the *econometric* approach vs the *deterministic* and *nonparametric* nature of the *mathematical programming* approach. An excellent survey of both these approaches can be found in Fried et al (1998) and Coelli et al (1998). Extensive surveys of practical applications of various productivity and efficiency measurement techniques are presented in Coelli et al (1998) and Avkiran (1999).

No attempt is made here to address Q#1-3. Rather, a Hyper Sensitivity Analysis framework is suggested that provides *methodological links between,* and *technical enchancements* (such as Composite Concave Programming) *to,* a variety of existing answers to abovementioned questions. This framework facilitates a collaboration between *Multiple Objective Mathematical Programming* (Steuer (1986)), *Composite Linear Programming* (Sniedovich (1986); Macalalag and Sniedovich (1996)), and *Data Envelopment Analysis* (Charnes et al (1986), Schrage (1997), and Avkiran (1999)).

The analysis relies on the *key assumption* that the decision maker is required to somehow quantify the "input-output" options available, in order to choose the "best" mix of "inputs" and "outputs" in some respect, but does not have *a priori* suitable means for doing so. Since this is indeed the case in the majority of productive efficiency measurement problems, the necessity of such a quantification may bring about several candidates for a real-valued *utility function* combining together the "input" and "output" terms. Note that the exact form of a "correct" utility function, i.e. one representing the decision maker's choice most accurately, is *unknown* - otherwise the difficulty of lack of means for quantification would not have arisen in the first place. In particular, situations are considered where the decision maker has a variety of objective functions rather than a single one. The problem is to be optimized with regard to each of them. Hence, sensitivity analysis is conducted with respect to changes in optimal solutions in response to changes in the objective function. On the other hand, in most productivity measurement applications, some properties of such a function may follow directly from the decision maker's analysis of the underlying economic model.

The *objective* of this paper is to present *Hyper Sensitivity Analysis* as a possible tool for handling such situations, that is as a tool for solving "input-output" productivity measurement problems in general, and efficiency measurement problems in particular, where, for various reasons, more than one utility function is under consideration.

This paper is organized as follows: Section 2 presents the formulation of the generalized "input-output" productivity measurement problem; Sections 3 and 4 are dedicated to an overview of known results on using *Composite Concave Programming* as a tool for conducting a *Hyper Sensitivity Analysis (HSA)* procedure (Section 3) and earlier developed various technical tools of Concave Composite Linear Programming nature that can be used for solving large-scale productivity and efficiency measurement problems (Section 4). In Section 5 some aspects of the relationship between Data Envelopment Analysis and Multiobjective Optimization are discussed from *HSA* perspective; while Section 6 summarizes the analysis as a whole.

2. PROBLEM FORMULATION

A variety of authors, including Sink (1989), Fried et al (1993), and Coelli et al (1993) provide a comprehensive analysis of a traditional "partial" productivity measurement approach based on single factor ratios. The single factor ratios approach is based on ratio measures of a combination of financial and throughput data, and is frequently encountered in accounting practice. The required data is usually collected for internal management purposes or as part of the requirement of the funding agency, which makes ratio measures comparatively easy to obtain. Despite their availability, single factor partial measures cannot be successfully used in the context of productivity measurement (see Sink (1989)). This is due to the fact that these measures are subjected to a variety of influences, such as factor substitutions, efficiency changes in the entire process, and demand for product and/or service fluctuations. In short, a single ratio measure cannot fully capture the *multi-input, multi-output* nature of the firm's business process. Fried et al (1993) also discuss these limitation but describe some situations where partial productivity and efficiency measures can be utilized.

The abovementioned shortcomings in productivity and efficiency measurement can be overcome by using the following "multiple - factor - productivity" approach. It is rooted in both "total-factor-productivity" measurements discussed in detail in Sink (1989) and *Multiple Objective Programming* (Steuer (1986)). The "total-factor-productivity" approach is based on the idea of calculating the set of *weights that represent the relative value of the inputs used and outputs produced.* Then the productivity score is introduced as some suitable combination of weighted inputs and weighted outputs as demonstrated further in this paper.

Consider the generalized version of the "input-output" productivity measurement problem. Let $m \in N$ be the number of inputs used by the system, $s \in N$ be the number of outputs produced. Let $c \in \Re_+^m$ be the vector of *inputs* used by the system, and let $r \in \Re_+^s$ be the vector of *outputs* produced. Let $\Re_+$ denote the nonnegative part of $\Re$, i.e. $\Re_+ := [0; \infty)$ and let $w \in \Re_+^{m+s}$ denote the vector of weights whose first m components represent the relative weights of inputs used, and last s components represent the relative weights of outputs produced by the system. Let total weighted input combination required by the system,

u_1, and total weighted output combination produced by the system, u_2, be defined as follows:

$$u_1(w) := \sum_{i=1}^{m} c_i w_i, \; u_2(w) := \sum_{j=1}^{s} r_j w_{m+j}. \tag{1}$$

Then, according to the "total-factor-productivity" approach, the productivity score for the system can be determined as some properly constructed real valued function of $u_1(w)$ and $u_2(w)$, say $g(u_1(w), u_2(w))$.

The role of the generalized "input-output" problem is to simplify the analysis for productivity measurement problems discussed in the previous section. The simplification is achieved by highlighting the essential features of the "input-output" family of the problems and leaving behind the details less relevant to this analysis.

The most important feature of any given "input-output" productivity problem is the decision maker's preference of a large total output ($u_2(w)$) and a small total input ($u_1(w)$). In other words, the decision maker is interested in "producing more for less". According to these preferences, some family G of utility functions $g(u_1(w), u_2(w))$ quantifying the options available, may be introduced. These functions have to satisfy the essential properties of the "input-output" productivity measurement problems, that is they are to be monotonically increasing with "total output" $u_2(w)$ and monotonically decreasing with "total input" $u_1(w)$.

Introduction of the family of functions $g \in G$ reflects the fact that there can be more than one natural choice for a combination of u_1 and u_2, depending on the underlying management objectives. For example, one of the most common nonparametric techniques used for measuring technical efficiency of firms is *Data Envelopment Analysis (DEA)* (Coelli et al (1993), Fried et al (1993), and Schrage (1997)). *DEA* compares the actual outputs produced and inputs used by a group of firms in a common time period. The *DEA* efficiency score is defined as the *ratio* of weighted sum of the outputs to the weighted sum of inputs, i.e. the inducing utility function is defined as

$$g(u_1(w), u_2(w)) := \frac{u_2(w)}{u_1(w)}. \tag{2}$$

Similar ratio type utility function applies in the case of the other important productivity score commonly used in management - *Malmquist index*. Caves et al (1982) demonstrated that the *Malmquist index* corresponds to the *ratio* of two distance functions. Examples of *Malmquist indexes* obtained using parametric linear programming distance functions can be found in Bjurek (1994) and Althin et al (1996).

Often in management and actuarial practice there is a necessity to apply various discounting procedures, such as discounting for time or discounting for the value of resources to ensure that the real value of resources to various parties is correctly identified. These issues also should be taken into account when deriving productivity measures, and, therefore in a variety of situations, one might derive the index based on, for example, standard output mix $u_2(w)$ and discounted input mix $a(1-e^{-\frac{u_1(w)}{a}})$, where $a > 0$ is a properly chosen scaling factor. In this case the following holds:

$$g(u_1(w), u_2(w)) := u_2(w) - a\left(1 - e^{-\frac{u_1(w)}{a}}\right) \tag{3}$$

Thus, the following is a natural framework for the analysis of the "input-output" productivity measurement problems:

Problem P_g, $g \in G$:

$$p(g) := \max_{w \in W} g(u_1(w), u_2(w)) \tag{4}$$

where W is some given set and G is a given family of real valued functions defined on some $U \subset \Re^2$. It is assumed that each g in G is monotonically increasing with $u_2(w)$ and monotonically decreasing with $u_1(w)$. Observe that the optimization is performed only with respect to the decision variable w, not functions $g \in G$. Thus, while conducting a Hyper Sensitivity Analysis, one is not concerned with "an optimal g ", but rather is interested in solving (4) for each g in G. The term *hyper* suggests that the sensitivity analysis is not with respect to numeric values, but rather with respect to functions.

As explained above, members of G have some common properties such as monotonicity or convexity. These properties may be used effectively in the analysis and allow to design the procedures doing better than solving the relevant optimization problem *independently* for each g in G.

This formulation highlights the important properties of the "input-output" productivity measurement model while allowing enough flexibility as far as particular features of the management situation under consideration are concerned.

3. PROPOSED APPROACH

The approach proposed in this paper for conducting Hyper Sensitivity Analysis of productivity measurement problems is based on C-Programming (Composite Concave Programming, Sniedovich (1986) and (1994)), which is a parametric method designed for global optimization of composite functions. The C-Programming format involves two problems: the *target problem* and its *parametric problem.* In this discussion, for each g in G, *Problem* P_g serves as a target problem and the associated parametric problem is as follows:

Problem $P(\lambda)$:

$$q(\lambda) := \max_{w \in W} \{\lambda_1 u_1(w) + \lambda_2 u_2(w)\}, \ \lambda = (\lambda_1, \lambda_2) \in \Lambda \subseteq \Re_- \times \Re_+ \quad (5)$$

where $\Re_+$ and $\Re_-$ are the strictly positive and negative parts of the real line respectively, and Λ is a suitable set to be discussed below. Observe that this family of problems does not depend on G in any explicit way except that the parametric space will naturally depend on G. In particular, for any fixed value of λ the optimal solution for *Problem* $P(\lambda)$ does not depend on G.

Let W_g^* denote the set of global optimal solutions to *Problem* P_g and let $W^*(\lambda)$ denote the set of global optimal solutions to *Problem* $P(\lambda)$. Also, provided that the function g is differentiable with respect to u, let $\nabla g(u(w))$ denote the gradient of g with respect to $u(w)$, namely define

$$\nabla g(u(w)) := \left(\frac{\partial}{\partial y_1} g(y), \frac{\partial}{\partial y_2} g(y) \right)_{y=(u_1(w),u_2(w))}, \ w \in W. \quad (6)$$

The link between the target problems and the parametric problem is established by the following *Fundamental Theorem of C-Programming* (Sniedovich (1986)):

Theorem 1 *Let* $g \in G$ *be* differentiable *and* pseudoconvex *with respect to* $u = (u_1(w), u_2(w))$ *on some open convex set containing*

$$\{u(w) : w \in W\}.$$

Then any optimal solution for Problem P_g *is also optimal with respect to parametric problem it generates, namely*

$$w \in W_g^* \Longrightarrow w \in W^*(\nabla g(u(w))). \quad (7)$$

Moreover, any optimal solution to this parametric problem is also optimal with respect to Problem P_g, *namely*

$$w \in W_g^* \Longrightarrow W^*(\nabla g(u(w))) \subset W^*. \quad (8)$$

There are some comments with regard to the Fundamental Theorem to be made at this point:

1 The significance of the *Fundamental Theorem* is that it guarantees that, subject to differentiability and pseudoconvexity conditions, there exists an optimal value of the parameter λ, namely a value λ^* such that any optimal solution for *Problem* $P(\lambda^*)$ is also optimal with respect to *Problem* P_g.

2 Note that although the Fundamental Theorem does not explicitly require monotonicity of g with respect to the components of u, the nonnegativity/nonpositivity of the components of $\nabla g(u(w))$ is often important for efficient solving the of associated parametric problem.

3 The solution for *Problem* P_g for any $g \in G$ can be obtained via solving *Problem* $P(\lambda)$ for a suitable range Λ of the parameter λ determined by the gradient of g over $u(W)$. Obviously, this strategy is useful only for cases where for every value of λ solving *Problem* $P(\lambda)$ is much easier than solving *Problem* P_g itself.

4 Utility function, in most cases, would be (pseudo)concave rather than (pseudo)convex with respect to outputs u_2. However, the *Fundamental Theorem* is still useful for the common case where utility function is linear with respect to u_2.

5 As was demonstrated in Sniedovich (1994), *C-Programming* framework can successfully handle instances of *Problem* P_g, where function g is additive and (pseudo)concave with respect to outputs u_2, while being (pseudo)convex with respect to inputs u_1.

The following example illustrates the way C-Programming approaches the productivity measurement problem.

Example 2 *Consider the following version of* Problem P_g *representing the discounted version of the standard productivity measurement problem:*

$$\max_{w \in W} \{g(u_1(w), u_2(w)) := \frac{u_2(w)}{a(1 - e^{-\frac{u_1(w)}{a}})}\}, \tag{9}$$

where $w \in W \subset \Re_+^{m+s}$, $u_1(w) := \sum_{i=1}^{m} c_i w_i$, $u_2(w) := \sum_{j=1}^{s} r_j w_{m+j}$ *as defined in (1). This problem belongs to the class of fractional programming problems and, generally speaking, is difficult to solve.*

Applying C-Programming techniques and taking into consideration that the function g is increasing with u_2 and decreasing with u_1, one gets the following parametric problem (5):

$$\max_{w \in W} \{\lambda_2 \sum_{j=1}^{s} r_j w_{m+j} + \lambda_1 \sum_{i=1}^{m} c_i w_i\},\ \lambda \in \Lambda \subset \Re_+ \times \Re_- \tag{10}$$

Since λ_2 is striclty positive, dividing the objective function of this problem by λ_2 yields the equivalent problem:

Problem P(μ):

$$\max_{w \in W} \{\sum_{j=1}^{s} r_j w_{m+j} + \mu \sum_{i=1}^{m} c_i w_i\},\ \mu \in M \subset \Re_- \tag{11}$$

where $\mu = \lambda_1/\lambda_2$.

For every fixed value of the parameter μ this problem is a standard Linear Programming *problem. Thus, this reduced one-parameter problem is a classical parametric linear programming problem (Hillier and Lieberman (1990), Gal (1995)). It can be solved by any commercial software capable of performing parametric analysis for linear programming problems (see Schrage (1997)).*

Note that the solution strategy suggested by C-Programming involves recovering an optimal solution for *Problem P_g* by solving the corresponding parametric problem for the suitable value of the parameter μ. Composite Concave Programming requires the ability to solve the parametric problem for *any* specified value of the parameter μ in a certain range M, which, in fact, means the ability of solving the parametric problem for *all* values of the parameter μ in that range. This is why the issue of identifying the lower and upper bounds $(\underline{\mu}, \overline{\mu})$ for the range M becomes very important.

To address this issue, recall that, subject to the conditions required by the *Fundamental Theorem of C-Programming*, if w^* is any globally optimal solution to *Problem P_g*, then w^* is also optimal with respect to its corresponding parametric problem, namely *Problem $P(\lambda)$*, $\lambda = \nabla g(u(w^*))$. Furthermore, any global optimal solution to this parametric problem is also globally optimal with respect to the corresponding *Problem P_g*. Therefore, there is a way to directly translate the bounds $\underline{\mu}$ and $\overline{\mu}$ into some bounds for the value of $\nabla g(u(w))$ where $w \in W$.

Assume that the family G in *Problem P* consists of two functions, for example g_1 and g_2. Let $\mu^{(1)} \in M$ and $\mu^{(2)} \in M$ denote the correspondingly optimal values of parameter in *Problem $P(\mu)$*. If, when looking for optimal values of μ, the associated C-Programming procedure covers the whole range M, then two different objective functions can be maximized using one such procedure. This means that if the interval M is wide enough, then by solving the same C-Programming parametric problem once, one can recover the solution for *Problem P* involving several different composite functions g in one go. The wider the interval M is taken, the more target problems with different composite functions g may be solved using the same parametric problem. The obvious trade-off here is that making M too wide may increase the number of parametric problems that must be solved. Typically, for C-programming problems, the efficient identification of a proper pair of bounds $(\underline{\mu}, \overline{\mu})$ does not present serious difficulties.

The following is then an outline of the *C-Programming approach* to the solution of *Problem* P_g for a given $g \in G$:

Step 1: Formulate the corresponding two-dimensional parametric *Problem P(λ)*;

Step 2: Reduce the dimensionality of parametric problem by formulating the corresponding one-dimensional *Problem P(μ)*;

Step 3: Solve *Problem P(μ)* for all μ in a suitable range M.

The third step is algorithmic/computational in nature: it employs C-Programming techniques to generate an optimal solution to *Problem* P_g. This step typically comprises the following sub-steps:

Step 3.1: Determine a suitable range of values, call it M, for the parameter μ as specified by (11) and discussed above.

Step 3.2: Find the optimal value of $g(u(w(\mu)))$ over all μ in M, where $w(\mu)$ denotes the optimal solution found for *Problem P(μ)*.

These substeps are typically fully automated in that they do not require any interaction with the decision maker. If g satisfies the differentiability and convexity conditions required by the *Fundamental Theorem,* then from the $w(\mu)$ optimizing *Problem P(μ)* we get an optimal solution for *Problem* P_g by simply evaluating the appropriate function g. It should be noted that from the decision-maker's point of view *Step 3* is a "Black-Box": it generates an optimal solution to *Problem P(λ)* without requiring any further interaction.

Note that although the formulation of *Problem* P_g given in (4) assumes the *linear* nature of input-output functions $u_1(w)$ and $u_2(w)$, thus restricting the class of corresponding optimization problems under consideration to *Composite Linear Programming*, the C-Programming solution methodology outlined above is capable of handling much wider classes of composite optimization problems. In particular, Churilov et al (1998) discuss composite quadratic programming problems, Churilov (1998) applies C-Programming to composite knapsack problems, while Domingo and Sniedovich (1993) investigate nonseparable dynamic programming problems using C-Programming. Such a wide scope of applications allows to use C-Programming for conducting Hyper Sensitivity Analysis of portfolio optimization problems (Churilov et al 1999) and problems with soft constraints (Byrne et al (1998)).

In the next section the discussion of algorithmic aspects of C - Programming as a tool for conducting Hyper Sensitivity Analysis of productivity and efficiency measurement problems is continued by focusing on *Composite Linear Programming* problems.

4. PRODUCTIVITY MEASUREMENT PROBLEM AS A COMPOSITE LINEAR PROBLEM

Macalalag and Sniedovich (1996) discuss the algorithmic and computational aspects of solving Composite Linear Programming problems, i.e. problems of the following form:

Problem CP:

$$f^* := \max_{x\in X} f(x) := \Phi(v_1(x), v_2(x)), \tag{12}$$

where both $v_1(x)$ and $v_2(x)$ are linear, i.e.

$$v_1(x) := dx + D,\ v_2(x) := ex + E, \tag{13}$$

and $D, E \in \Re$, $d, e \in \Re^n$, $A \in \Re^{l\times n}$, $b \in \Re^l$, $X := \{x \in \Re^n : Ax \leq b, x \geq 0\}$ and Φ is differentiable convex real-valued function on some open convex set $V \subseteq \Re^2$ containing the set $v(X) := \{v(x) : x \in X\}$.

As indicated earlier in our discussion, the following corresponding one-dimensional parametric problem can be derived for such cases:

Problem CP(μ):

$$f^*(\mu) := \max_{x\in X} \{f(x;\mu) := (e + \mu d)x\},\ \mu \in M := [\underline{\mu}; \overline{\mu}] \tag{14}$$

The results reported in Macalalag and Sniedovich (1996) indicate that the standard parametric sensitivity analysis techniques of the simplex method can be effectively used in conjunction with C-Programming approach. Based on standard Lagrangian results,Macalalag and Sniedovich (1996) demons trate that function $f^*(\mu)$ defined in (14) is continuous, piecewise linear, and convex with μ. Thus, the graph of $f^*(\mu)$ consists of several linear segments and the points where two neighboring linear segments of the graph meet. These points are usually referred to as *breakpoints* (Hillier and Lieberman (1990)). The *Fundamental Theorem of C-Programming* discussed in the previous section guarantees that one of the breakpoints corresponds to an optimal value of parameter μ.

Therefore, as indicated in Macalalag and Sniedovich (1996), in order to solve *Problem CP*, one has to solve *Problem CP(μ)* at subintervals of M over which the graph of $f^*(\mu)$ consists of line segments, and then identify the end points of these segments, i.e. the *breakpoints* of the graph.

Note that the parametric problem has the standard form of *parametric linear programming* problems, and therefore can be solved using the standard parametric analysis of the simplex method (see Hillier and Lieberman (1990)). Each iteration, taking place at each vertex of $f^*(\mu)$ requires one pivot operation and identifies the range, that is the linear segment of $f^*(\mu)$, upon which the current solution remains optimal, and thus the next *breakpoint* is also being identified.

Comprehensive computational results are reported in Macalalag and Sniedovich (1996), comparing the efficiency of several approaches for solving *Problem CP*. A summary of the results which estimate the average number of pivots required to solve a linear programming problem using the simplex method and the average number of pivots required to solve parametric problem is presented in Shamir (1987). The obvious conclusion is that the collaboration of C-Programming and *Parametric Linear Programming* techniques presents a very powerful solution approach for *Problem CP*.

Note that *Problem* P_g given in (4) presents a special case of *Problem CP*. In particular, recall that function $u_1(w)$ defined in (1) and representing the weighted some of inputs in *Problem* P_g has the following form:

$$u_1(w) := \sum_{i=1}^{m} c_i w_i, \tag{15}$$

while function $v_1(x)$ defined in (13) looks as follows:

$$v_1(x) := \sum_{i=1}^{n} d_i x_i + D. \tag{16}$$

Similarly, the form of functions u_2 and v_2 is mathematically identical. The only difference is of the *modelling* nature: both $u_1(w)$ and $u_2(w)$ apply to clearly defined and *separate* sets of inputs and outputs, while both $v_1(x)$ and $v_2(x)$ are defined as the weighted sum of possibly *all* components of the vector x, without subdividing these components into "input" and "output" parts. Thus, as discussed below, not only the C-Programming solution approach can be directly applied for solving generalized "input-output" productivity and efficiency measurement problems, but it also allows for interesting generalization of such class of problems. To illustrate the application of C-Programming approach to

Problem P_g, consider the following example of *DEA* problem discussed by Schrage (1997).

Example 3 *Below are four performance measures on six high schools: Bloom (BL), Homewood (HW), New Trier (NT), Oak Park (OP), York (YK), and Elgin (EL). Cost/pupil is the number of dollars spent per year per pupil by the school. "Percent not-low-income" is the fraction of the student body coming from homes not classified as low income. The writing and science scores are the averages over students in a school on a standard writing test and a standard science test. The first two measures are treated as inputs, over which teachers and administrators have no control. The test scores are treated as outputs.*

School	*Cost/Pupil*	*% not low income*	*Writing*	*Science*
BL	*89.39*	*64.3*	*25.2*	*223*
HW	*86.25*	*99*	*28.2*	*287*
NT	*108.13*	*99.6*	*29.4*	*317*
OP	*106.38*	*96*	*26.4*	*291*
YK	*62.40*	*96.2*	*27.2*	*295*
EL	*47.19*	*79.9*	*25.5*	*222*

(17)

The standard LP-transformed "ratio" DEA model for maximizing the score of New Trier appears below:

$$\max_w \quad 29.4w_3 + 317w_4 \tag{18}$$

s.t.

$$\begin{array}{rcl} 108.13w_1 + 99.6w_2 & = & 1 \\ 223w_4 + 25.2w_3 - 64.3w_2 - 89.39w_1 & \leq & 0 \\ 287w_4 + 28.2w_3 - 99w_2 - 86.25w_1 & \leq & 0 \\ 317w_4 + 29.4w_3 - 99.6w_2 - 108.13w_1 & \leq & 0 \\ 291w_4 + 26.4w_3 - 96w_2 - 106.38w_1 & \leq & 0 \\ 295w_4 + 27.2w_3 - 96.2w_2 - 62.40w_1 & \leq & 0 \\ 222w_4 + 25.5w_3 - 79.9w_2 - 47.19w_1 & \leq & 0 \\ w_1, w_2, w_3, w_4 & \geq & 0.0005 \end{array} \tag{19}$$

The equality constraint $108.13w1+99.6w2 = 1$ *is supposed to "normalize" the solution* w^*, *since by trivial observation if the vector*

$$(w_1^*, w_2^*, w_3^*, w_4^*)$$

is the solution, then so is the vector $(\alpha w_1^*, \alpha w_2^*, \alpha w_3^*, \alpha w_4^*)$, $\alpha \in \Re \setminus \{0\}$. *For simplicity reasons, Schrage (1997) presents the abovementioned LP transformation of the originally fractional programming problem, i.e. the problem with the objective function of the following form:*

$$g(u(w)) := \frac{u_2(w)}{u_1(w)} := \frac{29.4w_3 + 317w_4}{108.13w_1 + 99.6w_2} \tag{20}$$

Thus, in this example the ratio-type scoring function is used. On the other hand, DEA methodology does not necessarily assume the use of ratio function as the only candidate for scoring function (Schrage (1997)). This is precisely the point where C-Programming enters the scene: provided that function g satisfies the conditions stated in the Fundamental Theorem presented in the previous section, there is a well-defined link between the solution of this problem and its corresponding parametric problem given in (11), whose objective function has the following form:

$$\max_{w}\{29.4w_3 + 317w_4 + \mu(108.13w_1 + 99.6w_2)\}, \ \mu \in M \subset \Re_{-}. \quad (21)$$

If wide enough M is selected, by solving one parametric problem (21) several candidates for the scoring function such as $u_2(w) - \sqrt{u_1(w)}$ or $u_2(w)/(1 - e^{-u_1(w)})$ can be evaluated.

5. DISCUSSION

The basic motivation for *Hyper Sensitivity Analysis* discussed earlier in preceeding sections also sheds new light on the issue of the relationship between *Data Envelopment Analysis* and *Multiple Criteria Decision Analysis (MCDM)/Multiple Objective Linear Programming (MOLP)*. Several recent publications in this area discussed this relationship from a variety of perspectives. Stewart (1996) demonstrates the link between the ratio efficiency definition in DEA and a distance measure in input-output space based on linear value functions. Bouyssou (1999) discusses the equivalence between the concept of efficiency in *DEA* and that of convex efficiency in *MCDM.* The paper also contains some remarks on possible areas of interaction between *DEA* and *MCDM.* Joro et al (1998) show that structurally the DEA formulation to identify efficient units is quite similar to *Multiple Objective Linear Programming* model based on the reference point or the reference direction approach to generate efficient solutions. It is therefore concluded that *MOLP* provides interesting extensions to *DEA* and *DEA* provides new areas of application to *MOLP.*

As far as *Hyper Sensitivity Analysis* is concerned, these two areas cross-fertilize each other. Indeed, both *DEA* and *Multiple Objective Programming* provide one of the most natural cases demonstrating the usefulness and effectiveness of *Hyper Sensitivity Analysis.* Although in both cases it is sometimes difficult to obtain an exact mathematical representation for the utility function g, it is well known (Steuer (1986) and French (1988)) that, subject to certain conditions, one needs no knowledge about scoring utility function g provided that it is monotonic with respect to its arguments. In such cases an optimal solution to the single objective problem is also a non-dominated solution, in a Pareto sense, of the multiobjective problem. This fact is valid for any scoring function g satisfying the monotonicity conditions, hence the *same* set of non-dominated points applies to any such function. Thus, the idea of *Hyper Sensitivity Analysis* in this case suggests that one can find optimal solutions for many particular monotonic instances of scoring function g simply by recovering the set of all non-dominated points of the set $u(W) := \{u(w) : w \in W\}$. If the dimensionality of this set is not too large, the corresponding solutions can often be generated efficiently by standard optimization methods. When using C-Programming to conduct *Hyper Sensitivity Analysis*, instead of dealing with families of scoring functions based on their monotonicity properties, as is the case with multiobjective optimization, one requires the composite scoring utility function to be pseudoconvex and differentiable.

5. SUMMARY AND CONCLUSION

In this paper the use of C-Programming as a tool for performing *Hyper Sensitivity Analysis* in the area of productivity and efficiency measurement problems has been illustrated. A simple *Composite Linear Programming* - based approach for handling situations was presented, where several candidates for the function quantifying the utility of "input-output" pairs are available.

In particular, the generalized "input-output" productivity measurement problem was discussed and the C-Programming way of performing sensitivity analysis was demonstrated. Hyper Sensitivity Analysis for this class of problems can be carried out very efficiently, with very little extra computation, once solutions have been found for one utility function. Small size example have been used to illustrate the method. However, it should be borne in mind that the method can handle large-scale productivity measurement problems and allows for extra flexibility as far as various modelling issues are concerned.

References

Althin R., Fare R., and Grosskopf S. Profitability and Productivity Changes: An Application to Swedish Pharmacies. Annals of Operations Research, 1996; 66: 219-230

Avkiran N.K. *Productivity Analysis in the Services Sector with Data Envelopment Analysis.* Queensland, Australia: N.K.Avkiran Publishers, 1999

Bjurek H. Essays on Efficiency and Productivity Change with Applications to Public Service Production. Ekonomiska Studier 52, School of Economics and Commercial Law, University of Gothenburg, 1994

Bouyssou D. Using DEA as a Tool for MCDM: Some Remarks. Journal of the Operational Research Society, 1999; 50: 974-978

Byrne A., Sniedovich M., and Churilov L. Handling Soft Constraints via Composite Concave Programming. Journal of the Operational Research Society,1998; 49: 870-877

Caves D.W., Christensen L.R., and Diewert W.E. The Economic Theory of Index Numbers and the Measurement of Input, Output, and Productivity. Econometrica 1982; 50: 1393-1414

Charnes A., Cooper W.W., and Thrall R.M. Classifying and Characterizing Efficiencies and Inefficiencies in Data Envelopment Analysis. Operations Research Letters, 1986; 5(3):105-110

Churilov L. *Global Optimization through Composite Linearization.* PhD Thesis, Department of Mathematics and Statistics, The University of Melbourne, Australia, 1998

Churilov L., Ralph D., and Sniedovich M. A Note on Composite Concave Quadratic Programming. Operations Research Letters, 1998; 23: 163-169

Churilov L., Bomze I., Sniedovich M., and Ralph D. Hyper Sensitivity Analysis of Portfolio Optimization Problems. Submitted for publication. 1999

Coelli T., Prasada Rao D.S., and Battese G. *An Introduction to Efficiency and Productivity Analysis.* Boston, USA: Kluwer Academic Publishers,1998

Davis H. *Productivity Accounting.* USA: *The Wharton School Industrial Research Unit, University of Pennsylvania, 1951 (reprint 1978).*

Domingo A. and Sniedovich M. Experiments with Dynamic Programming Algorithms for Nonseparable Problems. European Journal of Operations Research 1993; 67(2): 172-187

French S. *Decision Theory - An Introduction to the Mathematics of Rationality.* New York: John Wiley and Sons, 1988

Fried H.O., Knox Lovell C.A., and Schmidt S.S. *The Measurement of Productive Efficiency: Techniques and Applications.* New York, USA: Oxford University Press, 1993

Fuentes H., Grifell-Tatje E., and Perelman S. "A Parametric Distance Function Approach for Malmquist Index Estimation". *Working Paper, CREPP 98/03, Centre de Recherche en Economie Publique et en Economie de la Population, Universite de Liege, Belgium, 1998*

Gal T. *Postoptimality Analyses, Parametric Programming and Related Topics.* Berlin, Germany: de Gruyter, 1995

Hillier F.C. and Lieberman G.J. *Introduction to Operations Research.* 5th edition. New York, USA: McGraw-Hill, 1990

Joro T., Korhonen P., and Wallenius J. Structural Comparison of Data Envelopment Analysis and Multiple Objective Linear Programming. Management Science 1998; vol.44, No.7: 962-970

Macalalag E. and Sniedovich M. Generalized Linear Programming and Sensitivity Analysis Techniques. Naval Research Logistics 1996; 43: 397-413

Schrage L. *Optimization Modeling with LINDO.* New York: The Scientific Press, 5th edition, 1997

Shamir R. The Efficiency of the Simplex Method, a Survey. Management Science 1987; 33: 301-334

Sink D.S. *Productivity Management: Planning, Measurement and Evaluation, Control and Improvement.* New York: John Wiley & Sons, 1989.

Sniedovich M. C-programming and the Minimization of Pseudolinear and Additive Concave functions. Operations Research Letters 1986; 5(4): 185-189

Sniedovich M. Algorithmic and Computational Aspects of Composite Concave Programming. International Transactions in Operations Research 1994; 1(1): 75-84

Steuer R. *Multiple Criteria Optimization: Theory, Computation, and Application.* New York: John Wiley & Sons, 1986

Stewart TJ. Relationship between Data Envelopment Analysis and Multicriteria Decision Analysis. Journal of the Operational Research Society 1996; 47: 654-665

REDUCED GAME PROPERTY OF LINEAR VALUES WITH EQUAL TREATMENT PROPERTY

Tsuneyuki Namekata[1] and Theo S.H. Driessen[2]

[1]*Department of Information and Management Science*
Otaru University of Commerce 3-5-21 Midori Otaru Hokkaido 047-8501 Japan

[2]*Faculty of Mathematical Sciences*
University of Twente P. O. Box 217, 7500 AE Enschede The Netherlands.

[1]*namekata@res.otaru-uc.ac.jp and* [2]*T.S.H.Driessen@math.utwente.nl*

Abstract A transferable utility game deals with problems of how to allocate the total benefit among players by considering contributions of various coalitions of players to the total benefit. A value for games is an allocation of the total benefit among them. Many values such as the Shapley value and the prenucleolus are introduced by their own equity consideration. Thus a consistency in terms of a reduced game is a useful tool to compare various solutions in a unified way, because it expresses the differences in solutions as those in associated reduced games. This paper axiomatizes linear values with Equal Treatment Property by the consistency in terms of a reduced game.

Keywords: TU-game, consistency, reduced game, Shapley value, Solidarity value.

1. INTRODUCTION

This paper deals with the axiomatic characterization of a class of values in terms of the reduced game property (RGP). Ruiz et al. (1998) characterize the Least Square (LS) values in terms of RGP in the case that their weights are consistent. The Shapley value is regarded as the LS value whose weights are consistent. The Shapley value has many reduced games for which it is characterized in terms of RGP (Driessen 1991, Namekata et al. 2000). The EN^{k}AC-value (Namekata et al. 1999) for k=2,…,n−2 is regarded as the LS value, but its weights are not consistent. Namekata et al. (2000) characterize the EN^{k}AC-value for k=1,…,n−1 in terms of RGP. Because the Solidarity value (Nowak et al. 1994) does not possesses inessential game property (IGP), it is not regarded as the LS value. The Solidarity value has not been characterized in terms of RGP. In this paper, reduced games are introduced for a linear value which possesses equal treatment property (ETP) and it is shown that this value is the unique value that is α-standard for two-person games and possesses RGP for those reduced games. The α-standardness for two-person games is a slight generalization of the usual standardness for two-person games. Thus we characterize the class of linear values that possesses ETP in terms of RGP. This class includes the LS values, the EN^{k}AC-value, and the Solidarity value.

An outline of the paper is as follows: Section 2 introduces linear values with equal treatment property in a very general form and shows that it is regarded as an Egalitarian Non-Individual Contribution Value. Section 3 considers their axiomatic characterization in terms of the reduced game property. Section 4 concentrates on the special form of reduced games which has an interesting meaning and compares the Shapley value with the LS prenucleolus and the Solidarity value through these reduced games. Section 5 presents a numerical example and illustrates how to use our result in a potential application area.

2. LINEAR VALUES WITH EQUAL TREATMENT PROPERTY

This section introduces linear values with equal treatment property, discusses the relationship between these values and Least Square values, and finally shows that a linear value with equal treatment property can be interpreted as an egalitarian non-individual contribution value. This interpretation will be used to characterize linear values with equal treatment property in terms of the reduced game property in Section 3.

A *transferable utility game* (or TU-game or cooperative game or coalitional game with side payments) is an ordered pair (N,v), where N is a finite set of *players* and $v{:}2^N \to \mathbf{R}$ is a *characteristic function* satisfying $v(\emptyset)=0$. A nonempty subset S of N (notation: $S\in 2^N$ or $S\subset N$ with $S\neq\emptyset$) is called *coalition*. The associated real number $v(S)$ is called the *worth* of coalition S and it represents the joint profits that the members of S can achieve due to their cooperative behavior in the (transferable utility) game (N,v). The cardinality of set S is denoted by $|S|$ or, if no ambiguity is possible, by s. Particularly, $n{:=}|N|$. Let G denote the class of all TU-games with a variable player set.

The (one-point or multi-valued) solution concepts for TU-games are concerned with the essential problem of how to divide the overall profits $v(N)$ of the grand coalition N among all the players of the game (N,v). A payoff vector $x\in\mathbf{R}^N$ is said to be *efficient* if it meets the efficiency principle $\sum_{j\in N} x_j = v(N)$. A *value* (or one-point solution concept) on G is a function σ that assigns an efficient payoff vector $\sigma(N,v)=(\sigma_i(N,v))_{i\in N}$ to every game

$(N,v)\in G$. The value $\sigma_i(N,v)$ of player i in the game (N,v) represents an assessment of his gains from participating in the game. In order to characterize some class of values we need some definitions of properties of a value as follows:

Definition 1. A value σ on G is said to

(i) be *linear* if $\sigma(N,v+w)=\sigma(N,v)+\sigma(N,w)$ for all $(N,v),(N,w)\in G$.

(ii) possess the *equal treatment property* (ETP) if $\sigma_i(N,v)=\sigma_j(N,v)$ for all $i,j\in N(i\neq j)$ in all game $(N,v)\in G$ satisfying $v(S\cup\{i\})= v(S\cup\{j\})$ for all $S\subset N-\{i,j\}$.

(iii) possess the *inessential game property* (IGP) if $\sigma_i(N,v)=v(\{i\})$ for all additive game $(N,v)\in G$ and all $i\in N$, where a game (N,v) is *additive* if $v(S)=\sum_{j\in S}v(\{j\})$ for all $S\subset N$.

This paper concentrates on a class of values that are linear and possess ETP. It has a simple form as the next lemma shows. See Ruiz et al. (1998) for proof.

Lemma 2. *A value σ on G is linear and possesses ETP if and only if there exists a collection of weights $\{(m_{n,s})_{s=1,\ldots,n-1}|n=2,\ldots\}$ such that*

$$\sigma_i(N,v)=\frac{v(N)}{n}+\sum_{S:i\in S\subset N,S\neq N} m_{n,s}v(S)-\sum_{S:S\subset N,S\neq\emptyset,N}\frac{s}{n}m_{n,s}v(S). \qquad (1)$$

Every weight $m_{n,s}$ takes any value in (1). Thus a value in Lemma 2 need not possess IGP. If we impose some condition on weights m, then the value possesses IGP as we show in the next proposition. Furthermore if all the weights are nonnegative, then the value is regarded as the Least Square value associated with the weights m. The Least Square (LS) value is defined in Ruiz *et al.* (1998) in the following way:

$$LS_i^m(N,v):=\frac{v(N)}{n}+\frac{1}{\sum_{s=1}^{n-1}\binom{n-2}{s-1}m_{n,s}}\left[\sum_{\substack{S:i\in S\subset N\\ S\neq N}} m_{n,s}v(S)-\sum_{\substack{S:S\subset N\\ S\neq\emptyset,N}}\frac{s}{n}m_{n,s}v(S)\right],$$

where all the weights be nonnegative and at least one of $m_{n,s}$ $(s=1,\ldots,n-1)$ is positive for every $n=2,\ldots$.

Proposition 3. *A linear value with ETP possesses IGP if and only if the weights satisfies*

$$\sum_{s=1}^{n-1}\binom{n-2}{s-1}m_{n,s}=1 \text{ for } n=2,\ldots. \qquad (2)$$

Further if all the weights $m_{n,s}(n=2,\ldots;s=1,\ldots,n-1)$ are nonnegative, then the value is the Least Square value associated with the weights m.

The proof is omitted since it is straightforward. We give some examples of linear values with ETP:

Example 4. The value in the form of (1) is equal to

(i). the Shapley value if $m_{n,s} = \dfrac{1}{n-s}\begin{pmatrix} n-1 \\ s-1 \end{pmatrix}^{-1}$,

(ii). the LS prenucleolus (Ruiz *et al.* 1996) if $m_{n,s} = \dfrac{1}{2^{n-2}}$,

(iii). the Solidarity value if $m_{n,s} = \dfrac{1}{(s+1)(n-s)}\begin{pmatrix} n-1 \\ s-1 \end{pmatrix}^{-1}$.

By checking the condition (2) directly we can show that the Shapley value and the LS prenucleolus are regarded as the LS values, whereas the Solidarity value is not.

This section is concluded by giving another useful representation of a linear value with ETP. An *Egalitarian Non-Individual Contribution Value* (ENIC-value) on G with an *Individual Contribution* (*IC*) (Driessen et al. 1997) is defined as

$$ENIC_i(N,v) := IC_i(N,v) + \frac{1}{n}\left[v(N) - \sum_{j \in N} IC_j(N,v) \right]. \qquad (3)$$

The ENIC-value has the following meaning: To allocate the total amount $v(N)$ to all members of N, (i) every player i firstly gets his individual contribution $IC_i(N,v)$, and (ii) every player secondly gets the rest of it equally, that is, $\frac{1}{n}\left[v(N) - \sum_{j \in N} IC_j(N,v) \right]$.

The next proposition shows that a linear value with ETP is regarded as an ENIC-value if the individual contribution of a player is given by the weighted sum of worths of coalitions including that player with respect to the weight m. The proof is easy, so is omitted.

Proposition 5. *A linear value with ETP can be regarded as an ENIC-value with Individual Contribution IC which is given by*

$$IC_i(N,v) := \sum_{S, i \in S \subset N, S \neq N} m_{n,s} v(S) \text{ for all } (N,v) \in G \text{ and all } i \in N. \qquad (4)$$

3. CONSISTENCY OF LINEAR VALUES WITH EQUAL TREATMENT PROPERTY

The authors consider axiomatic characterization of linear values with ETP on G by means of a consistency property in terms of reduced games. A *reduced game* is obtained from a given game (N,v), $n\geq 3$, by removing one player on the understanding that the removed player i will be paid according to a proposed payoff vector $x\in\mathbf{R}^N$. The remaining players form the player set $N-\{i\}$ of the reduced game; the characteristic function of which is composed of both the original characteristic function v and the proposed payoff vector x, denoted by $v^x:2^{N-\{i\}}\to\mathbf{R}$, satisfying $v^x(\emptyset)=0$. For the sake of efficiency of solutions in the reduced game, we impose $v^x(N-\{i\}):=v(N)-x_i$. The worths of other coalitions $S(\neq\emptyset,N-\{i\})$ in v^x are specified later in this section to satisfy the following reduced game property:

Definition 6. A value σ on G is said to

(i). possess the *reduced game property* (RGP) with respect to a reduced game $\left((N-\{i\},v^x)\right)_{i\in N,x\in\mathbf{R}^N}$ if $\sigma_j\left(N-\{i\},v^{\sigma(N,v)}\right)=\sigma_j(N,v)$ for all $(N,v)\in G$ with $n\geq 3$, all $i\in N$, and all $j\in N-\{i\}$,

(ii). be *standard for two-person games* (S2G) if, for all two-person games, $\sigma_i(\{i,j\},v)=v(\{i\})+[v(\{i,j\})-v(\{i\})-v(\{j\})]/2$,

(iii). be *α-standard for two-person games* (α-S2G) if, for all two-person games, $\sigma_i(\{i,j\},v)= v(\{i,j\})/2+\alpha[v(\{i\})-v(\{j\})]$.

See Figure 1 for the implication of standardness and α-standardness for two-person games. As α becomes smaller, the difference of the power of the players, that is, the difference of $v(\{i\})$ and $v(\{j\})$ plays the less part in the determination of the allocation of $v(\{i,j\})$. In particular the value assigns the equal share to two players regardless of their power, that is, their individual contribution $v(\{i\})$ and $v(\{j\})$ if $\alpha=0$. Note that 1/2-S2G is just a S2G.. Therefore α-S2G is regarded as a generalization of S2G.. Also note that the Solidarity value is 1/4-S2G.

Let us define a reduced game $(N-\{i\},v^x)$, $n\geq 3$, for the linear value with

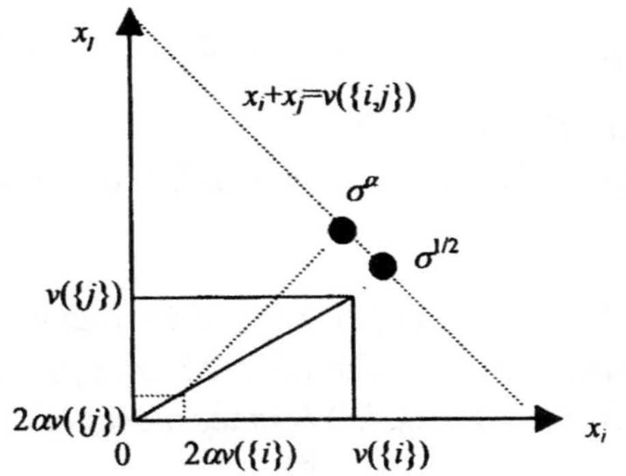

Figure 1. The implication of standardness and α-standardness for two-person games. $\sigma^{1/2}$ is standard for two-person games, while σ^α is α-standard for two-person games.

ETP in (1) as follows: For all $S \subset N-\{i\}$, $S \neq \varnothing, N-\{i\}$,

$$v^x(S) := a_s v(S) + b_s\left[v(S \cup \{i\}) - x_i\right] + c_s \sum_{l \in S}\left[v(S \cup \{i\} - \{l\}) - x_i\right] + d_s \sum_{l \in N-S\cup\{i\}} v(S \cup \{l\}). \tag{5}$$

The worth of a coalition S in the reduced game $(N-\{i\},v^x)$ is composed of four terms; the first and fourth terms are the worths of coalitions not including player i in the original game (N,v), the second and the third terms are the worths of coalitions including player i minus his payoff x_i which he gets when he leaves the game (N,v). We will give some interesting meaning of a special form of the reduced game later in Section 4. Coefficients $(a_s,b_s,c_s,d_s)_{s=1,\ldots,n-2}$ are determined by the following system of linear equations. Note that the coefficients $(a_s,b_s,c_s,d_s)_{s=1,\ldots,n-2}$ may not be determined uniquely by the equation (6) and that c_1 and d_{n-2} are arbitrary:

$$\begin{aligned}
&m_{n-1,1}a_1 = m_{n,1},\\
&m_{n-1,n-2}b_{n-2} = m_{n,n-1},\\
&m_{n-1,s}a_s + (s-1)m_{n-1,s-1}d_{s-1} = m_{n,s} \text{ for } s=2,\ldots,n-2,\\
&(n-s-1)m_{n-1,s}c_s + m_{n-1,s-1}b_{s-1} = m_{n,s} \text{ for } s=2,\ldots,n-2.
\end{aligned} \tag{6}$$

Some lemmata and a definition are needed to present the main Theorem 11. The following two lemmata play an important role to characterize the ENIC-value by RGP. The first lemma gives a sufficient condition that a value is characterized by RGP with respect to some reduced game. This sufficient condition holds for an ENIC-value if its individual contribution satisfies (8) as shown in the second lemma. Lemma 7 can be proved similarly to Lemma 3 in Namekata et al. (2000) and Lemma 8 is proved by simple calculation.

Lemma 7. *Let Φ be a value on G which is α-S2G and satisfies the following relation* (7) *for some reduced game* $\left((N-\{i\},v^x)\right)_{i\in N, x\in \mathbf{R}^N}$:
For all $|N|\geq 3$, all $i,j\in N(i\neq j)$, and all $x\in \mathbf{R}^N$

$$\Phi_j(N-\{i\},v^x) = \Phi_j(N,v) + \frac{1}{n-1}\left[\Phi_i(N,v) - x_i\right]. \tag{7}$$

(Note that (7) *means the RGP of Φ.)*
Then Φ is the unique value on G which is α-S2G and possesses RGP for that reduced game.

Lemma 8. *An ENIC-value satisfies* (7) *if its Individual Contribution IC satisfies the following property: For all $|N|\geq 3$, all $i,j\in N(i\neq j)$, and all $x\in \mathbf{R}^N$, the difference $IC_j(N-\{i\},v^x)-IC_j(N,v)$ is independent of j.*

(8)

Lemma 9. *If the coefficients $(a_s,b_s,c_s,d_s)_{s=1,\ldots,n-2}$ satisfy the relation* (6), *then the statement* (8) *holds.*

Proof. Sine the individual contribution *IC* for the linear value with ETP is given by (4) in Lemma 5, we have

$$IC_j(N-\{i\},v^x) = \sum_{S:j\in S\subset N-\{i\},S\neq N-\{i\}} m_{n-1,s}v^x(S)$$

$$= \sum_{\substack{S:j\in S\subset N-\{i\}\\ S\neq N-\{i\}}} m_{n-1,s}\begin{bmatrix} a_s v(S) + b_s\left[v(S\cup\{i\}) - x_i\right] \\ +c_s\sum_{l\in S}\left[v(S\cup\{i\}-\{l\}) - x_i\right] + d_s \sum_{l\in N-S\cup\{i\}} v(S\cup\{l\})\end{bmatrix}$$

$$= \sum_{S:j\in S\subset N-\{i\},S\neq N-\{i\}} m_{n-1,s}a_s v(S) + \sum_{S:\{i,j\}\subset S\subset N,S\neq N} m_{n-1,s-1}b_{s-1}v(S)$$

$$+ \sum_{S:j\in S\subset N-\{i\},S\neq N-\{i\}} m_{n-1,s}c_s\sum_{l\in S} v(S\cup\{i\}-\{l\})$$

$$+ \sum_{\substack{S:j\in S\subset N-\{i\}\\ S\neq N-\{i\}}} m_{n-1,s}d_s \sum_{l\in N-S\cup\{i\}} v(S\cup\{l\}) - x_i \sum_{\substack{S:j\in S\subset N-\{i\}\\ S\neq N-\{i\}}} m_{n-1,s}\left(b_s + sc_s\right). \quad (9)$$

Since $\sum_{S:j\in S\subset N-\{i\},S\neq N-\{i\}} m_{n-1,s}(b_s+sc_s) = \sum_{s=1}^{n-2}\binom{n-2}{s-1} m_{n-1,s}(b_s+sc_s)$, the last term in (9) is independent of j and can be ignored. The third term to the last and the second term to the last are rewritten in the following way:
The third term to the last in (9) is expressed as

$$\sum_{S:j\in S\subset N-\{i\},S\neq N-\{i\}} m_{n-1,s}c_s\sum_{l\in S} v(S\cup\{i\}-\{l\})$$

$$= \sum_{\substack{S:j\in S\subset N-\{i\}\\ S\neq N-\{i\}}} m_{n-1,s}c_s\left[v(S\cup\{i\}-\{j\}) + \sum_{l\in S-\{j\}} v(S\cup\{i\}-\{l\})\right]$$

$$= \sum_{S:i\in S\subset N-\{j\},S\neq N-\{j\}} m_{n-1,s}c_s v(S) + \sum_{l\in N-\{i,j\}}\ \sum_{S:\{i,j\}\subset S\subset N-\{l\},S\neq N-\{l\}} m_{n-1,s}c_s v(S) \quad (10)$$

$$= \sum_{S:i\in S\subset N-\{j\},S\neq N-\{j\}} m_{n-1,s}c_s v(S) + \sum_{S:\{i,j\}\subset S\subset N,s<n-1} (n-s)m_{n-1,s}c_s v(S)$$

$$= \sum_{S:\{i,j\}\subset S\subset N,s<n-1} (n-s-1)m_{n-1,s}c_s v(S) + \sum_{S:i\in S\subset N,s<n-1} m_{n-1,s}c_s v(S),$$

since the following relation holds

$$\{S|i\in S\subset N,\ s<n-1\}=\{S|\{i,j\}\subset S\subset N,\ s<n-1\}\cup\{S\subset N-\{j\}|i\in S,\ S\neq N-\{j\}\}.$$

The last term in (10) can be ignored by the independence of j. On the other hand, the second term to the last in (9) is expressed as

$$\sum_{S:j\in S\subset N-\{i\},S\neq N-\{i\}} m_{n-1,s}d_s \sum_{l\in N-S\cup\{i\}} v(S\cup\{l\})$$
$$= \sum_{l\in N-\{i,j\}} \sum_{S:\{l,j\}\subset S\subset N-\{i\}} m_{n-1,s-1}d_{s-1}v(S)$$
$$= \sum_{S:\{j\}\subset S\subset N-\{i\},S\neq\{j\}} (s-1)m_{n-1,s-1}d_{s-1}v(S).$$

Thus we have, by using (6),

$$IC_j(N-\{i\},v^x) = \sum_{\substack{S:j\in S\subset N-\{i\}\\ S\neq N-\{i\}}} m_{n-1,s}a_s v(S) + \sum_{\substack{S:\{i,j\}\subset S\subset N\\ S\neq N}} m_{n-1,s-1}b_{s-1}v(S)$$
$$+ \sum_{\substack{S:\{i,j\}\subset S\subset N\\ s<n-1}} (n-s-1)m_{n-1,s}c_s v(S) + \sum_{\substack{S:j\in S\subset N-\{i\}\\ S\neq\{j\}}} (s-1)m_{n-1,s-1}d_{s-1}v(S)$$
$$= m_{n-1,1}a_1 v(\{j\}) + \sum_{\substack{S:j\in S\subset N-\{i\}\\ S\neq\{j\},N-\{i\}}} \left[m_{n-1,s}a_s + (s-1)m_{n-1,s-1}d_{s-1}\right]v(S)$$
$$+ \sum_{\substack{S:\{i,j\}\subset S\subset N\\ s<n-1}} \left[(n-s-1)m_{n-1,s}c_s + m_{n-1,s-1}b_{s-1}\right]v(S) + \sum_{\substack{S:j\in S\subset N\\ s=n-1}} m_{n-1,n-2}b_{n-2}v(S)$$
$$+\left[(n-2)m_{n-1,n-2}d_{n-2}v(N-\{i\}) - m_{n-1,n-2}b_{n-2}v(N-\{i\})\right]$$
$$= m_{n,1}v(\{j\}) + \sum_{S:j\in S\subset N,1<s<n-1} m_{n,s}v(S) + \sum_{S:j\in S\subset N,s=n-1} m_{n,n-1}v(S)$$
$$+\left[\text{terms which are independent of } j\right]$$
$$= IC_j(N,v) + \left[\text{terms which are independent of } j\right],$$

since the following relations hold

$$\{S|j\in S\subset N,\ 1<s<n-1\} = \{S|\{i,j\}\subset S\subset N,\ s<n-1\}$$
$$\cup\{S\subset N-\{i\}|j\in S,\ S\neq\{j\},N-\{i\}\},$$
$$\{S|j\in S\subset N,\ s=n-1\} = \{S|\{i,j\}\subset S\subset N,\ s=n-1\}\cup(N-\{i\}).$$

Therefore we have proved the statement (8). Q.E.D.

We concentrate on the system of linear equations (6) for $(a_s,b_s,c_s,d_s)_{s=1,\dots,n-2}$. Let us consider the following linear equations for unknowns $(u_s,v_s)_{s=1,\dots,n-1}$:

$$m_{n-1,s}u_s + m_{n-1,s-1}v_s = m_{n,s} \quad \text{for } s=1,\dots,n-1, \tag{11}$$

where we define $m_{k,0}=m_{k,k}=0$ for $k=n-1$ and n for the sake of notational convenience. The equation (11) has at least one solution if and only if $m_{n-1,s}=m_{n-1,s-1}=0$ implies $m_{n,s}=0$. Let $(x_s,y_s)_{s=1,\dots,n-1}$ and $(x'_s,y'_s)_{s=1,\dots,n-1}$ be solutions of (11). Then $(a_s,b_s,c_s,d_s)_{s=1,\dots,n-2}$ in (6) are obtained by

$$\begin{cases} a_s = x_s & \text{for } s = 1,\dots,n-2, \\ b_s = y'_{s+1} & \text{for } s = 1,\dots,n-2, \\ c_s = \dfrac{x'_s}{n-s-1} & \text{for } s = 2,\dots,n-2, \\ d_s = \dfrac{y_{s+1}}{s} & \text{for } s = 1,\dots,n-3. \end{cases} \tag{12}$$

Note that that c_1 and d_{n-2} are arbitrary. We need the following definition:

Definition 10. The collection of weights $\{(m_{n,s})_{s=1,\dots,n-1} | n=2,\dots\}$ is said to be *RG-consistent*, if the weights satisfy, for n=2,...,

(i) $m_{n-1,1}=0$ implies $m_{n,1}=0$,
(ii) $m_{n-1,n-2}=0$ implies $m_{n,n-1}=0$, and
(iii) $m_{n-1,s}=m_{n-1,s-1}=0$ implies $m_{n,s}=0$ for $s=2,\dots,n-2$.

Thus the equation (6) has at least one solution given by (12) if and only if the weights $\{(m_{n,s})_{s=1,\dots,n-1} | n=2,\dots\}$ of the value is RG-consistent.

Now we present our main theorem that states the consistency of the linear value with ETP in terms of a reduced game.

Theorem 11. *The linear value with ETP in the form of* (1) *is the unique value on G which is $m_{2,1}/2$-S2G and possesses RGP for the reduced game* (5), *where the coefficients $(a_s,b_s,c_s,d_s)_{s=1,\dots,n-2}$ is given by* (12), *if the weights $\{(m_{n,s})_{s=1,\dots,n-1} | n=2,\dots\}$ of the value is RG-consistent.*

Proof. Note that the value is $m_{2,1}/2$-S2G. If the weights m is RG-consistent, then $(a_s,b_s,c_s,d_s)_{s=1,\dots,n-2}$ in (12) satisfy the relation (6). Therefore the theorem follows from Lemmata 7, 8, and 9. Q.E.D.

Let us compare our result with that of Ruiz et al. (1998). We firstly state the characterization of the LS value in terms of RGP in the next theorem. See Ruiz et al. (1998) or our main Theorem 11 and Remark 14 for proof. The weights m is *consistent* if it satisfies the following relation:

$$m_{n,s}=m_{n+1,s}+m_{n+1,s+1} \text{ for } n=2,\dots;s=1,\dots,n-1. \tag{13}$$

Theorem 12. *The LS value associated with weights m is the unique value on G which is S2G and possesses RGP with respect to the following reduced game; for $S \subset N-\{i\}$, $S \neq \emptyset, N-\{i\}$*

$$v^x(S) := \frac{m_{n,s+1}}{m_{n,s+1}+m_{n,s}}\left[v(S \cup \{i\}) - x_i\right] + \frac{m_{n,s}}{m_{n,s+1}+m_{n,s}} v(S), \tag{14}$$

if m is consistent.

The relationship between our RG-consistency and Ruiz et al.'s consistency (13) is given in the next proposition whose proof is omitted.

Proposition 13. *If the collection of nonnegative weights m is consistent, then it is also RG-consistent.*

We conclude that the result (in case of removing one player at a time) of Ruiz et al. (1998) is a special case of ours by Theorem 12, Proposition 13, and the following Remark 14.

Remark 14. If m is consistent, that is, $m_{n-1,s}=m_{n,s}+m_{n,s+1}$, then the reduced game (5) coincides with (14) if x, y, x', and y's in (12) are given by

$$\begin{cases} (x_s, y_s) = \begin{cases} \left(\dfrac{m_{n,s}}{m_{n-1,s}}, 0\right) & \text{for } s = 1,\dots,n-2, \\ \left(0, \dfrac{m_{n,n-1}}{m_{n-1,n-2}}\right) & \text{for } s = n-1, \end{cases} \\ (x'_s, y'_s) = \begin{cases} \left(\dfrac{m_{n,1}}{m_{n-1,1}}, 0\right) & \text{for } s = 1, \\ \left(0, \dfrac{m_{n,s}}{m_{n-1,s-1}}\right) & \text{for } s = 2,\dots,n-1. \end{cases} \end{cases}$$

4. SPECIAL FORM OF REDUCED GAMES THAT HAS AN INTERESTING MEANING

In this section we concentrate on the following form of a reduced game for the values in Example 4:

$$v^x(S) := r_s \left[\begin{array}{l} \dfrac{n-s-1}{n-1} p_s v(S) + \dfrac{1}{n-1}(1-p_s) \displaystyle\sum_{l \in N-S \cup \{i\}} v(S \cup \{l\}) \\ + \dfrac{1}{n-1} q_s \displaystyle\sum_{l \in S} \left[v(S \cup \{i\} - \{l\}) - x_l \right] + \dfrac{s}{n-1}(1-q_s)\left[v(S \cup \{i\}) - x_i \right] \end{array} \right] \tag{15}$$

where $0 \le p_s \le 1$, $0 \le q_s \le 1$, $r_s > 0$. The worth of S in the reduced game $(N-\{i\}, v^x)$ has the following meaning (see Figure 2).

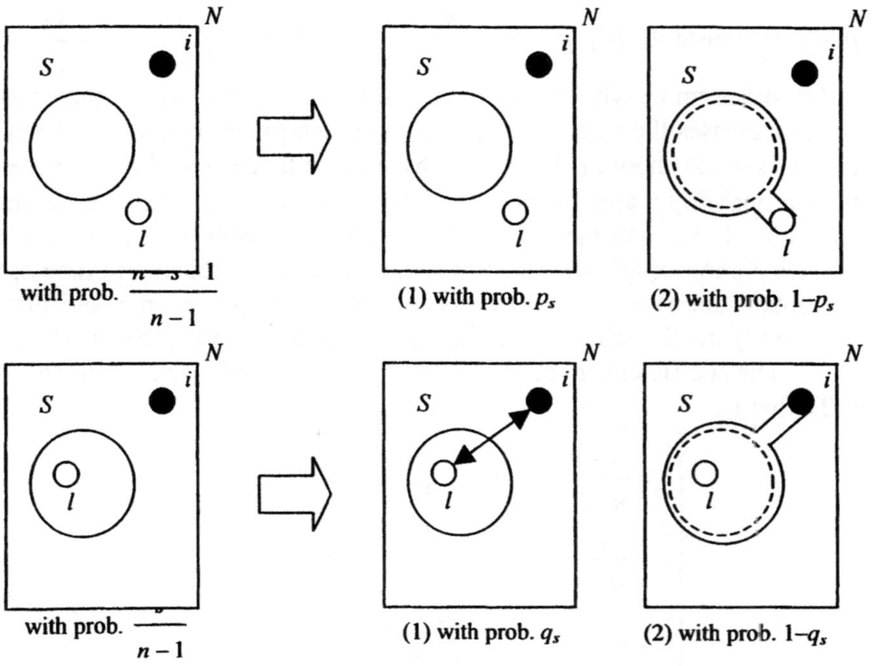

Figure 2. The way the coalition S tries to include player i.

Meaning: In order to estimate the worth of S in the reduced game $(N-\{i\},v^x)$, S tries to include player i, (1) by keeping the size of coalition, or (2) by increasing it by one player. But the candidate for (1) a replaced player or for (2) an adding player, say player l, is selected at random from $N-\{i\}$, that is, with probability $1/(n-1)$. If player l does not belong to S (see the upper diagram in Figure 2), then (1) the coalition S cannot include player i to form a coalition of the same size s with probability p_s and its worth in $(N-\{i\},v^x)$ is the same as $v(S)$, or (2) the coalition S cannot persuade player l to be replaced with player i to form a coalition of size $s+1$ with probability $1-p_s$ and its worth in $(N-\{i\},v^x)$ is $v(S\cup\{l\})$. If player l belongs to the coalition S (see the lower diagram in Figure 2), then (1) player i is replaced with player l to form the coalition $S\cup\{i\}-\{l\}$ with probability q_s, or (2) player i is directly added to form the coalition $S\cup\{i\}$ with probability $1-q_s$. When player i is removed from the game (N,v) with the payoff x_i, the coalition S in the reduced game $(N-\{i\},v^x)$ gets the rest of it, that is, (1) $v(S\cup\{i\}-\{l\})-x_i$ or (2) $v(S\cup\{i\})-x_i$. Thus the expected worth $v^x(S)$ of S in the reduced game $(N-\{i\},v^x)$ is given by (15), where r_s is some technical term. If the multiplicative term r_s is not equal to 1, then $v^x(S)$ is not an expected worth in the strict sense.

We give some interpretation of p_s and q_s for some cases, that is, (i)

$p_s=q_s=p$ (constant), (ii) $p_s = q_s = \dfrac{n-s-1}{n-1}$, and (iii) $p_s = q_s = \dfrac{n-s-1}{n}$. In order to determine whether (1) the coalition S keeps the size of the coalition or (2) increases the size by one player, a certain player is selected at random and is asked to choose (1) or (2). If we assume that every player chooses (1) with probability p and chooses (2) with probability $1-p$, then we obtain the case (i). If we assume that every player in S chooses (2), every player outside S chooses (1), and player i is excluded in this selection process, then we obtain the case (ii). If we assume that every player in S and player i choose (2) and the other player chooses (1), then we obtain the case (iii).

The coefficients p, q, and r's in (15) are derived from (12), that is,

$$\begin{cases} r_1 = \dfrac{n-1}{n-2}x_1 + (n-1)y_2, \\ r_s = \dfrac{n-1}{n-s-1}x_s + \dfrac{n-1}{s}y_{s+1} \quad \text{for } s = 2,\dots,n-3, \\ r_{n-2} = (n-1)x'_{n-2} + \dfrac{n-1}{n-2}y'_{n-1}, \\ p_s = \dfrac{n-1}{n-s-1}\dfrac{x_s}{r_s} \quad \text{for } s = 1,\dots,n-2, \\ q_s = 1 - \dfrac{n-1}{s}\dfrac{y'_{s+1}}{r_s} \quad \text{for } s = 1,\dots,n-2, \end{cases}$$

if x and y's satisfy $\dfrac{n-1}{n-s-1}x_s + \dfrac{n-1}{s}y_{s+1} = \dfrac{n-1}{n-s-1}x'_s + \dfrac{n-1}{s}y'_{s+1}$ for $s=2,\dots,n-3$.

By tedious but straightforward calculation, we can compute some interesting reduced games for values in Example 4. Table 1 summarizes the coefficients p, q, and r's of (15) for those reduced games. For convenience we present reduced games for the Solidarity value in the next corollary of our main Theorem 11 because they are not known yet:

$$v^x(S) = \begin{cases} \dfrac{n-s-1}{n-1}v(S) + \dfrac{1}{n-1}\displaystyle\sum_{l\in S}[v(S\cup\{i\}-\{l\}) - x_i] \quad (s \neq n-2), \\ \dfrac{n-s-1}{n-1}v(S) + \dfrac{1}{n}\displaystyle\sum_{l\in N-S\cup\{i\}} v(S\cup\{l\}) \\ + \dfrac{1}{n-1}\displaystyle\sum_{l\in S}[v(S\cup\{i\}-\{l\}) - x_i] \quad (s = n-2). \\ + \dfrac{n-2}{n}[v(S\cup\{i\}) - x_i] \end{cases} \tag{16}$$

$$v^x(S)=\begin{cases}\dfrac{n-2}{n-1}v(S)+\dfrac{s+1}{n(n-1)}\sum\limits_{l\in N-S\cup\{i\}}v(S\cup\{l\}) \\ +\dfrac{1}{n-1}\sum\limits_{l\in S}\left[v(S\cup\{i\}-\{l\})-x_i\right] & (s=1), \\ +\dfrac{s(s+1)}{n(n-1)}\left[v(S\cup\{i\})-x_i\right] \\ \dfrac{(n-s-1)^2}{n(n-1)}v(S)+\dfrac{s+1}{n(n-1)}\sum\limits_{l\in N-S\cup\{i\}}v(S\cup\{l\}) \\ +\dfrac{n-s-1}{n(n-1)}\sum\limits_{l\in S}\left[v(S\cup\{i\}-\{l\})-x_i\right] & (s\neq 1). \\ +\dfrac{s(s+1)}{n(n-1)}\left[v(S\cup\{i\})-x_i\right]\end{cases} \tag{17}$$

Corollary 15. *The Solidarity value is the unique value on G which is* 1/4-*S2G and possesses RGP for the reduced games given as follows, that is,* (16) *and* (17) *given above.*

Let us compare the differences between reduced games for various values. In Table 1 we list p, q, and r's of two reduced games for the Shapley value, those of one reduced game for LS prenucleolus, and those of two reduced games for the Solidarity value in this order. First we compare the Shapley value with the LS prenucleolus. Both of these two values satisfy S2G. If we set p=1/2 in the reduced game Sh_1 (the first row in Table 1) for the Shapley value, then we have $p_1=q_1=2/3$, $p_s=q_s=1/2$, $p_{n-2}=q_{n-2}=1/3$, and $r_1=r_{n-2}=3/2$, $r_s=1$. Thus, by comparing these p, q, and r's with those of the LS prenucleolus (the third row in Table 1), the difference between the Shapley value and LS prenucleolus is that in p_1, q_1, p_{n-2}, q_{n-2}, and r_s. The Shapley value and the LS prenucleolus differ mainly in the multiplicative term r_s; for the Shapley value it is equal to 1, whereas it is $1+1/(n-2)$ for the LS prenucleolus. If we set $\beta=n-2$ in the reduced game Sh_2 (the second row in Table 1) for the Shapley value, then we have $p_s=q_s=\dfrac{n-s-1}{n-1}$ and $r_s=\dfrac{n-1}{n-2}$. Thus, by comparing these p, q, and r's with those of the LS prenucleolus (the third row in Table 1), another interpretation of the difference between the Shapley value and LS prenucleolus is the difference in p_s and q_s; for the Shapley value, p and q's are interpreted as (ii) in the Meaning paragraph, while they are interpreted as (i) for the LS prenucleolus.

Table 1. Coefficients p, q, and r's for reduced games of various values.

	p_s	q_s	r_s
Sh_1	$p_1 = \dfrac{1}{2-p}$ $p_s = p(s \neq 1, n-2)$ $p_{n-2} = \dfrac{p}{1+p}$	$q_1 = \dfrac{1}{2-p}$ $q_s = p(s \neq 1, n-2)$ $q_{n-2} = \dfrac{p}{1+p}$	$r_1 = 2-p$ $r_s = 1(s \neq 1, n-2)$ $r_{n-2} = 1+p$
Sh_2	$p_1 = \dfrac{\beta}{3+2\beta-n}$ $p_s = \dfrac{n-s-1}{\beta+1}(s \neq 1)$	$q_1 = \dfrac{\beta}{3+2\beta-n}$ $q_s = \dfrac{n-s-1}{\beta+1}(s \neq 1)$	$r_1 = \dfrac{3+2\beta-n}{\beta}$ $r_s = \dfrac{\beta+1}{\beta}(s \neq 1)$
LS prenuc.	$p_s = \dfrac{1}{2}$	$q_s = \dfrac{1}{2}$	$r_s = \dfrac{n-1}{n-2}$
Sol (16)	$p_s = 1(s \neq n-2)$ $p_{n-2} = \dfrac{n}{2n-1}$	$q_s = 1(s \neq n-2)$ $q_{n-2} = \dfrac{n}{2n-1}$	$r_s = 1(s \neq n-2)$ $r_{n-2} = \dfrac{2n-1}{n}$
(17)	$p_1 = \dfrac{n}{n+2}$ $p_s = \dfrac{n-s-1}{n}(s \neq 1)$	$q_1 = \dfrac{n}{n+2}$ $q_s = \dfrac{n-s-1}{n}(s \neq 1)$	$r_1 = \dfrac{n+2}{n}$ $r_s = 1(s \neq 1)$

Next we compare the Shapley value with the Solidarity value. Note that the Shapley value satisfies S2G, that is, 1/2-S2G, while the Solidarity value satisfies 1/4-S2G. If we set p=1 in the reduced game Sh_1 (the first row in Table 1) for the Shapley value, then we have p_s=q_s=1, p_{n-2}=q_{n-2}=1/2, and r_s=1, r_{n-2}=2. By comparing these p, q, and r's with those in (16) (the fourth row in Table 1) for the Solidarity value, the difference between the Shapley value and the Solidarity value is that in p_{n-2}, q_{n-2}, and r_{n-2}; if we ignore this difference, then the Shapley value and the Solidarity value differ in α-standardness for two-person games (α=1/2 and α=1/4). If we set β=n−1 in the reduced game Sh_2 (the second row in Table 1) for the Shapley value, then we have $p_1 = q_1 = \dfrac{n-1}{n+1}$, $p_s = q_s = \dfrac{n-s-1}{n}$ and $r_1 = \dfrac{n+1}{n-1}, r_s = \dfrac{n}{n-1}$.

By comparing these p, q, and r's with those in (17) (the fifth row in Table 1) for the Solidarity value, the difference between the Shapley value and the Solidarity value is that in p_1, q_1, and r_s. Thus, in addition to the difference in α-standardness for two-person games, the Shapley value and the Solidarity value differ mainly in the multiplicative term r_s; the multiplicative term r_s for the Shapley is larger than that for the Solidarity value.

5. EXAMPLE AND APPLICATION

This section presents a numerical example and mentions potential application areas.

Let us consider a simple 4-person game given in Table 2. Player 2, 3, and 4 play the same role in the game. Player 1 has stronger position than them. Simple calculation shows that the Shapley value, the LS prenucleolus and the Solidarity value are given in Table 3. Which one of them is a reasonable solution in this situation? Consistency in terms of reduced game property will help us to select one solution as a reasonable one. Suppose, for example, that player 4 leaves the game with a payoff x. The reduced game (15) in this situation is given by

$$\begin{cases} v^x(123) = 100 - x, \\ v^x(12) = v^x(13) = \frac{r_2}{3}\left[80\left(1 - p_2\right) + 160\left(1 - q_2\right) - 2x\right], \\ v^x(23) = \frac{r_2}{3}\left[80\left(1 - p_2\right) + 100\left(1 - q_2\right) - 2x\right], \\ v^x(1) = v^x(2) = v^x(3) = -\frac{r_1}{3}x. \end{cases}$$

Table 2. Worth,

S	$v(S)$
1234	100
123	80
124	80
134	80
234	50
Otherwise	0

Table 3. Several solutions of the 4-person game.

	1	2	3	4
Sh	32.5	22.5	22.5	22.5
LS prenucleolus	30.625	23.125	23.125	23.125
Sol	26.875	24.375	24.375	24.375

If the players consider that the solution is 1/2-S2G and $p_2=q_2=1/3$ and $r_1=r_2=3/2$ ((ii) in the Meaning paragraph in Section 4) or $p_2=q_2=1/4$, $r_1=5/3$ and $r_2=4/3$ ((iii) in the Meaning paragraph in Section 4) as reasonable interpretation of them, then the solution should be the Shapley value. If the players consider that the solution is 1/2-S2G and $p_2=q_2=1/2$ and $r_1=r_2=3/2$ ((i) in the Meaning paragraph in Section 4) as reasonable interpretation of them, then the solution should be the LS prenucleolus. Finally if the players consider that the solution is 1/4-S2G and $p_2=q_2=1/4$, $r_1=3/2$ and $r_2=1$ ((iii) in the Meaning paragraph in Section 4) as reasonable interpretation of them, then the solution should be the Solidarity value.

Through a simple numerical example, we illustrate how to use our result in order to select a reasonable one among several solutions in the last paragraph. Other application areas of the consistency were discussed, for example, in Young (1987) and Thomson (1996). Young (1987) considered a taxation problem though it is not in a context of a transferable utility game. Thomson (1996) surveyed the applications of the consistency and converse consistency to a variety of problems in game theory, economics, and political science.

6. CONCLUSION

We considered the axiomatic characterization of linear values with ETP in terms of the reduced game property. We proposed reduced games and showed in our main Theorem 11 that the linear value with ETP is the unique value that is α-standard for two-person games and possesses RGP for those reduced games if it is RG-consistent. In particular we characterized the Solidarity value. We also compared the LS prenucleolus with the Shapley value and the Solidarity value with the Shapley value through our reduced games. Finally we discussed an example to illustrate how to use our result.

References

Driessen, T. S. H. and Funaki, Y. "Reduced Game Properties of Egalitarian Division Rules for TU-games." In eds T. Parthasarathy, B. Dutta, J. A. M. Potters, T. E. S. Raghavan, D. Ray and A. Sen. *Game Theoretical Applications to Economics and Operations Research*, Dordrecht: Kluwer Academic Publisher, 1997

Driessen, T. S. H. A Survey of Consistency Properties in Cooperative Game Theory. SIAM Review 1991; 33: 43-59

Namekata, T. and Driessen, T. S. H. The Egalitarian Non-k-Averaged Contribution (EN^{k}AC-) value for TU-Games. International Game Theory Review 1999; 1: 45-61

Namekata, T. and Driessen, T. S. H. Reduced Game Property of the Egalitarian Non-k-Averaged Contribution (EN^{k}AC-) Value and the Shapley Value. International Transactions in Operational Research 2000; 7: 365-382

Nowak, A. S. and Radzik, T. A Solidarity Value for n-Person Transferable Utility Games. International Journal of Game Theory 1994; 23: 43-48

Ruiz, L. M., Valenciano, F. and Zarzuelo, J. M. The Least Square Prenucleolus and the Least Square Nucleolus. Two Values for TU Games Based on the Excess Vector. International Journal of Game Theory 1996; 25: 113-134

Ruiz, L. M., Valenciano, F. and Zarzuelo, J. M. The Family of Least Square Values for Transferable Utility Games. Games and Economic Behavior 1998; 24: 109-130

Thomson, W. Consistent Allocation Rules. Rochester Center for Economics Research Working Paper No. 418 1996

Young, H. P. On dividing an amount according to individual claims or liabilities. Mathematics of Operations Research 1987; 12: 398-414

ALTERNATIVE TRANSFORMATIONS AND DUALITY OF LINEAR FRACTIONAL PROGRAMMING

G. C. O'Brien[1] and Lifen Wu[2]

[1] *Faculty of Law and Management, La Trobe University, Australia 3086*

[2] *Centre for Regional Economic Analysis, University of Tasmania, Australia 7005*

Abstract This paper discusses alternative solutions and dual formulations of linear fractional programming. These solutions lie between the two usual solutions obtained from denominator or numerator normalisation. Data Envelopment Analysis is presented as a special case of homogeneous linear fractional programming. The consequent analysis provide alternative solutions which lie between those from input- and output-oriented models for constant returns to scale DEA models and better solutions for variable returns to scale DEA models than those from input- and output-oriented models. A case study of metropolitan hospitals in Victoria, Australia is discussed.

Keywords: Linear fractional programming, duality, data envelopment analysis

1. LINEAR FRACTIONAL PROGRAMMING AND EQUIVALENT LINEAR PROGRAMMING MODELS

Fractional programming problems arise, for example, when maximising productivity, that is, maximising the ratio of output (of a combination of products) to labour input (heterogenous man hours), or when maximising profit to capital expended (rate of return) or when maximising expected value to the standard deviation of some measure of performance for an investment portfolio (return/risk) in circumstances where there are a number of incommensurate inputs and outputs. They occur in a diverse range of areas from financial analysis to sports performance assessment and location theory.

Special solution procedures have been developed for certain types of fractional programming problems to overcome the computational difficulties associated with optimisation in these circumstances. Pioneering work on fractional programming was carried out by Charnes and Cooper (1962). Stacu-Minasian (1997) provided a simplex algorithm which, unfortunately, does not apply to the case of homogeneous linear fractional programming problems. The most straightforward procedure is to transform the problem into an equivalent linear programming problem.

A linear fractional programming model has the form:

$$\text{Maximise } f(\mathbf{x}) = \frac{\mathbf{c}'\mathbf{x} + c_0}{\mathbf{d}'\mathbf{x} + d_0} \tag{1}$$

$$\text{subject to } \mathbf{Ax} \le \mathbf{b},$$

$$\text{and } \mathbf{x} \ge \mathbf{0},$$

where $\mathbf{c}$, $\mathbf{d}$ and $\mathbf{x} \in R^n$ and c_0 and d_0 are fixed scalars.

Provided that $\mathbf{d}'\mathbf{x} + d_0 \ne 0$, then the problem can be transformed into an equivalent *linear programming problem* by setting $\mathbf{y} = \dfrac{\mathbf{x}}{\mathbf{d}'\mathbf{x} + d_0}$ and $t = \dfrac{1}{\mathbf{d}'\mathbf{x} + d_0}$, so that $\mathbf{x} = \mathbf{y}/t$. This yields the equivalent *denominator normalised* LP problem

$$\text{Maximise } Z = \mathbf{c}'\mathbf{y} + c_0 t, \tag{2}$$

$$\text{subject to } \mathbf{Ay} - \mathbf{b}t \le \mathbf{0},$$

$$\mathbf{d}'\mathbf{y} + d_0 t = 1,$$

$$\text{and } \mathbf{y} \ge \mathbf{0}, \quad t \ge 0,$$

which can be solved by the simplex method.

Similarly, there is a *numerator normalised* equivalent linear program, for $\mathbf{c}'\mathbf{x} + c_0 \ne 0$,

$$\text{Minimise } Z = \mathbf{d}'\mathbf{y} + d_0 t, \tag{3}$$

$$\text{subject to } -\mathbf{Ay} + \mathbf{b}t \ge \mathbf{0}$$

$$\mathbf{c}'\mathbf{y} + c_0 t = 1,$$

$$\text{and } \mathbf{y} \ge \mathbf{0}, \quad t \ge 0.$$

These two equivalent linear programming problems have different duals.

The dual of (2) is

$$\text{Minimise } \theta \tag{4}$$

$$\text{subject to } \mathbf{z}'\mathbf{A} + \mathbf{d}\theta \ge \mathbf{c},$$

$$-\mathbf{z}'\mathbf{b} + d_0\theta \ge c_0,$$

$$\text{and } \mathbf{z} \ge \mathbf{0}, \mathbf{z} \in R^n, \theta \in R.$$

whereas, the dual of (3) is

$$\text{Maximise } \phi \tag{5}$$

$$\text{subject to } -\mathbf{z}'\mathbf{A} + \mathbf{c}\phi \le \mathbf{d},$$

$$\mathbf{z'b} + c_0\phi \le d_0,$$

and $\mathbf{z} \ge \mathbf{0}, \mathbf{z} \in R^n, \phi \in R.$

An alternative is to consider a simple affine transformation of (1). Set $\mathbf{y} = \dfrac{\mathbf{x}}{\phi(\mathbf{d'x} + d_0)}$ and $t = \dfrac{1}{\phi(\mathbf{d'x} + d_0)}$, so that $\mathbf{x} = \mathbf{y}/t$. Then (1) is equivalent to

$$\text{Maximise } Z = \phi(\mathbf{c'y} + c_0 t) \tag{6}$$

subject to
$$\mathbf{Ay} - \mathbf{b}t \le \mathbf{0}$$
$$\mathbf{d'y} + d_0 t = \frac{1}{\phi},$$
and $\mathbf{y} \ge \mathbf{0}$, $t \ge 0$,

Alternatively, set $\mathbf{y} = \dfrac{\mathbf{x}}{\theta(\mathbf{c'x} + c_0)}$ and $t = \dfrac{1}{\theta(\mathbf{c'x} + c_0)}$, so that $\mathbf{x} = \mathbf{y}/t$. Then (1) is equivalent to

$$\text{Minimise } Z = \theta(\mathbf{d'y} + d_0 t), \tag{7}$$

subject to
$$-\mathbf{Ay} + \mathbf{b}t \ge \mathbf{0}$$
$$\mathbf{c'y} + c_0 t = \frac{1}{\theta},$$
and $\mathbf{y} \ge \mathbf{0}$, $t \ge 0$.

The dual of (6) is

$$\text{Minimise } \frac{\theta}{\phi} \tag{8}$$

subject to
$$\mathbf{z'A} + \mathbf{d}\theta \ge \mathbf{c}\phi,$$
$$-\mathbf{z'b} + d_0\theta \ge c_0\phi,$$
and $\mathbf{z} \ge \mathbf{0}.$

and the dual of (7) is

$$\text{Maximise } \frac{\phi}{\theta} \tag{9}$$

subject to
$$-\mathbf{z'A} + \mathbf{c}\phi \le \mathbf{d}\theta,$$
$$\mathbf{z'b} + c_0\phi \le d_0\theta,$$
and $\mathbf{z} \ge \mathbf{0}.$

Obviously, (8) and (9) are identical and can be written as

$$\text{Minimise } \frac{\theta}{\phi} \tag{10}$$

subject to $\mathbf{z'A} - \mathbf{c}\phi + \mathbf{d}\theta \ge \mathbf{0},$

$$\mathbf{z'b} + c_0\phi - d_0\theta \leq \mathbf{0},$$

and $$\mathbf{z} \geq \mathbf{0}.$$

Because (8) is the dual of (6), we have $\phi(\mathbf{c'y} + c_0 t) \leq \frac{\theta}{\phi}$, i.e., $\frac{\mathbf{c'x} + c_0}{\mathbf{d'x} + d_0} \leq \frac{\theta}{\phi}$; Similarly because (9) is the dual of (7), we have $\theta(\mathbf{d'y} + d_0 t) \geq \frac{\phi}{\theta}$, i.e., $\frac{\mathbf{c'x} + c_0}{\mathbf{d'x} + d_0} \leq \frac{\theta}{\phi}$. Therefore, equation (10) is a fractional programming form of the dual of (1). The relationship of (1) through (10) can be demonstrated in the following Table.

Table 1. Relationship of linear fractional programming formulations

	$\theta = 1$	Numerator Normalisation		Denominator Normalisation	$\phi = 1$
Primal	(3)	(7)	(1)	(6)	(2)
Dual	(5)	(9)	(10)	(8)	(4)

2. CONSTANT RETURNS TO SCALE DEA MODEL

A special case of linear fractional programming is Data Envelopment Analysis, which is an application of homogeneous linear fractional programming. The following model is from Charnes, Cooper and Rhodes (1978, 1979):

$$\text{Maximize } \frac{\sum_{r=1}^{s} u_r y_{rk}}{\sum_{i=1}^{m} v_i x_{ik}} \tag{11}$$

$$\text{subject to } \quad \frac{\sum_{r=1}^{s} u_r y_{rj}}{\sum_{i=1}^{m} v_i x_{ij}} \leq 1, \ j = 1,\dots,n;$$

$$u_r \geq \varepsilon, \ r = 1,\dots,s;$$

$$v_i \geq \varepsilon, \ i = 1,\dots,m.$$

where x_{ij} and y_{rj} are the i-th input and r-th output of unit j, $\varepsilon > 0$ is a non-Archimedean number, and u_r and v_i are the (unknown) weights given to the inputs and outputs respectively.

Equation (11) is usually solved by being transformed into one of two equivalent linear programming models, ie. an input-oriented model by denominator normalisation or an output-oriented model by numerator normalisation.

Denominator normalisation involves setting $t = \dfrac{1}{\sum_{i=1}^{m} v_i x_{ik}}$, $\mu_r = tu_r$,

and $\nu_i = tv_i$, then $\sum_{r=1}^{s} \mu_r y_{rj} = \dfrac{\sum_{r=1}^{s} u_r y_{rj}}{\sum_{i=1}^{m} v_i x_{ik}}$ and $\sum_{i=1}^{m} \nu_i x_{ij} = \dfrac{\sum_{i=1}^{m} v_i x_{ij}}{\sum_{i=1}^{m} v_i x_{ik}}$.

The input-oriented equivalent model is as following,

$$\text{Maximise} \sum_{r=1}^{s} \mu_r y_{rk} \tag{12}$$

$$\begin{aligned} \text{subject to } & \sum_{i=1}^{m} \nu_i x_{ik} = 1 \\ & \sum_{r=1}^{s} \mu_r y_{rj} - \sum_{i=1}^{m} \nu_i x_{ij} \le 0, \ j = 1,\dots,n; \\ & \mu_r \ge \varepsilon t, \ r = 1,\dots,s; \\ & \nu_i \ge \varepsilon t, \ i = 1,\dots,m; \\ & t \ge 0. \end{aligned}$$

Numerator normalisation involves setting $t = \dfrac{1}{\sum_{r=1}^{s} u_r y_{rk}}$, $\mu_r = tu_r$, and

$\nu_i = tv_i$, then $\sum_{r=1}^{s} \mu_r y_{rj} = \dfrac{\sum_{r=1}^{s} u_r y_{rj}}{\sum_{r=1}^{s} u_r y_{rk}}$ and $\sum_{i=1}^{m} \nu_i x_{ij} = \dfrac{\sum_{i=1}^{m} v_i x_{ij}}{\sum_{r=1}^{s} u_r y_{rk}}$.

The output-oriented equivalent model is as follows,

$$\text{Minimise} \sum_{i=1}^{m} \nu_i x_{ik} \tag{13}$$

$$\text{subject to} \sum_{r=1}^{s} \mu_r y_{ik} = 1$$

$$\sum_{r=1}^{s} \mu_r y_{rj} - \sum_{i=1}^{m} \nu_i x_{ij} \le 0, \ j = 1,\dots,n;$$

$$\mu_r \geq \varepsilon t\ ,\ r = 1,\ldots,s\ ;$$
$$v_i \geq \varepsilon t\ ,\ i = 1,\ldots,m\ ;$$
$$t \geq 0\ .$$

(11), (12) and (13) are in multiplier form. The dual is called the envelopment form. The dual of (12) is,

Minimise θ_k (14)

$$\text{subject to} \sum_{j=1}^{n} y_{rj}\lambda_j - s_r^+ = y_{rk}\ , r = 1,\ldots,s\ ;$$
$$-\sum_{j=1}^{n} x_{ij}\lambda_j - s_i^- = -\theta_k x_{ik}\ , i = 1,\ldots,m\ ;$$
$$\varepsilon\left(\sum_{r=1}^{s} s_r^+ + \sum_{i=1}^{m} s_i^-\right) \geq 0$$
$$\lambda_j \geq 0\ , j = 1,\ldots,n\ ;$$
$$s_r^+ \geq 0\ , r = 1,\ldots,s\ ;$$
$$s_i^- \geq 0\ , i = 1,\ldots,m\ .$$

The dual of (13) is as following,

Maximise ϕ_k (15)

$$\text{subject to} \sum_{j=1}^{n} y_{rj}\lambda_j - s_r^+ = \phi_k y_{rk}\ , r = 1,\ldots,s\ ;$$
$$-\sum_{j=1}^{n} x_{ij}\lambda_j - s_i^- = -x_{ik}\ , i = 1,\ldots,m\ ;$$
$$-\varepsilon\left(\sum_{r=1}^{s} s_r^+ + \sum_{i=1}^{m} s_i^-\right) \leq 0$$
$$\lambda_j \geq 0\ , j = 1,\ldots,n\ ;$$
$$s_r^+ \geq 0\ , r = 1,\ldots,s\ ;$$
$$s_i^- \geq 0\ , i = 1,\ldots,m\ .$$

If, in (11), we set $t = \dfrac{1}{\phi_k \sum_{i=1}^{m} v_i x_{ik}}$, $\mu_r = t u_r$, and $v_i = t v_i$, then

$$\sum_{r=1}^{s} \mu_r y_{rj} = \frac{\sum_{r=1}^{s} u_r y_{rj}}{\phi_k \sum_{i=1}^{m} v_i x_{ik}} \text{ and } \sum_{i=1}^{m} v_i x_{ij} = \frac{\sum_{i=1}^{m} v_i x_{ij}}{\phi_k \sum_{i=1}^{m} v_i x_{ik}}.$$

The input-oriented equivalent model becomes,

$$\text{Maximise}\ \phi_k \sum_{r=1}^{s} \mu_r y_{rk} \tag{16}$$

$$\text{subject to} \sum_{i=1}^{m} v_i x_{ik} = \frac{1}{\phi_k}$$

$$\sum_{r=1}^{s} \mu_r y_{rj} - \sum_{i=1}^{m} v_i x_{ij} \leq 0\,,\ j = 1,\ldots,n\,;$$

$$\mu_r \geq \varepsilon t\,,\ r = 1,\ldots,s\,;$$

$$v_i \geq \varepsilon t\,,\ i = 1,\ldots,m\,;$$

$$t \geq 0\,.$$

If, in (11), we set $t = \dfrac{1}{\theta_k \sum_{r=1}^{s} u_r y_{rk}}$, $\mu_r = t u_r$, and $v_i = t v_i$, then

$$\sum_{r=1}^{s} \mu_r y_{rj} = \frac{\sum_{r=1}^{s} u_r y_{rj}}{\theta_k \sum_{r=1}^{s} u_r y_{rk}} \text{ and } \sum_{i=1}^{m} v_i x_{ij} = \frac{\sum_{i=1}^{m} v_i x_{ij}}{\theta_k \sum_{r=1}^{s} u_r y_{rk}}.$$

The output-oriented equivalent model becomes,

$$\text{Minimise}\ \theta_k \sum_{i=1}^{m} v_i x_{ik} \tag{17}$$

$$\text{subject to} \sum_{r=1}^{s} \mu_r y_{rk} = \frac{1}{\theta_k}$$

$$\sum_{r=1}^{s} \mu_r y_{rj} - \sum_{i=1}^{m} v_i x_{ij} \leq 0\,,\ j = 1,\ldots,n\,;$$

$$\mu_r \geq \varepsilon t\,,\ r = 1,\ldots,s\,;$$

$$v_i \geq \varepsilon t\,,\ i = 1,\ldots,m\,;$$

$$t \geq 0\,.$$

The dual of (16) and (17) are identical and can be written as,

$$\text{Minimise}\ \frac{\theta_k}{\phi_k} \tag{18}$$

$$\text{subject to} \sum_{j=1}^{n} y_{rj}\lambda_j - s_r^{+} = \phi_k y_{rk}\,, r = 1,\ldots,s\,;$$

$$-\sum_{j=1}^{n} x_{ij}\lambda_j - s_i^{-} = -\theta_k x_{ik}\,, i = 1,\ldots,m\,;$$

$$\varepsilon\left(\sum_{r=1}^{s} s_r^+ + \sum_{i=1}^{m} s_i^-\right) \geq 0$$

$$\lambda_j \geq 0, j = 1,\ldots,n;$$

$$s_r^+ \geq 0, r = 1,\ldots,s;$$

$$s_i^- \geq 0, i = 1,\ldots,m.$$

Equation (18) can be viewed as a mixed-oriented constant returns to scale model. Obviously, setting $\phi_k = 1$, (18) becomes the input-oriented CCR model; while setting $\theta_k = 1$, (18) becomes output-oriented CCR model.

If ϕ_k is selected to be a constant, set $\dfrac{\theta_k}{\phi_k} = \theta'_k$, $\phi_k y_{rk} = y'_{rk}$ and $\phi_k x_{ik} = x'_{ik}$, then $-\theta_k x_{ik} = -\theta'_k \phi_k x_{ik} = -\theta'_k x'_{ik}$, and input-oriented CCR model can be used.

If θ_k is selected to be a constant, set $\dfrac{\phi_k}{\theta_k} = \phi'_k$, $\theta_k y_{rk} = y'_{rk}$ and $\theta_k x_{ik} = x'_{ik}$, then $\phi_k y_{rk} = \phi'_k \theta_k y_{rk} = \phi'_k y'_{rk}$, and output-oriented CCR model can be used.

The minimum value of θ_k in input-oriented CCR model (θ_{kI}) when $\phi_k = 1$ and the maximum value of ϕ_k in output-oriented CCR model (ϕ_{kO}) when $\theta_k = 1$ generated from currently existed DEA codes are very useful. If $\theta_{kI} \leq \theta_k \leq 1$ in the output-oriented model or $1 \leq \phi_k \leq \phi_{kO}$ in the input-oriented model is set, either oriented model will fix $\dfrac{\theta_k}{\phi_k}$ to $\theta_{kI} = \dfrac{1}{\phi_{kO}}$ and an inefficient DMU can be projected to any interpolated point on a line formed by the two projected points of the two oriented models as follows,

$$\left(\theta_k x_{ik} - s_i^-, \phi_k y_{rk} + s_r^+\right) = \frac{1-\theta_k}{1-\theta_{kI}}\left(\theta_{kI} x_{ik} - s_{iI}^-, y_{rk} + s_{rI}^+\right) +$$

$$+\frac{\phi_k - 1}{\phi_{kO} - 1}\left(x_{ik} - s_{iO}^-, \phi_{kO} y_{rk} + s_{rO}^+\right) \text{ with } \frac{\theta_k}{\phi_k} = \theta_{kI} = \frac{1}{\phi_{kO}}$$

ie. $\left(\hat{x}_{ik}, \hat{y}_{rk}\right) = \dfrac{1-\theta_k}{1-\theta_{kI}}\left(\hat{x}_{ikI}, \hat{y}_{rkI}\right) + \dfrac{\phi_k - 1}{\phi_{kO} - 1}\left(\hat{x}_{ikO}, \hat{y}_{rkO}\right)$ with $\dfrac{\theta_k}{\phi_k} = \theta_{kI} = \dfrac{1}{\phi_{kO}}$

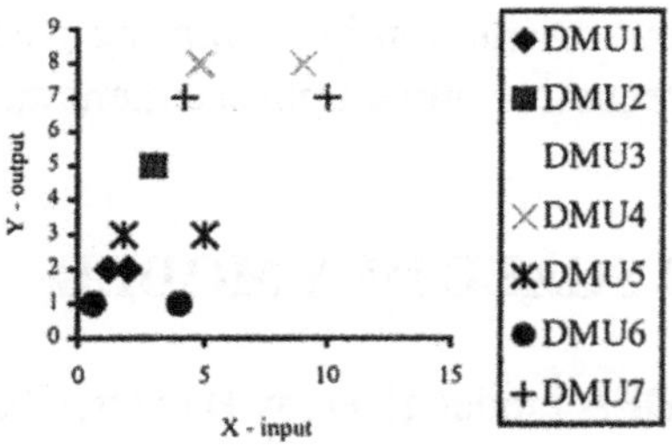

Figure 1. Input-oriented CCR model

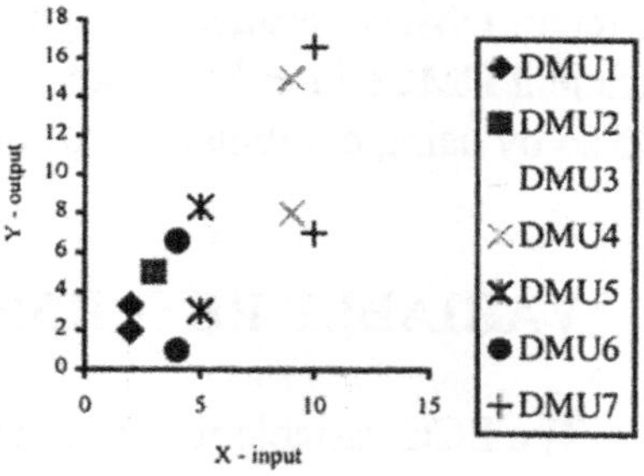

Figure 2. Output-oriented CCR model

Figures 1 and 2 demonstrate the two oriented CCR models[1]. DMU5 (5,3) is projected to (9/5,3) in input-oriented CCR model with $\theta_5 = \frac{9}{25}$ or to (5,25/3) in output-oriented CCR model with $\phi_5 = \frac{25}{9}$. If $\theta_5 = \frac{3}{5}$ is set, DMU5 can be projected to DMU2 (3,5) with $\phi_5 = \frac{5}{3}$. DMU6 (4,1) is projected to (3/5,1) in input-oriented CCR model with $\theta_6 = \frac{3}{20}$ or to (4,20/3) in output-oriented CCR model with $\phi_6 = \frac{20}{3}$. If $\phi_6 = 5$ is set, DMU6 can be projected to DMU2 (3,5) with $\phi_6 = \frac{3}{4}$, as shown in Figure 3.

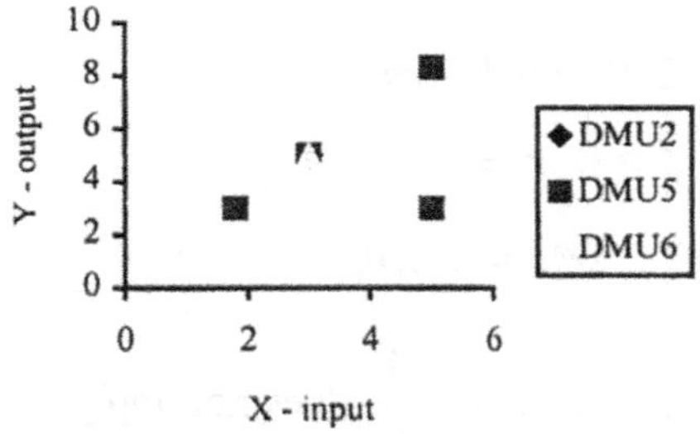

Figure 3. Mixed-orientation CRS model

(18) is the dual of (11) and provides more flexible projections for inefficient DMUs because it might be difficult or even impossible to project inefficient DMUs in either input- or output-orientation but might be practical to do so by using combinations of input reduction and output augmentation.

3. VARIABLE RETURNS TO SCALE DEA MODEL

The BCC variable returns to scale DEA model given by Banker, Charnes and Cooper (1984) imposes one more constraint, $\sum_{j=1}^{n} \lambda_j = 1$, in envelopment form of an oriented constant returns to scale DEA model. Similarly this convex constraint can be imposed as in (18) to obtain the envelopment form ofa mixed oriented variable returns to scale DEA model,

$$\text{Minimise} \frac{\theta_k}{\phi_k} \tag{19}$$

$$\text{subject to} \sum_{j=1}^{n} y_{rj}\lambda_j - s_r^+ = \phi_k y_{rk}, r = 1,\ldots,s;$$

$$-\sum_{j=1}^{n} x_{ij}\lambda_j - s_i^- = -\theta_k x_{ik}, i = 1,\ldots,m;$$

$$\varepsilon\left(\sum_{r=1}^{s} s_r^+ + \sum_{i=1}^{m} s_i^-\right) \geq 0$$

$$\sum_{j=1}^{n} \lambda_j = 1;$$

$$\lambda_j \geq 0, j = 1,\ldots,n;$$

$$s_r^+ \geq 0, r = 1,\ldots,s;$$

$$s_i^- \geq 0, i = 1,\ldots,m.$$

The constancy of the value of $\frac{\theta_k}{\phi_k}$ does not apply in the variable returns to scale DEA model, because θ_{kI} does not equal $\frac{1}{\phi_{kO}}$ since an inefficient DMU might be projected onto a different segment of the efficiency frontier by the output-oriented model than by the input-oriented model. In addition, nether the input-oriented model nor the output-oriented model might identify the optimal solution in the case where an inefficient DMU is projected onto an increasing returns to scale segment of the frontier by the input-oriented model and onto a decreasing returns to scale segment by the output-oriented model. In variable returns to scale DEA models, θ_k reaches its minimum value in the input-oriented model; while ϕ_k reaches its maximum value in the output-oriented model. However, the minimum $\frac{\theta_k}{\phi_k}$ value might not be achieved in either of the above two oriented models in that case. From the point of view of (19), the optimal solution obtained from either the input-oriented model or from the output-oriented model may therefore be suboptimal.

Similar to (18), if ϕ_k is selected to be a constant, set $\frac{\theta_k}{\phi_k} = \theta'_k$, $\phi_k y_{rk} = y'_{rk}$ and $\phi_k x_{ik} = x'_{ik}$, and input-oriented BCC model can be used; If θ_k is selected to be a constant, set $\frac{\phi_k}{\theta_k} = \phi'_k$, $\theta_k y_{rk} = y'_{rk}$ and $\theta_k x_{ik} = x'_{ik}$, and output-oriented BCC model can be used. $\frac{1}{\phi_{ko}}$ is the minimum value of $\frac{\theta_k}{\phi_k}$ in a case where both input- and output-oriented models project an inefficient DMU onto an increasing returns to scale segment frontier while θ_{kI} is the minimum value of $\frac{\theta_k}{\phi_k}$ in the case where both oriented models project that DMU onto a decreasing returns to scale segment. To obtain the minimum value of $\frac{\theta_k}{\phi_k}$ in (19) in the case where an inefficient DMU is projected onto increasing returns to scale segment of frontier by input-oriented model but onto decreasing returns to scale segment by output-oriented model, select some values of $\theta_{kI} \le \theta_k \le 1$ or $1 \le \phi_k \le \phi_{ko}$ and a series of $\frac{\theta_k}{\phi_k}$ values can be obtained. The minimum value of $\frac{\theta_k}{\phi_k}$ equals its value in constant returns to scale model and is located at the intersection of the different segments of the efficiency frontier. The ratio of increment of λ_j in comparison or reference set of DMUk over increment of θ_k or ϕ_k in different segment of efficiency frontier is different. Solving two simultaneous equations of that ratio in two segments intersecting the minimum $\frac{\theta_k}{\phi_k}$ value can obtain its θ_k or ϕ_k value.

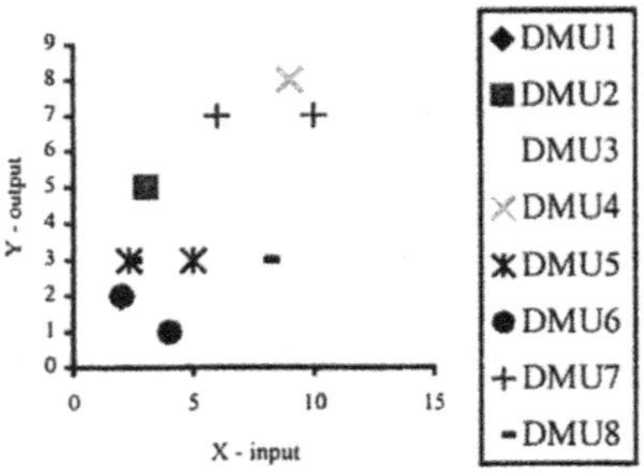

Figure 4. Input-oriented BCC model

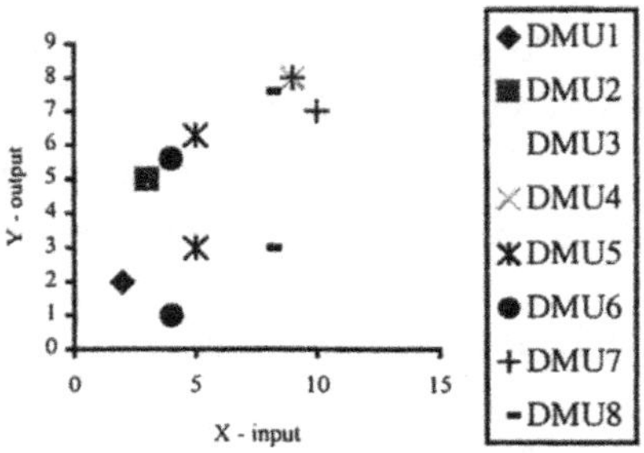

Figure 5. Output-oriented BCC model

As shown in Figures 4 and 5, DMU8 (8, 3) is projected to (7/3, 3) with $\theta_8 = \dfrac{7}{24}$ by input-oriented variable returns to scale model or to (8, 23/3) with $\phi_8 = \dfrac{23}{9}$ by output- oriented model. However, if $\theta_8 = \dfrac{3}{4}$ and $\phi_8 = \dfrac{7}{3}$ are set, DMU8 can be projected to DMU3 (6, 7) with $\dfrac{\theta_8}{\phi_8} = \dfrac{9}{28}$ and if $\theta_8 = \dfrac{3}{8}$ and $\phi_8 = \dfrac{5}{3}$ are set, DMU8 can be projected to DMU2 (3, 5) with $\dfrac{\theta_8}{\phi_8} = \dfrac{9}{40} < \dfrac{7}{24} < \dfrac{9}{28} < \dfrac{9}{23}$ as shown in Figure 6.

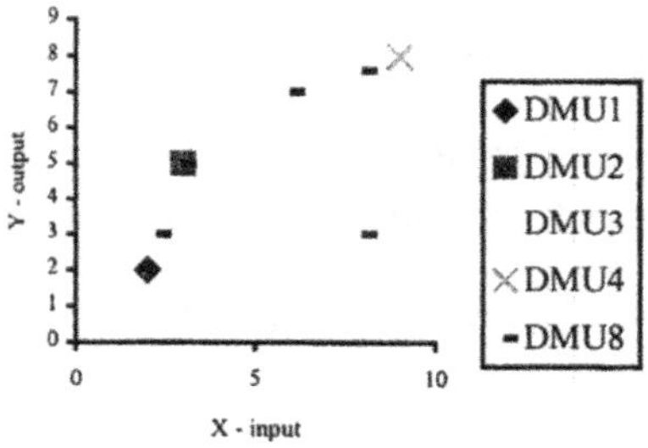

Figure 6. Mixed-orientation VRS model

Any projected point on the same segment as two given projected points satisfies the following relationships

$$\frac{\theta_k-\theta_{k2}}{\phi_k-\phi_{k2}}=\frac{\theta_{k1}-\theta_{k2}}{\phi_{k1}-\phi_{k2}},$$

$$\left(\theta_k x_{ik}-s_i^-,\phi_k y_{rk}+s_r^+\right)=$$

$$=\frac{\theta_k-\theta_{k2}}{\theta_{k1}-\theta_{k2}}\left(\theta_{k1}x_{ik}-s_{i1}^-,\phi_{k1}y_{rk}+s_{r1}^+\right)+\frac{\theta_1-\theta}{\theta_1-\theta_2}\left(\theta_2 x_{ik}-s_{i2}^-,\phi_2 y_{rk}+s_{r2}^+\right)$$

i.e. $\left(\hat{x}_{ik},\hat{y}_{rk}\right)=\dfrac{\theta_k-\theta_{k2}}{\theta_{k1}-\theta_{k2}}\left(\hat{x}_{ik1},\hat{y}_{rk1}\right)+\dfrac{\theta_{k1}-\theta_k}{\theta_{k1}-\theta_{k2}}\left(\hat{x}_{ik2},\hat{y}_{rk2}\right)$

or, $\left(\theta_k x_{ik}-s_i^-,\phi_k y_{rk}+s_r^+\right)=\dfrac{\phi_k-\phi_{k2}}{\phi_{k1}-\phi_{k2}}\left(\theta_{k1}x_{ik}-s_{i1}^-,\phi_{k1}y_{rk}+s_{r1}^+\right)+$

$$+\frac{\phi_{k1}-\phi_k}{\phi_{k1}-\phi_{k2}}\left(\theta_{k2}x_{ik}-s_{i2}^-,\phi_{k2}y_{rk}+s_{r2}^+\right)$$

i.e. $\left(\hat{x}_{ik},\hat{y}_{rk}\right)=\dfrac{\phi_k-\phi_{k2}}{\phi_{k1}-\phi_{k2}}\left(\hat{x}_{ik1},\hat{y}_{rk1}\right)+\dfrac{\phi_{k1}-\phi_k}{\phi_{k1}-\phi_{k2}}\left(\hat{x}_{ik2},\hat{y}_{rk2}\right)$,

where the subscripts 1 and 2 refer to the two given projected points.

(19) can reach different sub-optimal solutions depending on the value of θ_k or ϕ_k. Consider a multiplier form of (19) as following,

$$\text{Maximise } \frac{\sum_{r=1}^{s} u_r y_{rk}}{\sum_{i=1}^{m} v_i x_{ik}} - t o_k \tag{20}$$

$$\text{subject to } \sum_{r=1}^{s} u_r y_{rj} - \sum_{i=1}^{m} v_i x_{ij} - o_k \le 0,\ j=1,\ldots,n;$$

$$u_r \ge \varepsilon,\ r=1,\ldots,s;$$

$$v_i \ge \varepsilon,\ i=1,\ldots,m.$$

The input-oriented form of (20) can be obtained by setting $t=\dfrac{1}{\phi_k\sum_{i=1}^{m} v_i x_{ik}}$, $\mu_r = t u_r$, $v_i = t v_i$ and $\omega_k = t o_k$, then (20) becomes

$$\text{Maximise } \phi_k \sum_{r=1}^{s} \mu_r y_{rk} - \omega_k \tag{21}$$

$$\text{subject to } \sum_{i=1}^{m} v_i x_{ik} = \frac{1}{\phi_k}$$

$$\sum_{r=1}^{s}\mu_r y_{rj}-\sum_{i=1}^{m}v_i x_{ij}-\omega_k\leq 0,\ j=1,\ldots,n;$$
$$\mu_r\geq\varepsilon t,\ r=1,\ldots,s;$$
$$v_i\geq\varepsilon t,\ i=1,\ldots,m;$$
$$t\geq 0.$$

The output-oriented form of (20) can be obtained by setting $t=\dfrac{1}{\theta_k\sum_{r=1}^{s}u_r y_{rk}}$, $\mu_r=tu_r$, $v_i=tv_i$ and $\omega_k=to_k$.

The objective of (20) becomes

$$\text{Maximise}\frac{\sum_{r=1}^{s}u_r y_{rk}}{\sum_{i=1}^{m}v_i x_{ik}}+\text{Maximise}(-to_k)=\text{Minimise}\frac{\sum_{i=1}^{m}v_i x_{ik}}{\sum_{r=1}^{s}u_r y_{rk}}+\text{Minimise}\,to_k$$

$=\text{Minimise}\dfrac{\sum_{i=1}^{m}v_i x_{ik}}{\sum_{r=1}^{s}u_r y_{rk}}+to_k$, and (20) becomes

$$\text{Minimise}\,\theta_k\sum_{i=1}^{m}v_i x_{ik}+\omega_k \tag{22}$$

$$\text{subject to } \sum_{r=1}^{s}\mu_r y_{rk}=\frac{1}{\theta_k}$$
$$\sum_{r=1}^{s}\mu_r y_{rj}-\sum_{i=1}^{m}v_i xij-\omega_k\leq 0,\ j=1,\ldots,n;$$
$$\mu_r\geq\varepsilon t,\ r=1,\ldots,s;$$
$$v_i\geq\varepsilon t,\ i=1,\ldots,m;$$
$$t\geq 0.$$

4. CASE STUDY

Government funded services contribute about 10 per cent to Australia's gross domestic product and affect the daily lives of all Australians. For example, governments make available education and training, health and community services; they maintain law and order; and they provide subsidised care and shelter for people in need. Improvements in the performance of these government funded 'human' services have the potential to deliver major social and economic benefits and to improve our capacity to address social needs more effectively.

Human service delivery performance is receiving increased scrutiny as part of the ongoing process of microeconomic reform. In the absence of market contestability, comparative performance reporting provides a way of introducing competitive pressures for government service providers. Developing and reporting performance indicators is crucial to identifying performance improvements, and thus guiding decision making. Comparative performance reporting is typically undertaken cooperatively to assist all participants to improve their performance; it is not focused on determining if a specific level of performance is being achieved.

The Steering Committee for the Review of Commonwealth/State Service Provision, established in 1993, is developing and reporting jurisdiction based performance indicators for human services funded by governments. Their 1997 report[2] contains applications applying DEA to assess:

- acute care services in Victorian hospitals;
- Queensland oral health services for school students;
- NSW correctional centres;
- NSW police patrols; and
- NSW Roads and Traffic Authority (RTA) motor registries.

Many of their results may be sensitive because the optimal solution is attained in input or mixed orientation rather than in output orientation which they used to assess technical efficiency. The ranking and input targets of the assessed inefficient DMUs might be less accurate and could mislead these service delivery organisations rather than identifying efficient ones. For example, their Model 2[3] assessed Metropolitan/large country hospitals in Victoria. In 14 inefficient hospitals[4], only HP33 was correctly assessed in the output oriented model because both the input- and the output-oriented BCC model project it onto the increasing returns to scale segment of the efficiency frontier, 11 should be assessed in input oriented model because both the input- and the output-oriented BCC model project them onto the decreasing returns to scale segment, and HP23 and HP22 should be assessed in mixed orientation model because the input-oriented BCC model projects them onto the increasing returns to scale segment of the efficiency frontier and the output-oriented BCC model projects them onto the decreasing returns to scale segment. By such a mixed orientation projection, HP23 and HP22 are improved as efficient DMUs not only under variable returns to scale but also under constant returns to scale.

Table 2. Efficiency scores

	Constant returns to scale results					Variable returns to scale results			
	θ	ϕ	θ/ϕ	ϕ/θ	$\Sigma\lambda$	θ	ϕ	θ/ϕ	ϕ/θ
HP9	0.685	1	0.685	1.459	2.921	0.99	1	0.99	1.01
	1	1.459	0.685	1.459	4.261	1	1.002	0.998	1.002
HP17	0.908	1	0.908	1.101	5.457	0.993	1	0.993	1.007
	1	1.101	0.908	1.101	6.008	1	1.007	0.993	1.007
HP19	0.971	1	0.971	1.03	1.664	0.983	1	0.983	1.017
	1	1.03	0.971	1.03	1.714	1	1.016	0.984	1.016
HP13	0.907	1	0.907	1.102	1.702	0.944	1	0.944	1.06
	1	1.102	0.907	1.102	1.876	1	1.052	0.95	1.052
HP15	0.929	1	0.929	1.076	1.039	0.933	1	0.933	1.072
	1	1.076	0.929	1.076	1.118	1	1.054	0.949	1.054
HP23	0.916	1	0.916	1.092	0.994	0.916	1	0.916	1.092
						0.921	1.006	0.916	1.092
	1	1.092	0.916	1.092	1.086	1	1.055	0.948	1.055
HP27	0.851	1	0.851	1.175	1.948	0.932	1	0.932	1.073
	1	1.175	0.851	1.175	2.288	1	1.061	0.942	1.061
HP26	0.863	1	0.863	1.158	1.937	0.922	1	0.922	1.085
	1	1.158	0.863	1.158	2.243	1	1.073	0.932	1.073
HP33	0.918	1	0.918	1.089	0.765	0.936	1	0.936	1.069
	1	1.089	0.918	1.089	0.833	1	1.083	0.923	1.083
HP25	0.782	1	0.782	1.279	1.607	0.881	1	0.881	1.135
	1	1.279	0.782	1.279	2.056	1	1.095	0.913	1.095
HP34	0.849	1	0.849	1.178	1.278	0.87	1	0.87	1.15
	1	1.178	0.849	1.178	1.506	1	1.126	0.888	1.126
HP22	0.849	1	0.849	1.177	0.985	0.85	1	0.85	1.177
						0.863	1.016	0.849	1.177
	1	1.177	0.849	1.177	1.159	1	1.158	0.864	1.158
HP28	0.757	1	0.757	1.322	1	0.757	1	0.757	1.322
	1	1.322	0.757	1.322	1.322	1	1.264	0.791	1.264
HP37	0.661	1	0.661	1.514	1.726	0.688	1	0.688	1.453
	1	1.514	0.661	1.514	2.613	1	1.4	0.714	1.4

Table 3. Input and output targets

	X1	X2	X3	Y1	Y2	Y3	Y4
HP9	346.61	27.04	9577.41	3041.91	3477.79	2751.18	3.76
	345.91	27.30	9586.28	3047.69	3472.88	2756.37	3.75
HP17	1446.35	190.00	37940.64	7489.93	10921.25	15834.22	7.61
	1456.91	191.45	38216.48	7543.61	10997.02	15947.69	7.64
HP19	612.95	66.19	14386.53	4157.22	6423.28	5763.50	8.08
	623.60	67.34	14636.31	4223.57	6525.80	5855.51	8.10
HP13	734.76	84.58	16434.71	5598.74	7156.54	6422.79	9.29
	778.54	88.90	17414.07	5892.07	7532.83	6759.33	9.44
HP15	179.04	10.00	5013.49	1672.45	2073.79	1700.69	3.92
	191.99	10.72	5375.98	1762.09	2146.97	1791.85	3.72
HP23	204.70	3.78	5158.38	1648.62	2456.14	1682.25	4.07
	206.75	3.80	5187.78	1658.46	2481.02	1692.28	4.09
	211.96	4.13	5514.14	1739.92	2560.15	1775.41	4.09
HP27	322.40	15.84	9935.36	2406.52	3622.17	3351.43	5.87
	346.00	17.00	10641.04	2553.72	3859.61	3556.44	5.99
HP26	314.41	13.52	9470.89	2258.62	3476.86	3211.31	5.35
	341.17	14.67	10276.88	2423.31	3750.99	3445.47	5.48
HP33	151.19	1.40	3824.79	1462.39	1269.13	978.62	4.77
	164.09	2.22	4087.02	1584.24	1420.69	1046.77	5.25
HP25	393.66	23.20	11282.29	3340.40	4733.82	3567.89	7.15
	461.43	26.33	12805.81	3656.85	5159.78	3905.88	6.89
HP34	209.06	6.91	6017.12	1665.12	2414.46	2045.04	4.34
	240.43	7.95	6919.97	1872.90	2763.38	2302.07	4.33
HP22	234.33	7.37	5614.27	1996.57	2638.54	1792.72	5.59
	239.60	7.49	5700.03	2027.63	2714.82	1820.60	5.49
	277.77	8.68	6608.02	2311.90	3412.52	2075.84	4.77
HP28	190.41	5.52	4633.63	1146.76	1907.23	1731.84	7.08
	250.18	7.30	6123.94	1450.00	2490.55	2189.81	6.40
HP37	436.26	36.40	11007.91	2578.66	4330.64	4483.68	6.57
	655.61	52.87	15989.77	3609.28	5698.64	6275.66	5.87

5. CONCLUSION

The alternative transformations discussed in this paper provide a general solution rather than a representative one to the linear fractional programming problem and the results presented here apply more generally to any fractional programming problem with concave numerator, convex denominator and convex constraint.

As an added bonus these results and the DEA models described in this paper provide alternate constant returns-to-scale solutions and improved variable returns-to-scale solutions for DEA of the standard linear fractional programming model.

Notes

1. The Seven DMU model which we use to demonstrate the various DEA models is from page 25, Charnes, A., Cooper, W.W., Lewin, A.Y. and Seiford, L.M., *Data Envelopment Analysis: Theory, Methodology, and Application*. Boston, Kluwer Academic Publishers, 1994.

2. Steering Committee for the Review of Commonwealth/State Service Provision 1997, *Data Envelopment Analysis: A technique for measuring the efficiency of government service delivery*, AGPS, Canberra. Available at *http://www.pc.gov.au/service/gspindex.html* or *http://www.pc.gov.au/gsp/dea/index.html*

3. Pages 72 - 73, ibid.

4. Instead of 13, HP9 should be assessed as inefficient.

References

Banker, R.D., Charnes, A., Cooper, W.W. Some models for estimation of technical and scale inefficiencies in data envelopment analysis. Management Sciences 1984; 30:1078-1092.

Charnes, A., Cooper, W. W. Programming with linear fractional functionals. Naval Research Logistics Quarterly 1962; 9: 181-186.

Charnes, A., Cooper, W.W., Lewin, A.Y., Seiford, L.M. *Data Envelopment Analysis: Theory, Methodology, and Application*. Dordrecht: Kluwer Academic Publishers, 1994.

Charnes, A., Cooper, W.W., Rhodes, E. Measuring the efficiency of decision making units. European Journal of Operational Research 1978; 2: 429-444.

Charnes, A., Cooper, W.W., Rhodes, E. Short communication: Measuring the efficiency of decision making units. European Journal of Operational Research 1979; 3: 339.

Cooper, W.W., Seiford, L.M., Tone, K. *Data Envelopment Analysis: A Comprehensive Text with Models, Applications, References and DEA-Solver Softwar*. Dordrecht: Kluwer Academic Publishers, 2000

Hillier, F.S., Lieberman, G.J. *Introduction to Operations Research*, 5th edition, New York: McGraw-Hill Publishing Company, 1990.

O'Brien, G.C., Wu, L. Bound Restrictions on DEA Slack and Its Feasibility. 5th International Conference of Decision Sciences Institute, 1999; Athens, Greece.

O'Brien, G.C., Wu, L. DEA Duality and Infinitude of Primal Optimal Solutions. INFORMS-KORMS Seoul 2000 Conference, 2000; Korea.

Stancu-Minasian, I.M. *Fractional Programming: Theory, Methods and Applications*. Dordrecht: Kluwer Academic Publishers, 1997.

MAX-MIN OPTIMIZATION OF THE MULTIPLE KNAPSACK PROBLEM: AN IMPLICIT ENUMERATION APPROACH

Takeo Yamada

Department of Computer Science, The National Defense Academy, Yokosuka, Kanagawa 239-8686, Japan

yamada@nda.ac.jp

Abstract The binary knapsack problem is fundamental in combinatorial optimization, and algorithms to solve problems with nearly a million items are now available through the Internet. This paper is concerned with a variation of the problem, where there are n items to be packed into m knapsacks. Our problem is to find the assignment of items into knapsacks such that the minimum of the knapsack profits is maximized. This problem is referred to as the max-min multiple knapsack problem ($\mathbf{M^3KP}$). First, some upper bounds and a heuristic algorithm are presented, and based on these, we explore algorithms to solve the problem to optimality. Then, we make use of a novel pruning method to develop an implicit enumeration algorithm that can solve $\mathbf{M^3KP}$s with up to a few hundred items exactly.

Keywords: Multiple knapsack problem, max-min optimization, implicit enumeration.

1. INTRODUCTION

In this article we are concerned with a variation of the binary *knapsack problem* (**KP**, Martello and Toth, 1990), where we have n items to be packed into m knapsacks. Let w_j and p_j denote the weight and profit of the jth item respectively, and the knapsack capacities are $c_1, c_2, \ldots, c_m$. Without much loss of generality we assume the following.

- $\mathbf{A}_1$. Problem data w_j, p_j $(j = 1, \ldots, n)$ and c_i $(i = 1, \ldots, m)$ are all positive integers.
- $\mathbf{A}_2$. Items are arranged in non-increasing order of profit per weight, that is,

$$p_1/w_1 \geq p_2/w_2 \geq \cdots \geq p_n/w_n. \qquad (1)$$

A$_3$. Knapsacks are numbered in non-increasing order of capacity, that is,

$$c_1 \geq c_2 \geq \ldots \geq c_n. \tag{2}$$

The problem is to fill the knapsacks with items such that the capacity constraints are all satisfied and the *minimum* of the knapsack profits is maximized, where by *knapsack profit* we mean the total profit of items in each knapsack. Let x_{ij} be the decision variable such that $x_{ij} = 1$ if item j is included in knapsack i, and $x_{ij} = 0$ otherwise. Then, the problem is formulated as the following *max-min multiple knapsack problem*.

M^{3}KP:

$$\text{maximize} \quad z(x) := \min_{1 \leq i \leq m} \{\sum_{j=1}^{n} p_j x_{ij}\} \tag{3}$$

$$\text{subject to} \quad \sum_{j=1}^{n} w_j x_{ij} \leq c_i, \qquad i = 1, 2, \ldots, m \tag{4}$$

$$\sum_{i=1}^{m} x_{ij} \leq 1, \qquad j = 1, 2, \ldots, n \tag{5}$$

$$x_{ij} \in \{0, 1\}, \qquad \forall i, j. \tag{6}$$

By $z(x)$ we denote the objective function for $x = (x_{ij})$, and for convenience, we introduce knapsack 0 to accept items that are rejected from all other knapsacks. The capacity of this knapsack is $c_0 = \infty$, and (5) may be replaced by

$$\sum_{i=0}^{m} x_{ij} = 1, \; j = 1, 2, \ldots, n. \tag{7}$$

Throughout the paper $x^\star$ denotes an optimal solution with the corresponding objective value $z^\star := z(x^\star)$. **M^{3}KP** is $\mathcal{NP}$-hard (Garey and Johnson, 1979), since for $m = 1$ it reduces to **KP** which is already $\mathcal{NP}$-hard.

M^{3}KP is closely related to the *multiple knapsack problem* (**MKP**, Martello and Toth, 1980), except that the objective function is the *minimum*, as opposed to the *sum* in **MKP**, of the knapsack profits. Such a max-min approach aims at obtaining a 'balanced' solution, since in **MKP** we often get a solution with very unbalanced knapsack profits. Similar max-min (or min-max) formulation has been considered for various combinatorial optimization problems such as the mini-max spanning forest problem (Yamada *et al.*, 1997), the knapsack sharing problem (Yamada et al., 1998), and others (Du and Pardalos, 1995).

Such an assignment problem with a max-min type objective function may be encountered by a city security authority, where several emergency brigades should be assigned to corresponding bases. Each base has budget constraint associated with it, and each emergency worker has a cost and a defined level of competence assigned. The level of competence of an emergency brigade is a sum of individual levels. If an accident happens, the emergency team from the nearest base has to respond. The accident is equally likely to happen at any area. The problem for the authority is to assign brigades so as to make sure that the worst competent brigade is still capable of fixing the situation, that is, the authority desires to maximize the lowest brigade competence subject to all the budget constraints.

In this article we derive some upper bounds, develop heuristic algorithms that give lower bounds, and present algorithms to solve the problem to optimality. Since conventional methods are unable to solve problems except for very tiny instances, we introduce a novel *pruning* method, and based on this develop an *implicit enumeration* (see, for example, Wolsey, 1998) algorithm. Combined with an *interval reduction* (Gill et al., 1981) method, the developed algorithm is able to solve $\mathbf{M^3KP}$s with up to a few hundred items to optimality.

2. UPPER AND LOWER BOUNDS

2.1 Upper Bounds

An upper bound to $\mathbf{M^3KP}$ can be found by replacing (6) with

$$x_{ij} \geq 0, \quad \forall i, j. \tag{8}$$

The resulting problem can be converted into a linear programming problem that is easy to solve, and the optimal objective value to this *continuous relaxation* is denoted as $\bar{z}^1$.

Alternatively, we may consider the following *auxiliary* problem, which is obtained by setting

$$y_j := \sum_{ij}^{m} x_{ij} \tag{9}$$

and summing all the knapsacks into a large one.

Aux:

$$\text{maximize} \quad \frac{1}{m}\sum_{j=1}^{n} p_j y_j \tag{10}$$

$$\text{subject to} \quad \sum_{j=1}^{n} w_j y_j \le \sum_{i=1}^{m} c_i, \tag{11}$$

$$0 \le y_j \le 1, \qquad j = 1, 2, \ldots, n. \tag{12}$$

Aux is a continuous knapsack problem, and under condition $\mathbf{A}_2$ it is easily solved as follows. Let W^j and P^j be, respectively, the accumulated weight and profit of items 1 through j, that is,

$$W^j := \sum_{k=1}^{j} w_k, \; P^j := \sum_{k=1}^{j} p_k. \tag{13}$$

By $\phi_0(\cdot)$ we denote the piece-wise linear function connecting the points $(0,0)$, (W^1, P^1), $\ldots$, (W^n, P^n), (∞, P^n) (See Figure 1). Clearly, $\phi_0(\cdot)$ is non-decreasing and concave. Then the optimal objective value to **Aux** is given by $\phi_0(\sum_{i=1}^{m} c_i)/m$ (Martello and Toth, 1990), and thus

$$\bar{z}^2 := \lfloor \frac{1}{m}\phi_0(\sum_{i=1}^{m} c_i) \rfloor \tag{14}$$

gives another upper bound to $\mathbf{M^3KP}$.

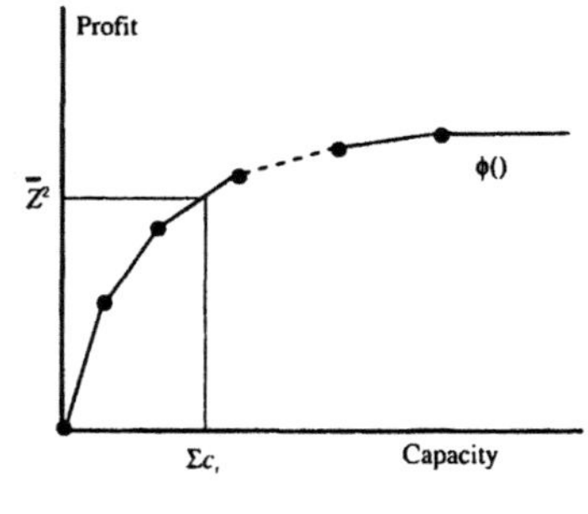

Figure 1. Function $\phi_0(\cdot)$.

Proposition 1

$$\bar{z}^2 \ge \bar{z}^1 \ge z^\star. \tag{15}$$

Proof: Since $\bar{z}^1 \ge z^\star$ is obvious, we show $\bar{z}^2 \ge \bar{z}^1$. Let (x_{ij}) be a feasible solution to the continuous relaxation of $\mathbf{M^3KP}$. Then, clearly

(y_j) is feasible to **Aux**, and we have

$$\frac{1}{m}\sum_{j=1}^{n} p_j y_j = \frac{1}{m}\sum_{i=1}^{m}\{\sum_{j=1}^{n} p_j x_{ij}\} \geq \min_{1\leq i\leq m}\{\sum_{j=1}^{n} p_j x_{ij}\},$$

which completes the proof.

Thus $\bar{z}^2$ is *weaker* than $\bar{z}^1$ as an upper bound. However, the former is much easier to calculate and often very close to $\bar{z}^1$.

2.2 Greedy Algorithm

Associated with an arbitrary feasible solution $\underline{x}$ to **M³KP** we have a lower bound $\underline{z} := z(\underline{x})$. Here we give a greedy method to obtain a good initial feasible solution. The greedy algorithm consists of n steps, with jth step corresponding to the decision for item j. That is, we do the following for $j = 1$ to n: at step j items $1, 2, \ldots, j-1$ have already been assigned to knapsacks (including knapsack 0), and we put item j into the feasible knapsack of the smallest *penalty*, where the penalty of item j to knapsack i is defined by

$$\xi_i(j) := \Delta c_i(j) + P_i(j). \tag{16}$$

Here $\Delta c_i(j)$ and $P_i(j)$ denote, respectively, the remaining capacity and the total profit of items included in knapsack i at the beginning of step j.

The output $\underline{x}^1$ and $\underline{z}^1$ from this algorithm is referred to as the *greedy* solution.

2.3 Local Search

Let x be a feasible solution to **M³KP**. We introduce a *2-opt neighborhood* of x as the set of solutions obtained from x by changing assignment of at most two items to knapsacks. These are obtained from x either by (i) moving item j_1 from knapsack i_1 to knapsack i_2, (ii) exchanging items j_1 and j_2 between knapsacks i_1 and i_2, (iii) moving item j_1 from knapsack i_1 to knapsack i_2, and item j_2 from knapsack i_2 to knapsack 0, or (iv) moving item j_1 from knapsack 0 to knapsack i_1, and item j_2 from knapsack i_1 to knapsack i_2. Here i_1, i_2, j_1, j_2 are arbitrary integers satisfying $1 \leq i_1 \neq i_2 \leq m$ and $1 \leq j_1 \neq j_2 \leq n$.

The *local search* (Aarts et al., 2000) algorithm starts with the greedy solution, and repeats improvements by moving to a better solution as long as such a neighbor can be found. The output from this algorithm, $\underline{x}^2$ and $\underline{z}^2$, is referred to as the *2-opt* solution.

3. EXACT ALGORITHMS

3.1 Branch-and-Bound Method

With the lower and upper bounds given in the previous section, we can construct a branch-and-bound algorithm (see, for example, Wolsey, 1998). Here we use $\underline{z}^2$ and $\bar{z}^2$ as the lower and upper bounds, and the *branching strategy* is as follows. We take **M^{3}KP** as the initial subproblem, and at each subproblem evaluate its lower and upper bounds. Here we obtain a feasible solution $\underline{x}$, and use this to generate n children in the following way. That is, the kth child is defined with the first $k-1$ items assigned to the same knapsacks as in $\underline{x}$, and the kth item is *forbidden* to be so assigned ($k = 1, 2, \ldots, n$). Thus, we have a tree of subproblems, which is then traversed in the *depth-first* (Sedgewick, 1998) fashion in our implementation.

Unfortunately, the developed algorithm is able to solve only very tiny instances. For example, we were unable to solve problems with $m = 2, c_1 = 100, c_2 = 75$, and $n > 40$, mainly due to excessive memory requirements incurred by the recursive structure of the implemented program.

3.2 Integer Programming

Since **M^{3}KP** is a 0-1 integer programming problem, it can be solved using softwares such as LINDO, CPLEX, or SOPT (Fourer, 1999). We tried XPRESS-MP (Dash Optimization Inc., 2000). Table 1 compares the methods of this section for randomly generated instances. Here B&B stands for the branch-and-bound method described in Subsection 3.1, #sub is the number of subproblems generated, and CPU is the time in seconds on an HP9000-B132L workstation. Also shown are the initial lower and upper bounds ($\underline{z}^2$ and $\bar{z}^2$), and the optimal objective value $z^\star$. XPRESS-MP solved larger problems than the branch-and-bound algorithm, but was unable to solve instances with $n > 140$, either.

4. IMPLICIT ENUMERATION

4.1 Complete Enumeration and Pruning

In this and subsequent sections we consider primarily the case of two knapsacks ($m = 2$). Although extension to the case of $m > 2$ is straightforward, the developed algorithm is usually inefficient. With

Table 1. Comparison of exact solution methods.

n	$\underline{z}^2$	$\overline{z}^2$	z^*	B&B #nodes	B&B CPU	XPRESS-MP #nodes	XPRESS-MP CPU
20	149	176	164	2462	4.83	159	1.0
40	249	264	253	147283	326.11	573	4.0
60	320	342	340	-		1672	11.6
80	414	428	421	-		4660	32.4
100	455	462	457	-		5795	48.9
120	493	511	510	-		3683	31.4
140	540	567	563	-		4660	41.8

$m = 2$, we have three alternatives for each item: (i) put item j into knapsack 1, (ii) put it into knapsack 2, or (iii) put it neither into knapsack 1 nor 2 (put it into knapsack 0).

Thus, in *complete enumeration* **M^{3}KP** is solved in $O(3^n)$ steps. Since this is hopeless unless $n < 20$, we introduce some pruning methods. Let (j, W_1, W_2, P_1, P_2) be the *node* representing the state where items $1, 2, \ldots, j-1$ have been put into knapsacks (including knapsack 0), and the accumulated weight (profit, resp.) of knapsack i is W_i (P_i, resp.). Here items $j, j+1, \ldots, n$ remain to be included into knapsacks.

From node (j, W_1, W_2, P_1, P_2) three nodes $(j+1, W_1+w_j, W_2, P_1+p_j, P_2)$, $(j+1, W_1, W_2+w_j, P_1, P_2+p_j)$, $(j+1, W_1, W_2, P_1, P_2)$ are generated depending on the decision (i) - (iii) for item j. Then, starting with node $(0,0,0,0,0)$ we have a tree consisting of $\sum_{j=0}^{n} 3^j$ nodes. To cut off some of the nodes (and their descendents) we introduce some conditions for pruning.

Before stating the conditions, however, we prepare some terminology. Consider two nodes (j, W_1, W_2, P_1, P_2) and $(j, W_1', W_2', P_1', P_2')$. We say the former is *dominated* by the latter if

$$W_i' \leq W_i, \;\; P_i' \geq P_i, \;\; i = 1, 2.$$

Next, let $\bar{z}(j, W_1, W_2, P_1, P_2)$ be an upper bound *conditioned* at (j, W_1, W_2, P_1, P_2), that is, an upper bound to the problem with respect to items $j, j+1, \ldots, n$, where items $1, 2, \ldots, j-1$ have been put into knapsacks with accumulated profit P_i and weight W_i ($i = 1, 2$). The derivation of this bound is discussed later, but for the moment we assume this is available.

Then, the following pruning conditions are straightforward. That is, Node (j, W_1, W_2, P_1, P_2) can be terminated if either one of the following is satisfied.

C$_1$. $W_1 > c_1$ or $W_2 > c_2$.
C$_2$. (j, W_1, W_2, P_1, P_2) is dominated by some other node.

C$_3$. $\bar{z}(j, W_1, W_2, P_1, P_2) < \underline{z}$ for some lower bound $\underline{z}$ to $\mathbf{M^3KP}$.

These conditions eliminate weight-over nodes, dominated nodes, and unpromising nodes.

4.2 Conditional Upper Bound

We now discuss how to evaluate $\bar{z}(j, W_1, W_2, P_1, P_2)$. Let us define $\phi_j(\cdot)$, similarly to $\phi_0(\cdot)$ shown in Figure 1, as the piece-wise linear, concave and non-decreasing function obtained by connecting points $(0,0), (W^{jj}, P^{jj}), \ldots, (W^{jn}, P^{jn}), (\infty, P^{jn})$, where

$$W^{jk} := \sum_{l=j}^{k} w_l, \ P^{jk} := \sum_{l=j}^{k} p_l. \tag{17}$$

By ΔP_i we denote the increase of profit from knapsack i by putting the remaining items into this knapsack. Since the remaining knapsack capacity is $\Delta c_i := c_i - W_i$, the maximum profit obtainable from this problem is at most $\phi_i := \phi_j(\Delta c_i)$. Thus, we have $\Delta P_i \leq \phi_i$ $(i = 1, 2)$. Furthermore, if two knapsacks are combined together, the increase in profit is at most $\phi_0 := \phi_j(\Delta c_1 + \Delta c_2)$. So we have $\Delta P_1 + \Delta P_2 \leq \phi_0$, and $\bar{z}(j, W_1, W_2, P_1, P_2)$ can be evaluated by solving the following.

$$\begin{aligned} \text{maximize} \quad & \min\{P_1 + \Delta P_1, P_2 + \Delta P_2\} && (18)\\ \text{subject to} \quad & \Delta P_1 + \Delta P_2 \leq \phi_0 && (19)\\ & 0 \leq \Delta P_i \leq \phi_i, \qquad i = 1, 2. && (20) \end{aligned}$$

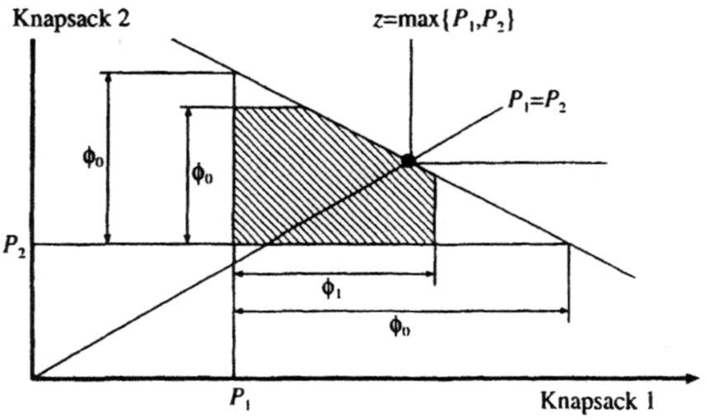

Figure 2. Derivation of $\bar{z}(j, W_1, W_2, P_1, P_2)$.

From Figure 2, the solution to this problem, and therefore $\bar{z}(j, W_1, W_2, P_1, P_2)$, is given as follows.

$$\bar{z}(j, W_1, W_2, P_1, P_2) := \begin{cases} P_1 + \phi_1, & \text{if } P_1 + \phi_1 \leq \theta, \\ \theta, & \text{if } P_1 + \phi_0 - \phi_2 \leq \theta \leq P_1 + \phi_1, \\ P_2 + \phi_2, & \text{if } \theta < P_1 + \phi_0 - \phi_2, \end{cases} \tag{21}$$

where

$$\theta := (P_1 + P_2 + \phi_0)/2. \tag{22}$$

4.3 Implicit Enumeration Algorithm

With the pruning conditions $\mathbf{C}_1$-$\mathbf{C}_3$, the implicit enumeration algorithm can be constructed as follows. Here Ω represents the set of *active* nodes.

Algorithm IMPLICIT_ENUMERATION($\underline{z}$).
input: A lower bound $\underline{z}$.
begin
  $\Omega := \{(0,0,0,0,0)\}$; (* Initial set of active nodes *)
  for $j = 0$ **to** n **do**
    for $(j, W_1, W_2, P_1, P_2) \in \Omega$ **do**
      if neither $\mathbf{C}_1$, $\mathbf{C}_2$ nor $\mathbf{C}_3$ holds **then**   (* Generate sub-nodes *)
        $\Omega := \Omega \cup \{(j+1, W_1 + w_j, W_2, P_1 + p_j, P_2), (j+1, W_1, W_2 + w_j, P_1, P_2 + p_j), (j+1, W_1, W_2, P_1, P_2)\}$;
      $\Omega := \Omega \setminus \{(j, W_1, W_2, P_1, P_2)\}$;   (* Remove current node *)
  find $(n+1, W_1, W_2, P_1, P_2) \in \Omega$ that maximizes $z := \min\{P_1, P_2\}$;
  output $(n+1, W_1, W_2, P_1, P_2)$ and the corresponding z^*;
end.

This algorithm may be thought of as another implementation of the branch-and-bound method, where we have exloited the novel upper bound (21). It also differs from the algorithm of Subsection 3.1 in branching strategy, that is, the way the subproblems are generated, and the subproblems are scanned in the breadth-first manner.

For an $\mathbf{M}^3\mathbf{KP}$ with $m = 2, n = 16$, $c_1 = 100$, $c_2 = 75$ and profits and weights generated randomly over [1,100], the complete enumeration needs to examine more than $3^{16} = 43,046,721$ nodes, while IMPLICIT_ENUMERATION found the optimal solution after generating only 420 nodes.

4.4 Interval Reduction Method

In the implicit enumeration algorithm we assumed that we have a lower bound $\underline{z} \leq z^*$. However, IMPLICIT_ENUMERATION runs with an arbitrary value for $\underline{z}$, not necessarily a lower bound. Then, we may try with an arbitrary 'estimation' of a lower bound. Such an estimation is denoted simply as z. If an estimation is wrong, that is, $z > z^\star$, the algorithm eventually eliminates all the active nodes, and terminates without finding an optimal solution. Figure 3 shows the number of nodes generated for various values of z in the $\mathbf{M^3KP}$ with $n = 16$. Here black bars show the case where an optimal solution is found, and white ones stand for the other case. These are divided into two disjoint groups depending on the correctness of estimation, that is, $z \le z^\star$ or $z > z^\star$.

Obviously the number of nodes generated in the implicit enumeration algorithm is a non-increasing function of z, as we see in Figure 3. Then, instead of using the initial lower bound $\underline{z}$, we may try, for example,

$$z := \lfloor \alpha\bar{z} + (1-\alpha)\underline{z} \rfloor, \tag{23}$$

where α is a parameter satisfying $0 < \alpha < 1$. If z is actually a lower bound, we will obtain an optimal solution after examining (much) fewer nodes than doing with $\underline{z}$. If the estimation was wrong, the algorithm fails, and we repeat again with a reduced z. This leads us to the following *interval reduction* (see Gill et al., 1981) method .

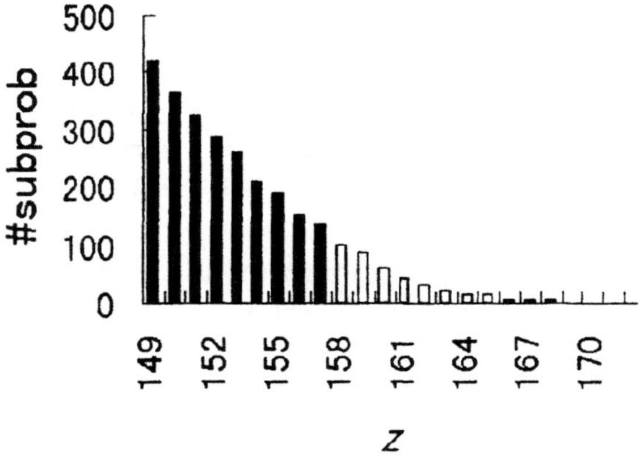

Figure 3. The number of nodes generated for various z.

Algorithm INTERVAL_REDUCTION.
input: An upper bound $\bar{z}$ and a lower bound $\underline{z}$.
begin
  repeat (* Infinite loop *)
    $z := \lfloor \alpha\bar{z} + (1-\alpha)\underline{z} \rfloor$; (* Estimated lower bound *)
    IMPLICIT_ENUMERATION(z);
    if an optimal solution is found **then**

output the solution and **stop**;
else $\bar{z} := z$; (* Reduce interval *)
end.

Here, from some preliminary experiments we take $\alpha = 0.8$. As for the initial bounds, we may take $\bar{z} := \bar{z}^2$ and $\underline{z} := \underline{z}^2$ discussed in Section 2. Then, in each iteration of the algorithm the length of $[\underline{z}, \bar{z}]$ shrinks by the factor of α, and an optimal solution is found in finite steps.

5. NUMERICAL EXPERIMENTS

5.1 Design of Experiments

The number of knapsacks is fixed at $m = 2$, and n ranges from 100 to 500. The weights and profits are assumed random and independent over $[1, 100]$, and we tried the following three cases.

Case a : $c_1 = 100, c_2 = 75$
Case b : $c_1 = 200, c_2 = 150$
Case c : $c_1 = 400, c_2 = 300$

5.2 Result of Experiments

Table 2 gives the result of experiments with the implicit enumeration algorithm combined with the interval reduction method. Here shown are the initial lower and upper bounds ($\underline{z}_2$ and $\bar{z}_2$), the optimal objective value (z^*), the number of nodes generated (#nodes), the number of repetitions in interval reduction (#rep.), and CPU time in seconds on an HP9000-B132L. Each row is the average of 10 independent runs.

From the table, we observe the following:

1 The problem becomes harder to solve, with respect to both memory requirement and computation time, as n and/or knapsack capacities increase. We were unable to solve larger problems than those shown in Table 2, mainly due to insufficient memory.

2 The number of iterations in interval reduction remains almost constant for various values of n and knapsack capacities.

Table 2. Result of Experiments.

Case	n	$\underline{z}_2$	$\bar{z}_2$	z^*	#nodes	#rep.	CPU sec.
	100	459.20	475.60	471.60	4771.1	2.1	0.4
	200	676.60	691.40	689.30	22325.1	1.5	9.1
a	300	847.30	857.50	855.20	39985.6	2.1	11.2
	400	957.00	966.30	964.10	44607.6	1.5	20.2
	500	1099.00	1109.50	1108.10	61240.2	1.2	65.2
	100	682.90	695.00	692.00	32937.7	1.7	16.1
	200	978.80	990.20	987.80	91913.9	1.2	142.5
b	300	1231.40	1241.80	1240.00	169381.3	1.1	728.5
	400	1403.80	1413.30	1411.70	217314.8	1.2	850.1
	100	1000.60	1010.90	1007.90	263849.1	1.8	336.6
c	200	1411.50	1420.80	1416.20	579865.5	1.3	1382.9
	300	1764.10	1769.80	1768.10	1122895.9	1.6	2719.1

6. CONCLUSION

We have formulated $\mathbf{M^3KP}$ and developed both heuristic and exact algorithms. Especially, using the implicit enumeration algorithm combined with an interval reduction method we were able to solve much larger problems than those solvable by other conventional methods. To solve yet larger problems to optimality, however, more sophisticated algorithms equipped with novel pruning methods are needed.

References

Aarts, E. and Lenstra, J. K. ed. Local Search in Combinatorial Optimization. Chichester:John Wiley & Sons, 1997.

Dash Optimization Inc. XPRESS-MP Release 11. 2000.

Du D.-Z. and Pardalos, P.M. Minimax and Applications. Boston:Kluwer Academic Publishers, 1995.

Fourer. R., Software survey: linear programming. OR/MS Today 1999, 26:64-71.

Garey, M. R. and Johnson, D. S. Computers and Intractability: A Guide to the Theory of NP-Completeness. New York:Freeman and Company, 1979.

Gill, P. E., Murray, W., Wright, M. H. Practical Optimization. New York:Academic Press, 1981.

Martello, S. and Toth, P. Knapsack Problems: Algorithms and Computer Implementations. Chichester:John Wiley & Sons, 1990.

Martello, S. and Toth, P. Solution of the zero-one multiple knapsack problem. European Journal of Operational Research 1980; 4:276-283.

Sedgewick, R. Algorithms in C, 3rd Edition. Reading:Addison-Wesley, 1998.

Wolsey, L. A. Integer Programming. Chichester:John Wiley & Sons, 1998.

Yamada, T., Takahashi, H. and Kataoka, S A branch-and-bound algorithm for the mini-max spanning forest problem. European Journal of Operational Research, 1997; 101:93-103.

Yamada, T., Futakawa, M. and Kataoka, S. Some exact algorithms for the knapsack sharing problem. European Journal of Operational Research, 1998; 106:177-183.

A PLANNING SYSTEM FOR PRECISION FARMING BASED ON AN AUTONOMOUS TRACTOR

Keiji Suzuki[1], Kazuki Takamatsu[2], Taku Okuno[3], Azuma Ohuchi[2], Yukinori Kakazu[2]

[1]*Future University-Hakodate*
Kameda-Nakano 116-2, Hakodate City, Hokkaido, 040-8655, Japan
Phone: +81-138-34-6229, Fax:+81-138-34-6301
e-mail: suzukj@fun.ac.jp

[2]*Research Group of Complex Systems Eng.,*
Graduate School of Eng., Hokkaido University
North-13 West-8, Kita-ku, Sapporo, Hokkaido, 060-8628, , Japan
Phone: +81-11-716-6496, Fax: +81-11-706-7834
e-mail: {takamatu, ohuchi, kakazu}@complex.eng.hokudai.ac.jp

[3]*Information Science Research Center Co., Ltd.*
North-9 West-3, Sapporo, Hokkaido, , Japan
Phone: +81-11-737-9157, Fax: +81-11-737-9159
e-mail: okuno@isrc.co.jp

Abstract: This paper describes the planning system developed for precision farming. Particularly, the system aims to support the cooperative works between global positioning system (GPS) based autonomous tractor and its applicators. The autonomous tractor can run precisely in fields according to GPS based navigation. The applicators can be controlled to perform variable outputs according to the precise position in fields. Using the tractor and its applicators, we can realize precision farming according to several conditions in fields. In order to use the autonomous tractor system, the planning system is developed for storing the field information, navigating the tractor and planning the variable farm-work. In this paper, we show the outline of the total systems, planning methods of navigation and optimizing variable farm-work. The results of experiments applied to the planning system will be shown.

Keywords: planning, precision farming, management.

1. INTRODUCTION

Precision farming or precision agriculture can be defined as the use of information technologies to obtain knowledge about important crop parameters, and to optimize the management of the production system on an appropriate scale (Blackmore, 1994). It is a concept that is gaining a great deal of attention recently. In order to realize site-specific management such as precision farming, computer controlled applicators, for example, fertilizer applicators and sprayers, are required. Furthermore, autonomous tractors that can trace precise positions will be preferable to increase the effectiveness of the precision farming (Hara, 1999). Here, we assume an autonomous tractor can receive the global positioning system (GPS) signals to take precise positions itself. Namely, the precise navigation of an autonomous tractor without winding like human operators will create advantage utilizing the field more effectively. Furthermore, we can expect to decrease several losses related to the amount of fertilizer or agricultural chemicals by precisely controlled tractors and applicators.

To use these machines effectively, Field Information Systems (FIS) are required. The FIS must store sensing data of fields and create site-specific working plans from those data. Ordinary, GISs (geographic information systems) are used as FISs because functions of these FISs are remaining to store the sensing data and create the site-specific working plans (Blackmore et al, 1994). However, to increase the effectiveness of the precision farming, the autonomous tractor will be required and the functions of the FIS will be increased to manage the tractor. Thus, FIS should create the plan not only site specific working plans but also make the control plans of the autonomous tractor combined with the site specific working plans. In this paper, we report the developing FIS that includes the navigation planning. Our developing system can create both of the working maps and the navigation plans. The navigation plans must include attributes for the control plans of the applicators. In this paper, we introduce the functions of the FIS and experimental results of using the output plans of the FIS. It will be expected for the optimization problems in FISs to become a new application field of Operations Research area.

2. PRECISION FARMING

Today, precision farming is motivated by the economic and environmental concern of excessive applications of chemical and other agricultural inputs regardless of within-field variability. The concept of precision farming is to treat within-field variability. Since soil within a field varies in productivity, optimum use of resources cannot be achieved by conventional farming, uniform applications of seed, fertilizer, pesticides, etc., across an entire field. The standard cycle of general precision farming is

1. sensing the data of within-field variability,
2. creating field information maps from sensed data,
3. planning application maps based on the field information maps,
4. operations in the field based on application maps.

Implementation becomes feasible only with the development of modern electronic tools. These tools include sensors to measure the many variables affecting crop production, GPSs for geo-referencing of the data, GISs for handling the geo-referenced data, and variable rate applicators able to vary application rates as the applicator traverses a field.

In addition to the above standard precision farming strategy, autonomous tractors are expected to increase the preciseness and optimality of several working plans. In the next section, we will introduce the schematic view of precision farming with an autonomous tractor and a field information system.

3. PRECISION FARMING WITH AUTONOMOUS TRACTOR

Autonomous tractors and autonomous vehicles for farm working based on GPS have been developed recently to decrease efforts of the human operators (Iida et al, 1999)(Yukumoto et al, 1997). However, these tractors aren't developed to apply to the precision farming. If the autonomous tractors are applied to the precision farming, the effectiveness of the precision farming will be increased. To increase the effectiveness, however, the tractors are required not only to run in the field without operator, but also to cooperate with variable applicators.

We developed the most advanced autonomous tractor, which can be applied to precision farming, based on a KUBOTA MD-77. As computer controlled variable applicators, the sprayer and the fertilizer applicator are developed for equipped the autonomous tractor (Hara, 1999) (see Figure 1). This tractor has the control area network (CAN) (Freimann, 2000) to communicate the map based controller, the navigation controller, and the variable applicators for achieving the site-specific working.

Supported by the CAN and several control systems, cooperative work can be achieved between the tractor and these applicators (Okuno et al, 2000). For example, the precise navigation will enable to precise planting of crops. The precise planting will increase the utilization of a field. According to the recording position of planting, the fertilizer applicator and the sprayer can work only in the planting area. Because the work area can be decided correctly, the required amount of fertilizer and chemicals can be determined precisely. Thus, economical and effective farming will be realized. To support these works based on the autonomous tractor, the field information system that can make the cooperated plans is required. In the next section, we introduce the developed FIS that can support the activity of the autonomous tractor.

Figure 1: The computer controlled fertilizer applicator (left figure) and the sprayer (right figure) attached to the autonomous tractor. The fertilizer applicator developed by Circle Tekko Co., LTD. The sprayer developed by Toyonoki Co., LTD. The tractor developed by KUBOTA Co., LTD. and Hokkaido University. The CAN based control systems developed by ISRC Co., LTD.

4. DEVELOPMENT OF FIELD INFORMATION SYSTEM

Currently, developed field information systems (FISs) can store sensing data of field status and process it to convert application maps for applicators (ESRI, 1997). In order to treat the properties of the cooperative works between the autonomous tractor and the variable applicators, our developing FIS has additional functions for creating navigation plan of the autonomous tractor and calculating required amount of chemicals and fertilizer according to the navigation plan. First, we will show the configuration of the system.

4.1 Configuration of the System

The FIS is designed as the client-server system to suitable flexible configuration. The client user interface is based on a web browser and provides unitary GUI keeping reasonable guideline (Takamatsu et al, 2000). The field information server is constructed with Java on UNIX platform and mounts a GIS kernel for management of geographical data, a database management system (DBMS) for management of data files via JDBC, and a mechanism for connecting the precision farming (PF) application modules.

The schematic view of the configuration of the system is illustrated in Figure 2.

As precision farming application modules, two planning modules are provided as follows:

1. **PF Applicator Planning Engine**: To achieve the site-specific variable field works, control plans of the applicator are needed for the fertilizer applicator, the sprayer, and so on.
2. **Path Planning Engine**: For the autonomous tractor, the system should generate the required paths and coordination of the activities between the tractor and the applicators.

These planning functions will be presented in the next section.

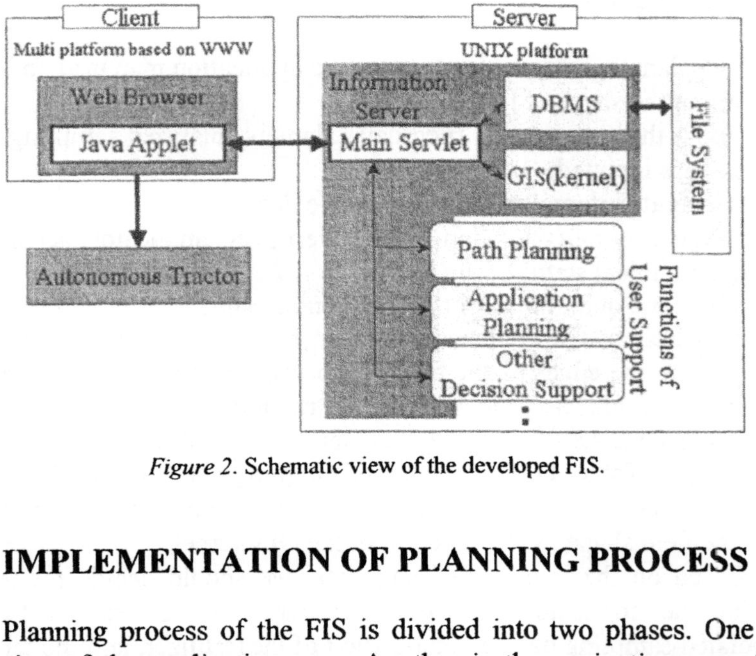

Figure 2. Schematic view of the developed FIS.

5. IMPLEMENTATION OF PLANNING PROCESS

Planning process of the FIS is divided into two phases. One is for generating of the application map. Another is the navigation map for the autonomous tractor. The navigation map is loaded into the control system of the tractor. The application map is used in the map management system on the tractor. The tractor control system and the map management system on the tractor are connected to each other by the CAN and exchange the information of the GPS, the velocity, and the status of the tractor and the applicators. According to the exchanged information, appropriate outputs of the applicator can be achieved. The navigation map has not only reference position data but also the control sequences of the tractor itself and the applicator. Because the tractor controller is required to control the position of the lift arms which connect the applicator, and the degree of the revolution of the PTO (power take off) which drives the applicator. Therefore, the coordination between the tractor activity and the applicator condition should be planned in the navigation phase.

The details of the planning processes are explained according to the implementation to a real field. The location of the field is in Memuro town, where is in the east part of Hokkaido in Japan. The field size is about 200m long and 100m wide. The objective of planning is to plant young beet. Before planting young beet, we have to make ridges and distribute the fertilizer along the ridges. Then planting of young beet must be on the same ridges precisely. In this work, the quantities of the fertilizer and the position of the ridges are quite important.

5.1 Application Map Planning

The planning process for creating the application map is as follows:

1. Input the shape of fields.
2. Input the status of the filed data. Namely, analyzed sampling data of soil are inputted.
3. Generate interpolation map of above data.
4. Decide output function of the fertilizer applicator according to geographical status of the field.
5. Generate output plan of the application map.

The data for the field shape is taken from the trace of the tractor. By running along the edges of the field, the precise position of the field can be taken. Next, in this case, the soil data of the field is inputted. The analyzed data of soil consists of pH, P (phosphorus), K (potassium), Lime and N (nitrogen) (see Table. 1). The number of sampling points is 84. The sampling points are associated with global positions. The sample data are interpolated within the field shape by the function of the GIS kernel.

Based on the interpolated soil data, we should decide the fertilizer map for the applicator. The fertilizer map can be decided from the interpolated sampling data, a composition of the fertilizer, and the target soil condition. This making process of the fertilizer map is important but unknown factors still exist (Blackmore, 1996). In this case, we take a strategy that N and P values are unified over the field to be close to the recommended rates of N and P for beet. The fertilizer we use is a composite type consisting of P and N. The rate of composition can be selected from a few certain grades. For example, we use the fertilizer called S182 that consists of 11 % of N, 18 % of P and 12 % of K. It means that we can't achieve the recommended rate from the hybrid fertilizer without error. Therefore, we optimize the following function to minimize the sum of error.

$$E_r = \sum_L \left(w_1 \left| P_L + R_P F_L - P_{std} \right| + w_2 \left| N_L + R_N F_L - N_{std} \right| \right) \quad (1)$$

where, L denotes the index of grid area in the field,

P_L and N_L are amount of P and N in the grid area L in the field,

F_L is amount of input fertilizer to the grid area L,

R_P and R_N are composite rate of the fertilizer,

P_{std} and N_{std} are standard amount of P and N respectively,

w_1 and w_2 are coefficients.

Table 1. A part of analyzed data of soil.

Sample No.	pH (H_2O)	P (phosphorus) (*mg*/100g)	K (potassium) (*mg*/100g)	Lime (*mg*/100g)	N (nitrogen) (*mg*/100g)
51	5.9	4.0	17.0	157.9	2.9
52	5.9	2.6	17.8	153.5	3.2
53	5.8	1.8	14.5	97.7	1.8
...	...	...	...	...	...
80	5.6	3.2	20.4	114.0	2.4
81	5.7	3.2	28.2	97.1	2.1
...	...	...	...	...	...

To minimize the error, we have to tune the coefficients, w_1 and w_2. The coefficients can be decided by the mean square error. In this implementation, not only the mathematical decision of the coefficients, but also the opinion of the expert of the farmer is considered. Thus, the coefficients are decided as $w_1 = 0.9$ and $w_2 = 0.1$. The final fertilizer map is shown in Figure 3.

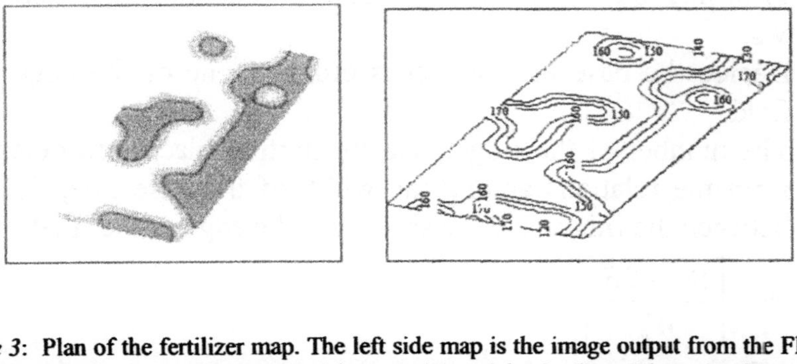

Figure 3: Plan of the fertilizer map. The left side map is the image output from the FIS. The right side map is counter plot image of the quantities of the fertilizer. The unit of the value is kg/10a.

5.2 Navigation Planning

The process of the navigation planning for the autonomous tractor is as follows:

1. Input the parameters, like as a width between ridges, side road areas for the tractor, and etc.
2. Decide the direction of the ridges in the field to maximize the total path length.
3. If needed, the positions to be supplied are determined.
4. Make the working process along to the paths including the revolution of tractor's engine, the status of the lift arms, the PTO status, and etc.
5. Decide the required paths according to the working areas of the applicators.
6. Publish the order of the paths and the paths that are embedded the attributes for realizing the working process.

In this planting case, we start to create the ridges to maximize the planting area. To determine the planting area, the parameters of the width of the ridges, the side road areas for the tractor should be determined. The side road areas for the tractor include the areas for rotating actions of the tractor. The tractor can rotate autonomously without guidance or operations of a human. The required area of rotation will change affected by the field condition. Thus the required rotating areas will be decided from the past records of the rotations of the tractor. The side road areas are set to include the rotating areas.

To determine the positions of the ridges, the direction of the ridges must be determined because we assume to make the parallel ridges. Because the shape of the field is the distorted rectangle, we evaluate the direction to maximize the total length of the ridges that means to maximize the planting area. The maximizing process is as follows:

1. Decide the base line, which is ordinary one of the edge of the field.
2. The number of the ridges and its starting places are determined from the relations among the width of the base line, the width between the ridges and the shift from the edge of the filed.

$$N = \left\lceil \frac{W_B - 2S}{W_R} \right\rceil \tag{2}$$

$$R_j^S = j \times W_R + S\ ,\ R_j^S \in BaseLine \tag{3}$$

where, W_B is the width of the base line. W_R is the width between the ridges. S denotes the shift value from the edge of field. R_j^S is start point of *j*-th ridge on the base line.

3. To maximize the length of the ridges, the direction of the ridges against the base line is optimized.

$$\max.\sum_{j}^{N} L(R_j^S, R_j^E) \tag{4}$$

where, $L(R_j^S, R_j^E)$ denotes the length between the start point, R_j^S, and the end point, R_j^E, of the *j*-th ridge. R_j^E is decided from the direction and the cross point of the edge of the planting area.

Along to the ridges, the fertilizer is distributed by the applicator. Thus we required determining the replenishment point of the fertilizer. The position of the replenishment will affect the total working time span but we decide the position without the optimization in this case.

The real working process requires the cooperation between the tractor and the applicator. The cooperation includes the setting of the revolution of the engine, the status of the PTO, the lift arm position, and the status of the applicator. Namely, we have to decide the approach condition to starting the fertilizing, the working condition while fertilizing, how to leave the end point of the plating area with lifting the applicator or not, and etc.

After the decision of the places of the ridges and above conditions of the system, the required paths are planned. The covering area of the applicator will determine the number of paths. In this case, the fertilizer applicator can fertilize 4 ridges at once and the planting applicator can plant at two ridges at once. Thus, the planning modules suggest making 152 ridges in the field and making 38 paths in the navigation plan for fertilizing and 76 paths for planting.

After the order of the paths is determined by the operator, the navigation planning module will publish the whole navigation plan that includes the attributes to realize the cooperative activities between the tractor and the applicators. The final output path image for the fertilizing is shown in Figure 4.

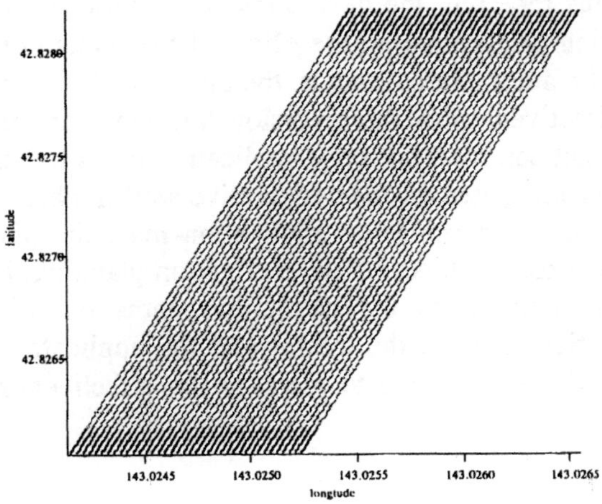

Figure 4: Navigation paths for fertilizing. The dark part of the lines denotes the side road areas and light gray area means working area for fertilizing.

5.3 Review of Plan

Before applying the plans to the real world, the operator will need to review the plans. One such reviewing function, we prepare the evaluation of total quantities of the fertilizer and the chemicals in each path. By using the autonomous tractor, the prediction of the quantities can be decided precisely. Thus, in the real field works, the operator will be able to reduce the redundant cost of the fertilizer and the chemicals. In this implementation case, the evaluation of the quantities of the fertilizer is shown in Table 2.

Table 2. Required quantity of the fertilizer in each path.

Path No.	Path Length (m)	Quantity (kg/10a)
1	221.68	99.55
2	242.83	108.01
3	240.68	107.18
...	...	...
12	243.37	101.09
13	245.37	99.61
14	242.72	98.62
...	...	...

6. CONCLUSION

In this paper, we describe the developing system as the Field Information System (FIS) that can support the activity of the autonomous tractor. In previous research, the development of autonomous tractors only focus on decreasing or eliminating the effort of the human operator. In this development of the autonomous tractor, the ability of the tractor is used for increasing the effectiveness of the precision farming. For this purpose, we develop the FIS that can treat not only application maps but also navigation plans including the attributes for the cooperative working between the tractor and the applicators. Concerning the application map, the planning process for the fertilizer is explained. As for the navigation planning, it is shown that the FIS not only determines the navigation paths but also creates the cooperative activities between the tractor and the applicator for fertilizing. These functions will be expected to increase the effectiveness of precision farming.

A subject of increasing the effectiveness of precision farming is related to the agricultural research area. However, the precision farming can't achieve without the information technologies. Particularly, the required techniques are related to the operations research area. Several types of optimization problems and data mining problems are still remaining. For example, one of the critical issues is that knowledge about important parameters is still unclear. Thus, we can't decide what kind of fertilizing policy is correct or not. If we apply several data mining techniques based on collections of stored data, this knowledge will be obtained. Therefore, according to increasing the stored data, the developed FIS will become platform to apply the advanced techniques of operations research to real field management problems.

7. ACKNOWLEDGMENTS

This research was supported by Regional Consortium Project of New Energy and Industrial Technology Development Organization (NEDO) of Japan.

References

Blackmore, S.. Precision Farming: An Introduction. *Outlook on Agriculture* 1994; 23(4): 275-280.

Blackmore, S., P.N. Wheeler., P.N, Morris, R.M., Morris, J., Jones, R.J.A.. The Role of Precision Farming in Sustainable Agriculture: A European Perspective. *In the Proc. of the 2nd International Conference on Site-Specific Management for Agricultural Systems* 1994; 777-793

Blackmore, S.. AN INFORMATION SYSTEM FOR PRECISION FARMING. *In the Proc. of Brighton Crop Protection Conference: Pests and Diseases* 1996; 3:1207-1214.

ESRI GIS Solutions for Production Agriculture, *A white paper of Environmental Systems Research Institute, Inc.* 1997.

Hara, Y.. Research on Precision Farming in Hokkaido. *Journal of the Japanese Society of Agricultural Machinery* 1999; 61(4): 19-23 (in Japanese).

Iida, M., Maekawa, T., Kudo, M., Umeda, M.. Autonomous Follow-up Vehicle System for Agriculture (Part 2). *Journal of the Japanese Society of Agricultural Machinery* 1999; 61(6): 141-147 (in Japanese).

Freimann, R.. Autonavigation and Implement Controlled Process Automation on CAN. *In the Proc. of Int. Symposium on Electronic Farm Communication with LBS* 2000; 64-87. Okuno, T., Takamatsu, K., Suzuki, K., and Kakazu, Y.. Map-based Control of Tractors and Implements for Precision Farming. *In the Proc. of the 10th Int. Conf. on Flexible Automation and Intelligent Manufacturing* 2000; 1:26-28.

Takamatsu, K., Suzuki, K., Okuno, T., Ohuchi, A., Kakazu, Y.. Management System with Web-based Interface for Precision Farming. *In the Proc. of the 10th Int. Conf. on Flexible Automation and Intelligent Manufacturing* 2000; 1:271-278.

Yukumoto, O., Matsuo, Y.. Navigation technology for tilling robots. *In the Proc. of mobile Int. symposium on agricultural Bus-system LBS and PA* 1997; 59-94.

SPOT PRICE FORECASTING MODELS FOR RISK AVERSION IN VICTORIA'S ELECTRICITY MARKET

Patrick Tobin and Alan Brown
Centre for Mathematical Modelling
Swinburne University of Technology
PTobin@groupwise.swin.edu.au

Abstract The bidding process in deregulated spot markets calls for effective prediction of the spot price in helping to frame suitable bid prices for the offer stack. Spot prices also feed schedules for plant operation. Demand forecasting is well established and demand is a major factor in determining a spot price. However supply is also a factor and effective price forecasting needs to take account of both of these.

Keywords: Forecasting,electricity market.

1. INTRODUCTION

Spot markets for electric power are becoming increasingly common as deregulation of utilities progresses (see e.g. Kliman, 1994). These markets rely on effective forecasting methods of demand and spot pricing to minimise risks associated with bidding offer stacks and the related power contracts used for hedging. The spot price is a major input to modeling the so-called *unit commitment problem* which is the process whereby scheduling in power stations is arranged to minimise generation cost while meeting required load (Takriti et al, 2000). Victoria's spot market is now well established, having operated since 1994 – one of the first spot markets for power created in Australia. Power can be drawn from the national grid as well as generated locally.

All major power stations (greater than a certain capacity) which are connected to the national grid must sell their electricity through the National Electricity Market (NEM). Market participants bid their capacity into the market. A bid array is entered for each half-hour period of a day for each generating unit in a power station.

During each five-minute period the bids from all stations are aggregated and sorted in ascending order of price. Then the cumulative amount of electricity for each price band in the market is calculated to construct a supply curve. The system demand then enters into the process and is used to determine the pool price and which units will be operating. All units at the pool price or below will operate up to the level determined by their bids.

The five-minute prices are then averaged to half-hourly prices and cash-flows are settled based on the half hourly price and the total energy generated by each unit in each half-hour period. Each generating unit gets paid the clearing price (sometimes known as the Spot Market Price (SMP) or Marginal Price), multiplied by their output generated. The basic notion of a spot price was initially outlined by Schweppe et al (1988) although there is ongoing research on how to best determine this spot price (see e.g Xie et al, 2000).

The fact that a low bid can still get paid a high price opens the prospect for players in the market to compete for volume and rely on others to set high prices.

The pattern of demand is generally well established and day-ahead forecasts of demand are generally reliable with the usual caveats on unexpected weather phenomena or unusual idiosynchratic occasions (see e.g. Charytoniuk et al, 1999; Taylor et al, 2000; Connor, 1996). As an illustration of the latter, we observe that the minimal demand for power recorded in Victoria in the period July 1, 1999 to June 30, 2000 occurred on New Year's eve – a time when widespread fear about possible Y2K problems caused general shutdown of all non-essential activity. This occurred on top of the usual seasonal lull.

Spot prices are far more variable than demand – they certainly follow similar patterns, but they are seriously affected by dispatch shortfalls, which arise predictably and unpredictably from plant downtime due to routine maintenance, breakdown or industrial action. Various approaches have been taken to forecast these prices. Szukata et al (1999) applied artificial neural networks to forecast the spot price in Victoria's electricity market with some success. We will apply a regression approach to forecast in the same market.

2. DEMAND FORECASTING

As noted demand forecasting for the electricity market is well established and a variety of processes can be applied to achieve this with substantial accuracy.

Examination of a demand pattern in Figure 1 over a (typical) three-week period from Sunday April 30 to Sunday May 14 in 2000 reveals the regularity. The initial time chosen was 4.30 am – a typical start time for offer stacks.

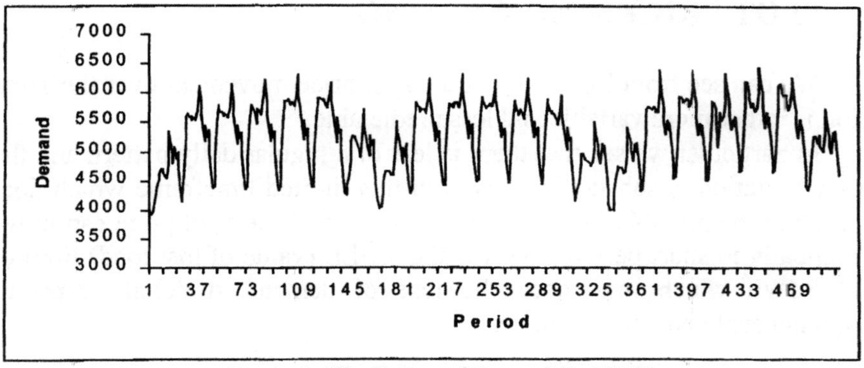

Figure 1: The Pattern of Demand (in MWh).

An ARIMA forecast model for the subsequent week compared actual values to predictions obtained from the original fortnight. The residuals from this are shown in Figure 2. In general the fit tracks the data well with this ARIMA model. Of the three weeks of hourly data two were used to create the model and a third week of data was used for validation.

The graph of the relative errors of this fit is in Figure 2 .

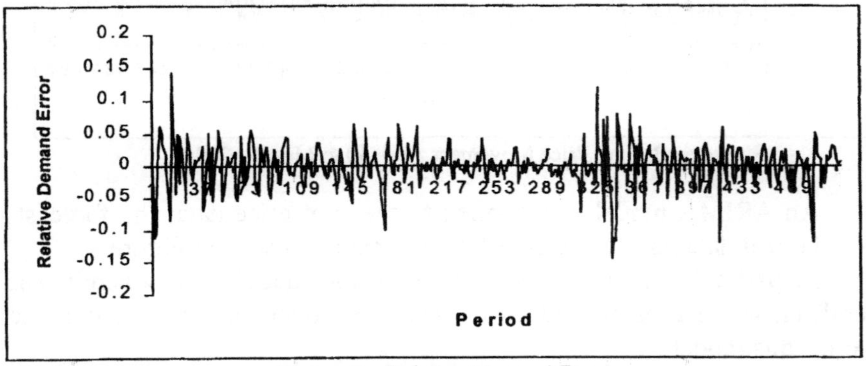

Figure 2: Relative Demand Errors.

From the ARIMA model we see that demand forecast accuracy averages around ± 2% here – an acceptable level. Combining ARIMA forecasts with some ad hoc procedures to accommodate holiday loads etc enables reasonable short-term forecasts of demand. In fact we have forecast some way ahead whereas for many purposes including spot price forecasts themselves, day-ahead forecasts would suffice. Longer forecasts can feed decisions on planned maintenance and plant upgrades.

Connor (1996) applied a neural network to successfully predict power demand on a US utility. He noted that nonlinear models like this generally outperform the classical linear ARIMA models – especially if they are fine-tuned with filters to treat outliers. The key conclusion is that demand itself is relatively predictable.

3. SPOT PRICE FORECASTING

We can see from Figure 3 that the spot price movement over the same period is rather more variable and less predictable.

In particular we see that there is less of a regular daily pattern and the weekly variation is substantial – even in this limited timeframe which does not capture the possible extreme situations. In fact the Spot price can move dramatically in short periods towards the VOLL (value of lost load) limit of $5000/MWh which may occur because of demand stretching capacity though unusual demand or outages.

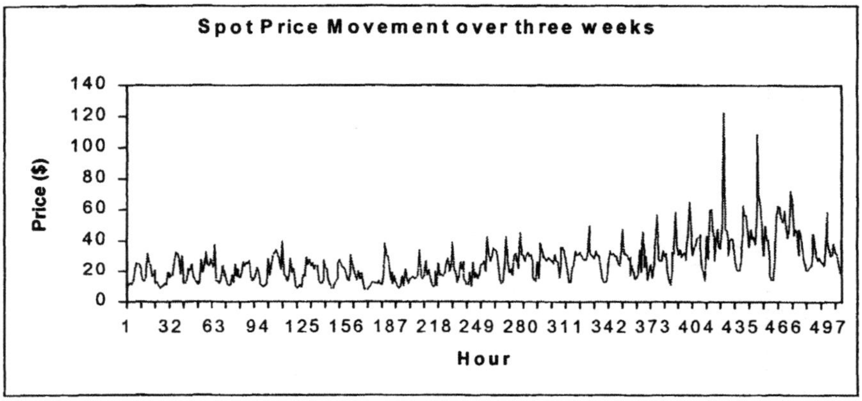

Figure 3: Spot Price Movement April 30 to May 21, 2000.

An ARIMA model was applied to the spot price data; the forecasts track the real data poorly compared with demand, as seen in Figure 4.

In particular, in the validation week the model fails severely and overall error averages around 20%. The MSE over three weeks is about 94.6 for this model

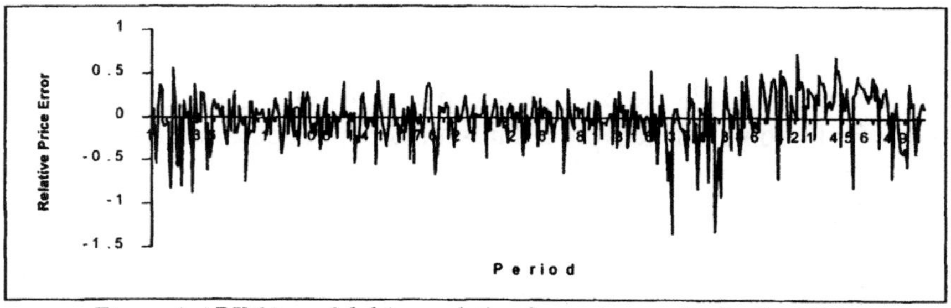

Figure 4: ARIMA model driven relative forecast errors for the Spot Price.

Clearly the variability is too great to be captured just in a time series model.

Spot price variation is so substantial that a log transformation of the price data is a reasonable option. Using a log to base 2 we see that a unit change in the log price now corresponds to a doubling of the price and that log prices will be effectively confined between 0 and 13. (Negative prices are possible of course – it is often desirable to pay to shed load rather than close plant down because of costs associated with ramping and cold starts. As a first pass, we can ignore prices below \$1 / MWh as these are relatively rare.)

The log price data for this period is shown in Figure 5 together with the log demand data.

The degree of variability in the log price is suppressed graphically but modelling using ARIMA models reveals all the same problems in forecasts as the original demand data.

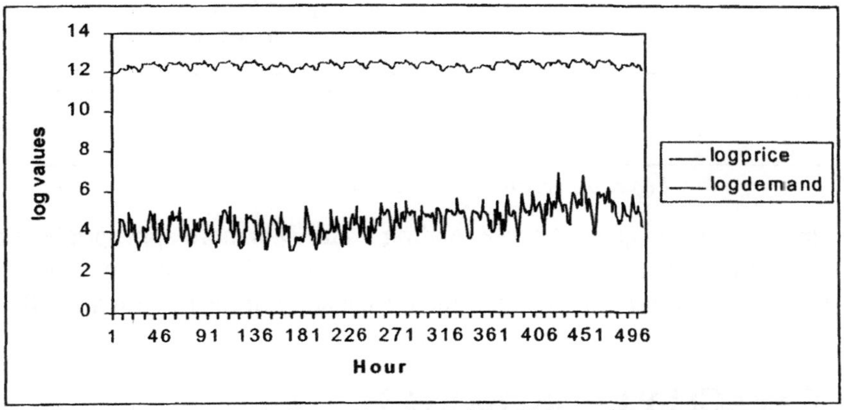

Figure 5: Comparing logs of Spot Price and Demand.

We need to consider what features affect the spot price. Clearly prices must be demand driven to an extent in a deregulated market. However the high predictability of demand is not reflected in forecasts for spot prices. We see this comparing the log values of spot price and demand over this period in Figure 5. One problem is the related issue of spare capacity. The Victorian supply is not limitless and when there are plant outages either planned or unplanned then the prospect of supply shortfalls arises which may not even be able to be met by drawing on the national grid.

A natural first step in relating demand and price is a regression model. As prices are highly volatile and values are in radically different ranges, it is more convenient to regress the logs of these variables leading to the model

$$\log P = a \log D + b$$

relating price P and demand D.

As a first step we look at a scatterplot of log demand and log price at hourly intervals over the three weeks starting April 30 at 4.30am. This is shown in Figure 6.

We see that the linear relationship is moderate here. Only 36% of variability in log price is accounted for by variation in log demand. Examining this model over a few weeks suggests that a ≈ 3 and b ≈ -30. This would generate a cubic rule $P = kD^3$ locally – a relationship which is not justifiable by any obvious physical models, and which cannot be extrapolated.

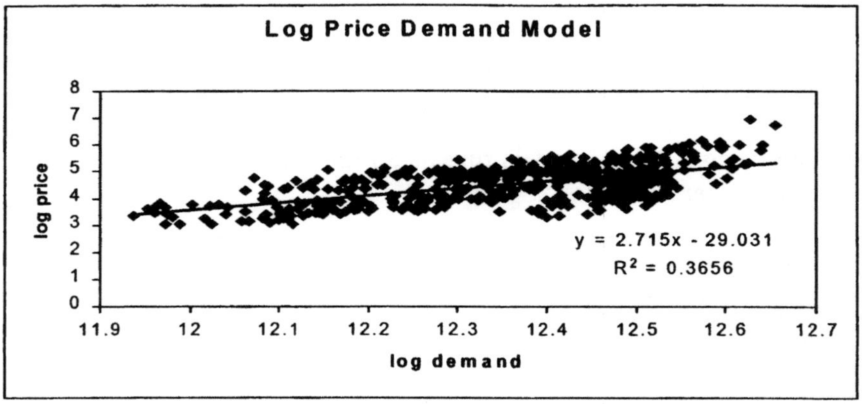

Figure 6. Variation of demand with spot price.

In fact, a more general curve relating *P* and *D* is more appropriate and may yield this 'cubic' as a local approximation. However, in order to properly model this we need to examine the capacity of the system as well.

4. CAPACITY CONSTRAINTS

The Victorian generation market includes a large number of plants of differing types and size although the principal source of power is brown coal. The plants and their nominal maximum capacities are given in the Appendix. In practice these capacities may be exceeded for short periods of time. The data on the available power generating capacity of local power generators over the three weeks from April 30 2000 is shown in Figure 7.

The graph in Figure 7 does not include some of the generating capacity nor the access to the grid. Hence it understates the available power which we will need to correct. The exact amount of co-generator capacity is not established – this includes the power put back into the system by industrial and domestic users who have an excess. The Figures also did not incorporate the black coal facility at Anglesea or the local hydroelectric plant capacity.

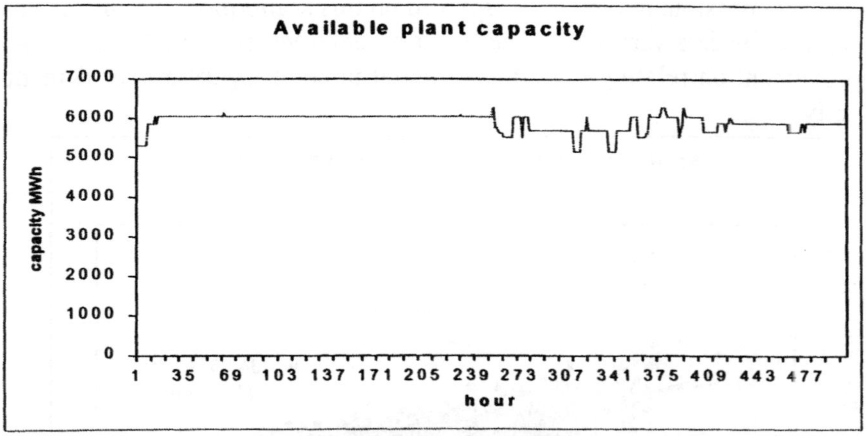

Figure 7. Available Generating Capacity in Victorian Plants.

The grid provides for some power both from the NSW and Snowy River scheme and from South Australia but these may actually be negative(Victoria providing grid supply) and the hydro-electicity is peak demand linked and related to water supplies.

In Figure 8 we see the graph of excess capacity over the same period. This reveals that there were several shortfalls over the period but in fact these were made up from the missing local capacity and the grid from South Australia and the Snowy River plant.

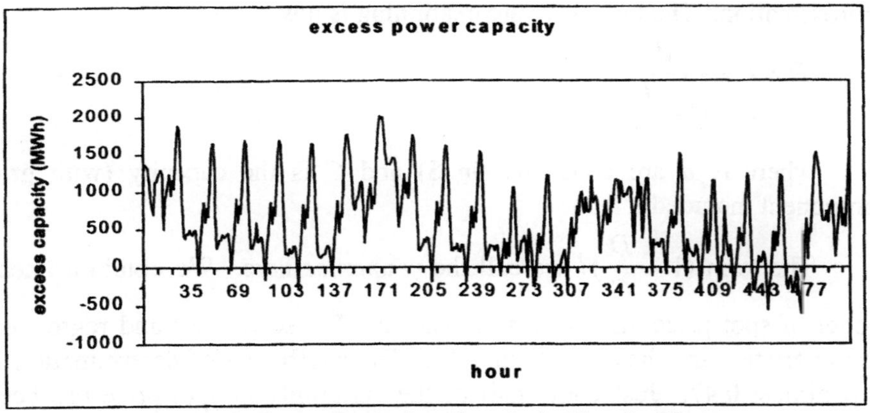

Figure 8. Excess Generating Capacity in Victorian Plants.

There were more stresses on generating capacity in the final week and this is reflected in the rising spot prices. It seems that spot prices vary in some inverse relationship with this excess capacity. To examine this hypothesis we look at a scatterplot of spot price against the available excess power for the period of three weeks from April 30. This is displayed in Figure 9.

Two possible models are displayed on this scatterplot. Neither seems to capture the data very well. The linear model gives rise to $R^2 = 0.43$ – an improvement on relating logarithms of spot price to logarithms of demand directly.

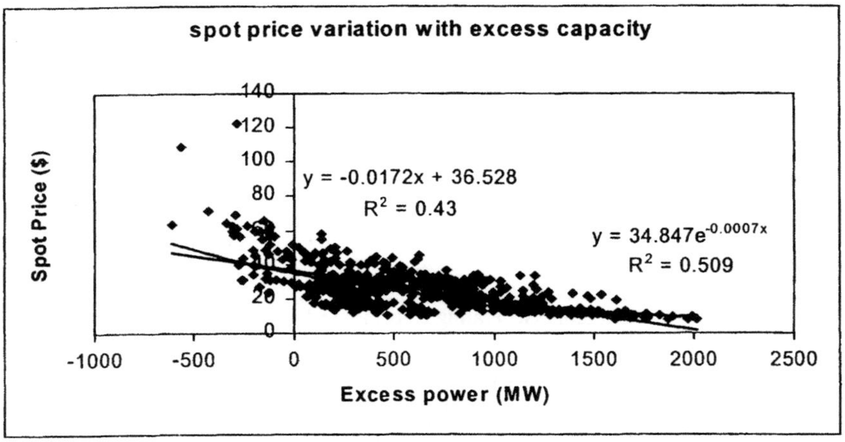

Figure 9. Price variation with excess capacity.

A model of inverse variation is suggested by the scatterplot. We may examine the available proportion of capacity, which brings the numbers back into scale. To include the grid component we will add 1750 MW. This is almost certainly too high a value in practice as it matches the available total from South Australia and the Snowy River Scheme but will serve as a first approximation. The model we are examining is now

$$P = c + \frac{d}{(1 - \frac{D}{C})}$$

where *c, d* are constants (in \$) and *C* is the capacity (with grid component included).

The term $(1 - \frac{D}{C})^{-1}$ we call the *resonance* here. We obtain a linear model of spot price against this resonance. The scatterplot and regression characteristics are shown in Figure 10. The coefficient of determination is now almost 0.47. Just under half of the variability in spot price has been accounted for by this resonance, confirming the expected strong dependence on both demand and available capacity. Here we see $c \approx -15$ and $d \approx 16$.

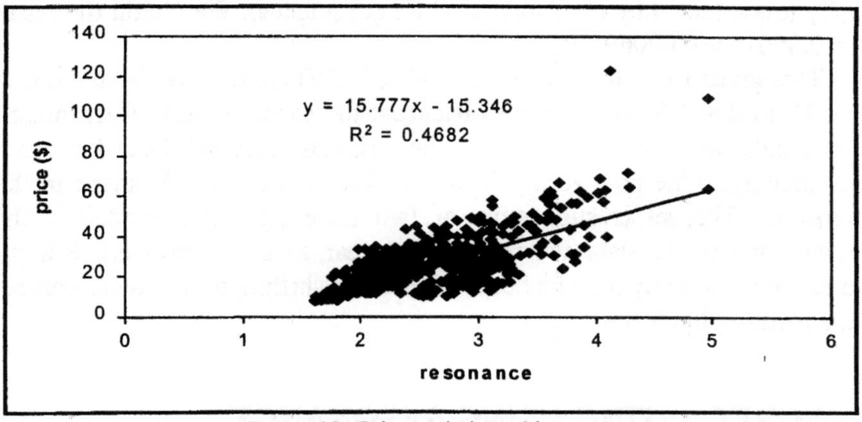

Figure 10: Price variation with resonance.

This model itself may understate the degree to which larger values of resonance will escalate the spot price. Outlier values are suggestive but too few to be conclusive here. It is noteworthy however that a linear model on this scatterplot yields a slightly higher coefficient of determination than exponential or polynomial regression models do – and this linear model is more easily explained.

It is interesting that we have almost a case where c = - d here. In such a case we have $\frac{C}{D} = 1 + \frac{d}{P}$ suggesting that in short periods where C may be held constant, linear regression of the reciprocals of demand and spot price is reasonable.

If we look at our capacity data it seems that we have effectively constant C for about half the three-week period. Obtaining a scatterplot of reciprocals of demand and spot price gives the result shown in Figure 11.

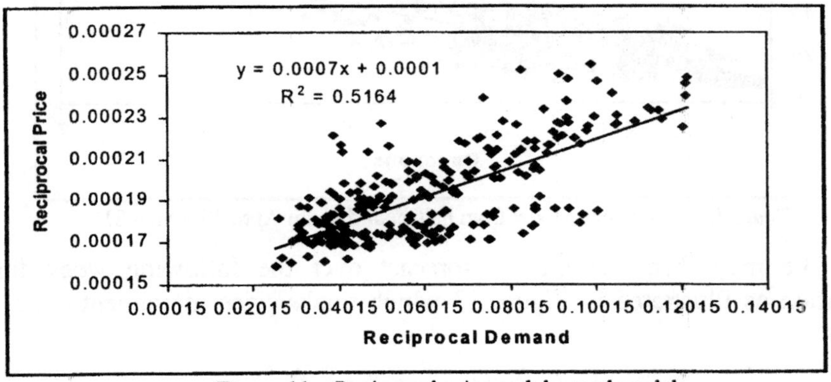

Figure 11: Reciprocal price and demand model

The coefficient of determination is fairly high here (0.51), although for the full three- week data set it drops to about 0.38. The linear model is actually less affected by the data period length selected. We obtain the result $1/D = 0.00015 + 0.000695/P$.

This gives 1/C = 0.00015 so C = 6700 (MW) approximately and d/C = 0.000695 so d = 4.6 ($). This does backcast the capacity quite well though our rationale for the process from the original data set had d = 15.5 approximately. One noteworthy feature of Figure 11 is the V-shape in the scatterplot. The set is suggestive of two differing linear models – the dominant one tracks data where prices appear to be suppressed. Pricing policies for a company may be better directed at shifting to the lower part set of data forming the V.

5. FORECASTING USING RESONANCE

We can now apply some of this analysis to generating forecasts of prices using known demand and capacity. We will use the available half-hourly data to model this as more data improves on the resonance model. Using half-hourly data over a three-week period enables the coefficient of determination for regressing price on resonance to lift to 0.49 from 0.37 for half hour data.

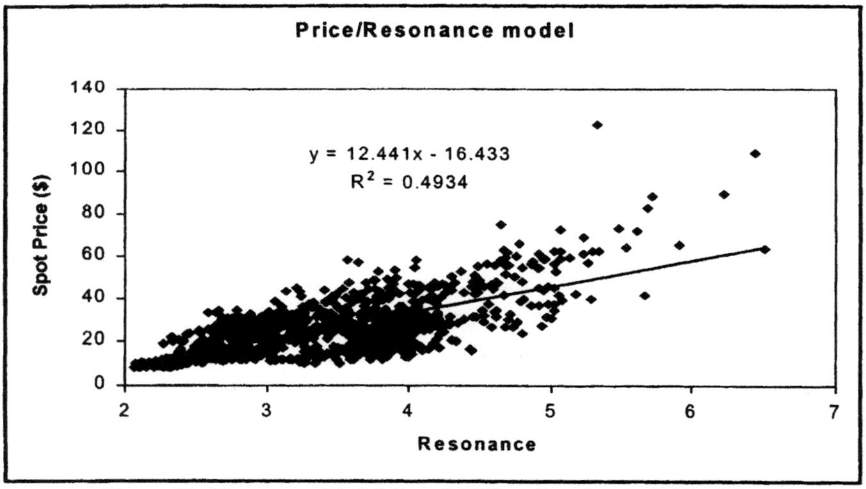

Figure 12: Resonance Model from half-hour data for April 30 – May 21.

We may then use this to forecast over the following week for validation as illustrated in Figure 13 which tracks price movement over 4 weeks.

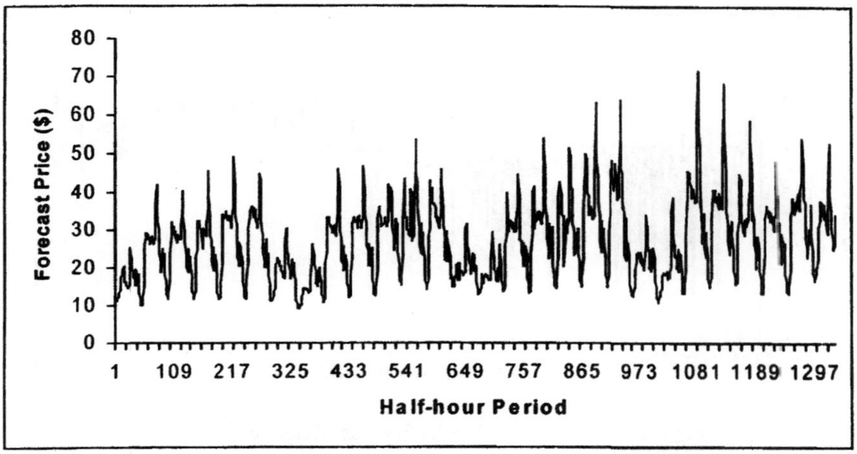

Figure 13: Resonance Model Forecasts for April 30 – May 28.

The more detailed half-hourly model appears to track the true spot price better but it imposes a degree of regularity not really observed in the data. The errors are still substantial as Figure 14 reveals and the relative error averages 30%. Of course the prices are fluctuating more in the fourth week – errors appear more stable early on but are generally higher in relative terms as Figure 15 shows!

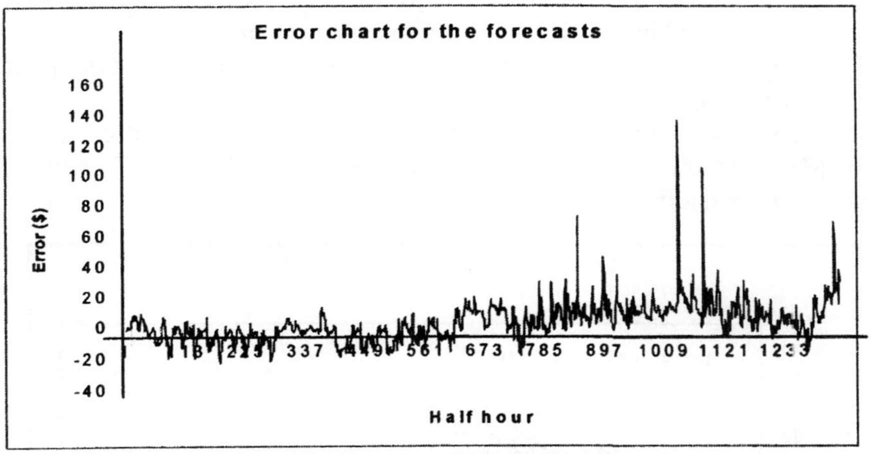

Figure 14: Resonance Model Forecast Errors.

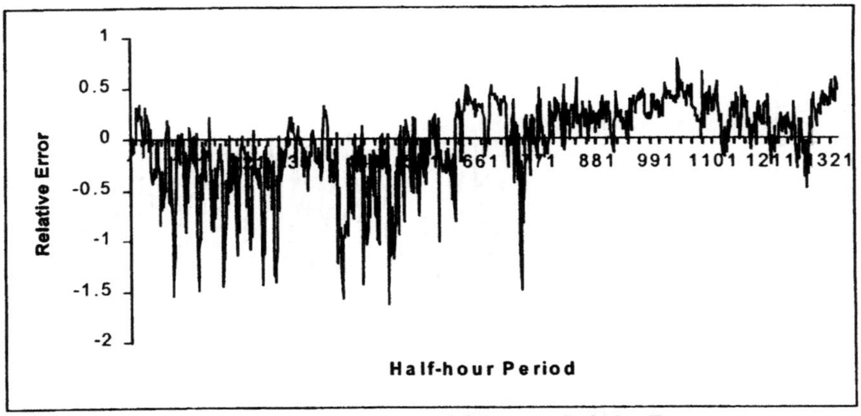

Figure 15: Resonance Model Forecast Relative Errors.

It is appropriate to question the choice of adding in the grid capacity as 1750 MW over all time periods. In fact this is certainly not available all the time and may mislead us to the true capacity and associated resonance pressure on demand.

We have seen that the excess is sometimes negative in peak demand time if we do not allow for grid access to South Australia and the Snowy River scheme. We may improve the model in a number of ways:

(a) arrange to add in only sufficient capacity to ensure resonance is positive
(b) arrange to add in a value which optimises the error
(c) arrange to add in a different amount on a regular basis for each of peak and off peak demand periods.

We look at the first option now reducing the added grid capacity to 950 MW. Results initially are encouraging (see Figure 16) with the determination coefficient now quite good.

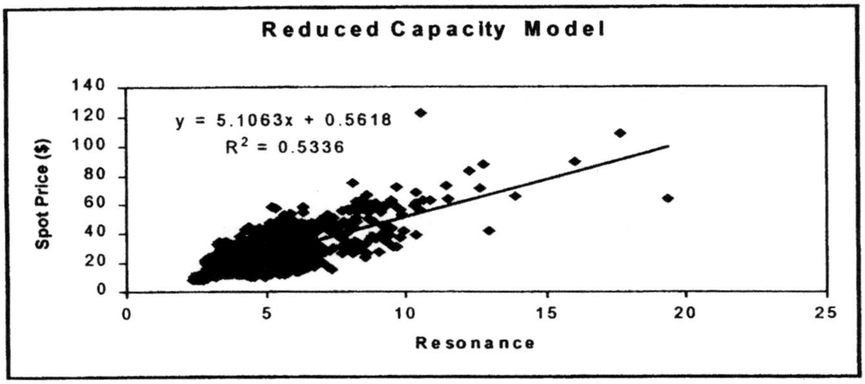

Figure 16: Resonance Model with reduced grid input for April 30 – May 21.

Using this model to build forecasts through to the next week as well for validation we get the results in Figure 17. These look much improved over existing forecasts.

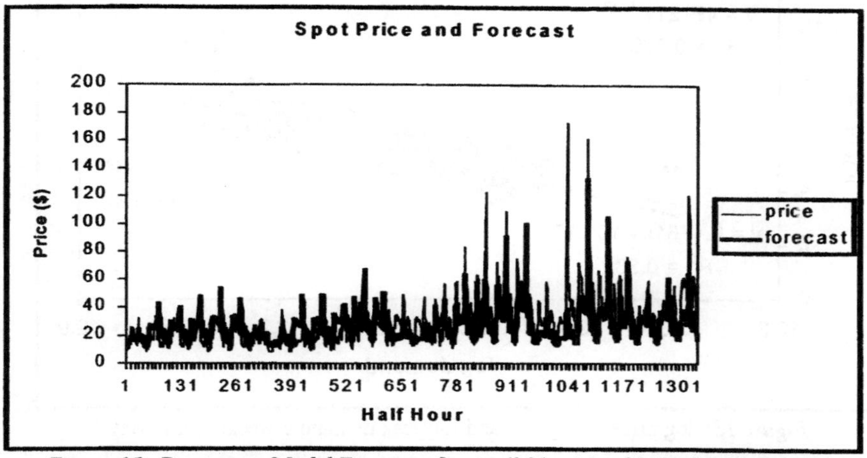

Figure 17: Resonance Model Forecasts for April 30 – May 28: Reduced Grid Input.

The MSE over the first three weeks has dropped to 84.2 from 146 confirming our view that this process of forecasting can be successful and that the available grid supply is rather less than initially modelled! We see that the MSE gives improvement over the ARIMA model used. However errors are still frequently very high and relative error still runs about 30% over the three-week period.

An informal analysis of the effect of varying the grid component suggests that 950 MW is near prediction-optimal as each of 900 MW and 1000 MW increase the MSE marginally. However the third option suggested here may be more reasonable as the available power from the Snowy River scheme follows an orthodox pattern connected to water use. This invites further research.

6. PEAK DEMAND MODELLING

The models may be improved by separating out the analysis for weekends and for peak and off peak demand in the weekdays. This automatically ensures we are forecasting like with like. In Figure 18, we can see some common regression models for log price and log demand over half hour periods in the five weeks of peak demand (7.30 am to 7.30 pm) starting Monday May 1. In the first instance the plot has been modelled for linear models as well as exponential and power laws. The simpler linear model proves best here and is also better than we have seen for regressing log price over log demand in the full weekly data (see Figure 6).

It is apparent that focussing on a particular subset of data is useful.

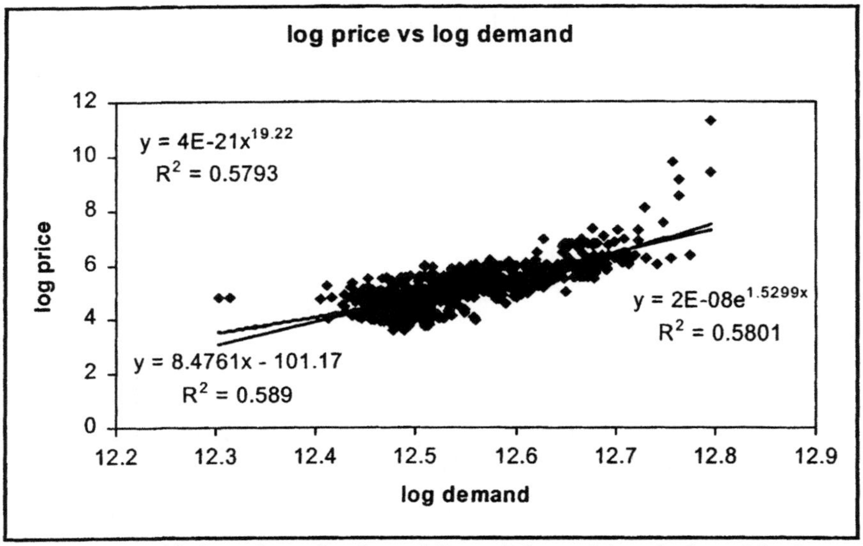

Figure 18: log price vs log demand for peak demand 5 weeks from May 1.

We can now examine these data in a resonance model as previously. We will examine log price as a function of resonance over the first four weeks and use this to build a model for the full five weeks, validating it in the fifth week. This is shown in Figure 19.

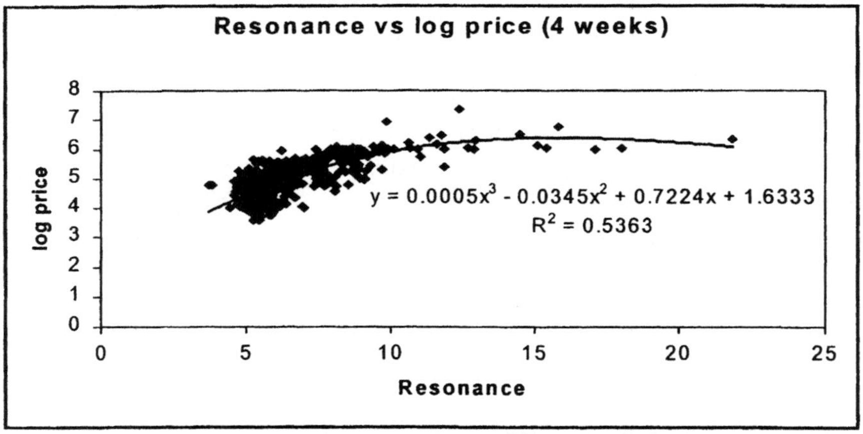

Figure 19: log price vs resonance for peak demand 4 weeks from May 1.

It seems that a cubic polynomial fit works quite well for these data. We try and check this by generating forecasts over the full 5 weeks. The errors from this log forecast model are shown in Figure 20.

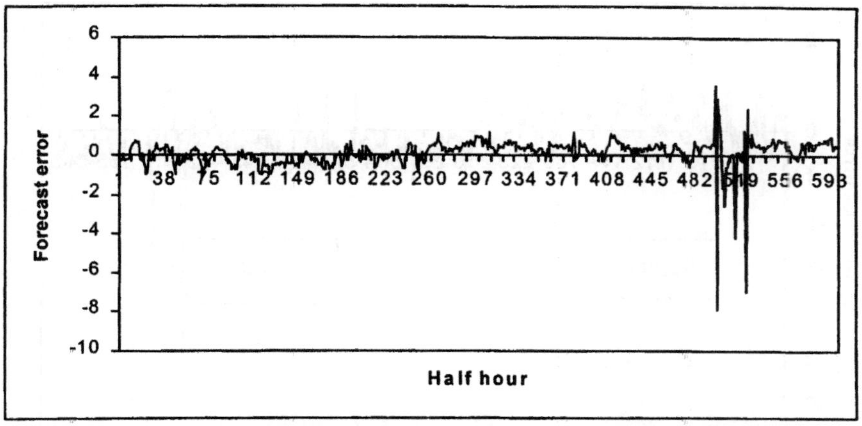

Figure 20: Tracking log forecast errors over 5 weeks from May 1.

Clearly the model fails in extreme cases. The more extreme spot prices possible did not occur in weeks 1 to 4. The model overreacts to the resonance pressure. We look at a cubic model generated over the full five weeks (see Figure 21).

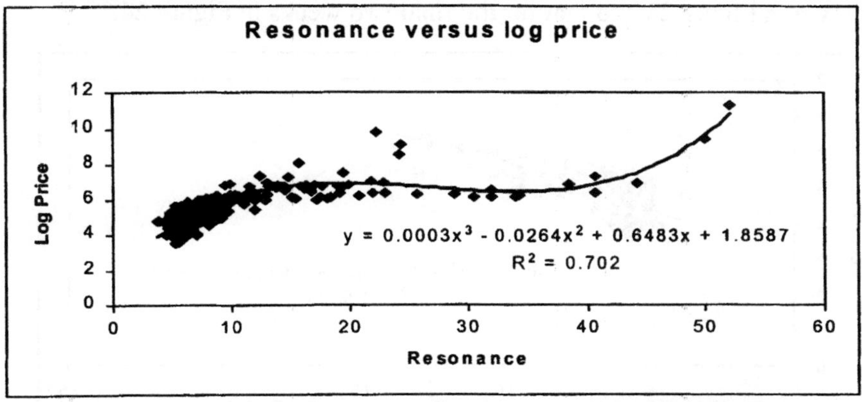

Figure 21: log price vs resonance for peak demand 5 weeks from May 1.

This performs much better as can be seen in Figure 22. It seems any model, which must cover the full range of prices, needs to be built from data including that range. It is noticeable that the forecasts here tend to be worse than before for the 'normal' range of prices however. This reminds us that the spot prices appear to be operating within different 'regimes' in each part of the data set. Peak prices may still be following the usual unconstrained demand or they may be set by stretched resources. These regimes are identified in Figure 22.

To forecast prices in each regime we may use data only from within that regime so that extreme values do not distort the forecasts.

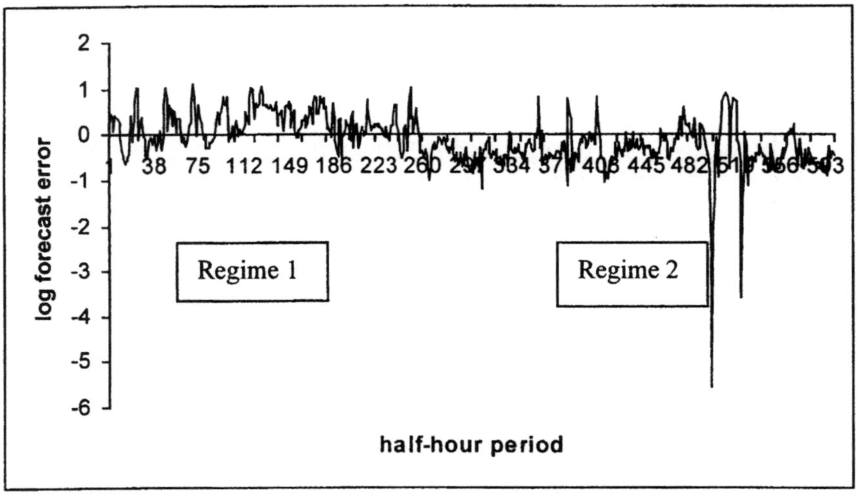

Figure 22: Forecast errors over peak demand 5 weeks from May 1.

Here we will split the five weeks of data into two groups - one of three weeks and one of two weeks. The regression model for the first three weeks is shown in Figure 23 and that for the final two weeks in Figure 24.

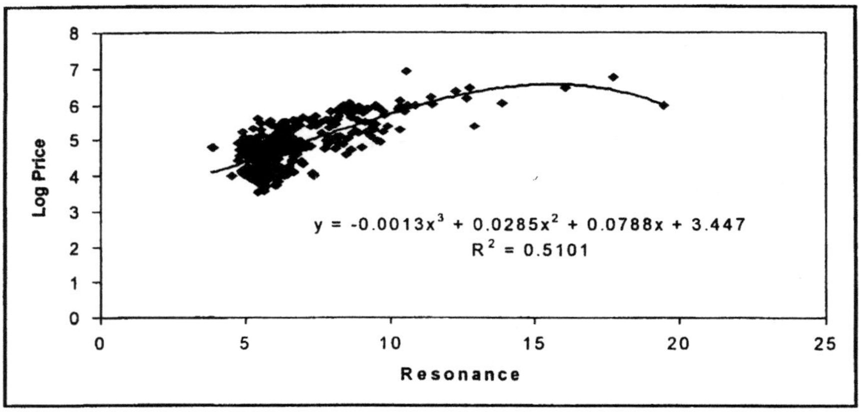

Figure 23: Regression model of resonance and log price in regime 1.

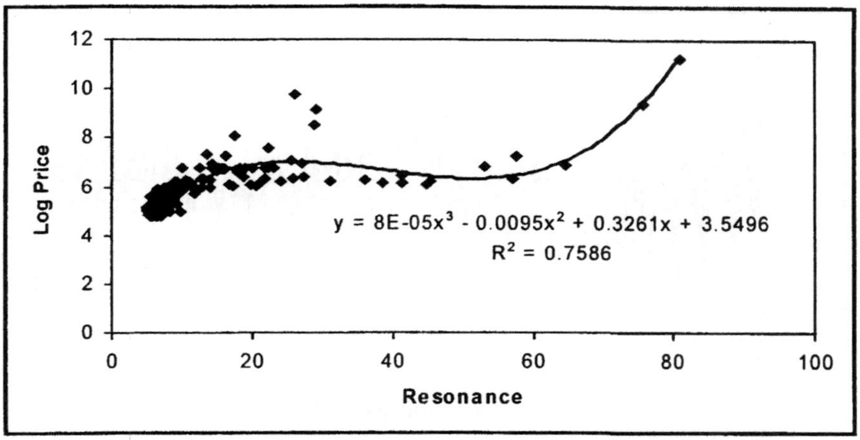

Figure 24: Regression model of resonance and log price in regime 2.

The equations generated were used to create the combination forecast in Figure 25. The relative errors for this model are displayed in Figure 26. The errors seem more balanced between regimes than evident in either Figure 15 or Figure 22.

The error analysis of this combination strategy is instructive. The MSE found earlier for the half hourly data was 84.3 using a covering grid capacity of 950 MW but taken over the first three weeks of data. This was actually raised to 11593.8 over the five weeks of data due to occasional extreme errors in the last two weeks! The mean absolute deviation of forecasts (MAD) was 20.026 over that period. When we apply a combination model to the peak demand period over that five weeks the MSE is reduced to less than half - 4576 – and the MAD dropped to 14.33. But the mean absolute relative error remains high at 23% and the relative error appears to retain some cycling.

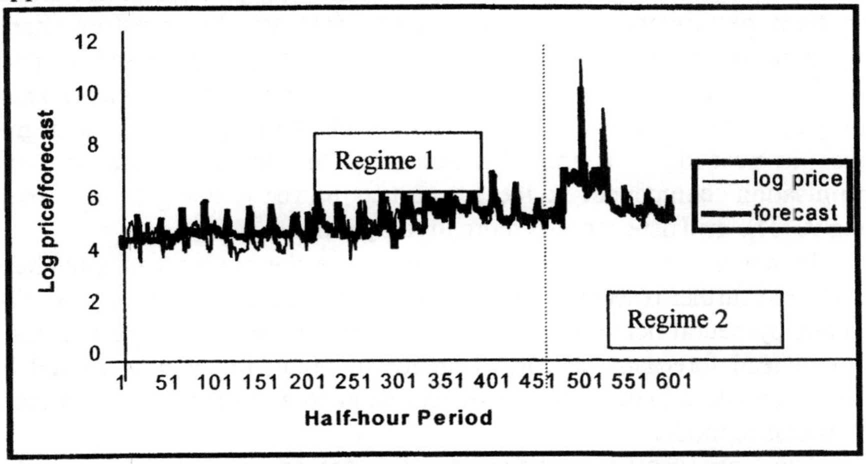

Figure 25: Log Forecasts for peak demand 5 weeks from May 1.

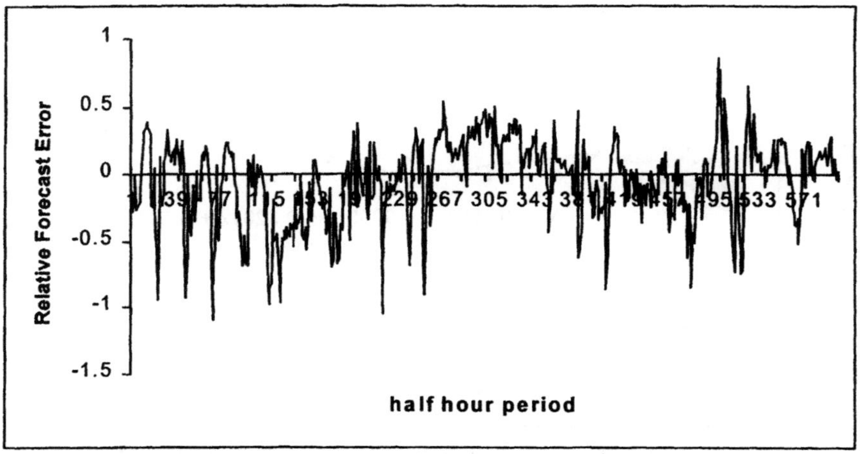

Figure 26: Relative Errors from the combined forecast over five weeks from May 1.

The forecasting issue then becomes one of identifying which regime the peak demand period is operating in. In fact we have identified regimes rather crudely here as cases where supply is under pressure by demand or is plentiful. We can fine tune this regime concept to tie it more to the price levels at which the various contributing power sources elect to enter the market. This is a subject of ongoing research.

7. CONCLUSION

It is possible to combine demand and supply into an effective predicting tool for the spot price which outperforms any time series based forecast models. The results of this forecasting enable more effective bidding strategies to be implemented, which will help to minimise risk in long term performance. We can improve forecasts by transformations, polynomial regression and applying models separately to peak and off peak times in weekdays as well as to weekends. Clearly the forecast dependence on supply means that accurate information on the co-generators and grid access is needed. It may be necessary to include a national grid 'optimisation' component in the forecast as power is being traded freely along it. The grid aspect needs more investigation.

In practice we may do better restricting the analysis to day-ahead models and further research should include direct comparisons with existing methods by neural networks and other techniques. When we are focussing on day-ahead forecasts the correct choice of predicting regime should be easier as should a better estimate of real demand and capacity which feeds the resonance model.

There are also risk aversion issues in the analysis of downtimes for generators and in the frequency of high spot prices. The study of extreme values in time series data sets is based on various speculative distribution theories (see e.g. Broussard and Booth) but may help in establishing some long-term risk policies.

8. ACKNOWLEDGEMENTS

The authors wish to thank Brad Smith of Yallourn Energy for his generous supply of relevant data and comments on structural issues in the Australian power industry. Patrick Tobin completed part of this research while on Industrial leave at CSIRO Division of Mathematical and Information Sciences.

References

Broussard, J.P. and Booth, G.G. (1998) The behaviour of extreme values in Germany's stock index futures: An application to intradaily trading *European Journal for Operations Research* 104 393-402

Charytoniuk, W., Chen, M.S, Kotas, P. and Van Olinda, P. (1999) Demand Forecasting in Power Distribution Systems using Nonparametric Probability Density Estimation *IEEE Transactions on Power Systems* 14, 4, 1200- 1206, November.

Connor, J.T. (1996) A Robust Neural Network Filter for Electricity Demand Prediction *Journal of Forecasting* 15, 437-458

Kliman, M (1994) Competitive Bidding for Independent Power *Energy Policy* January, 41-56.

Schweppe, F.C., Caramanis, M.C., Tabors, R.D. and Bohn, R.E. (1988) *Spot Pricing of Electricity*. Kluwer. Boston, MA.

Szukata, B.R., Sanabria, L.A. and Dillon, T.S. (1999) Electricity Price Short Term Forecasting using Artificial Neural Networks *IEEE Transactions on Power Systems* 14, 3, 851-857, August.

Takriti, S., Krasenbrink, B. and Wu, L. (2000) Incorporating Fuel Constraints and Electricity Spot Prices into the Stochastic Unit Commitment Problem *Operations Research* 48, 2, 268-280

Taylor, J.W. and Majithia, S. (2000) Using combined forecasts with changing weights for electricity demand profiling *Journal of the Operational Research Society* 51,1,72-82, January.

Xie, K, Song, Y-H, Stonham, J. Yu, E. and Liu, G. (2000) Decomposition Model and Interior Point Methods for Optimal Spot Pricing of Electricity in Deregulation Environments *IEEE Transactions on Power Systems* 15, 1, 39- 50 February.

Appendix Plant Nominal Capacity and Type in Victoria

Generator	Units	rating	total	fuel	Register Party
Anglesea	1	150	150	black coal	SECV
Morwell	1	22	22	brown coal	Energy Brix Australia
	3	32	96	brown coal	Energy Brix Australia
	1	65	65	brown coal	Energy Brix Australia
Newport	1	500	500	natural gas	Ecogen
Jeeralang A	4	56.6	226	natural gas	Ecogen
Jeeralang B	3	79.6	239	natural gas	Ecogen
Hazelwood	8	205	1640	brown coal	Hazelwood Power
Loy Yang A	4	500	2000	brown coal	Loy Yang Power
Loy Yang B	2	500	1000	brown coal	Edison Mission Energy
Yallourn W	2	350	700	brown coal	Yallourn Energy
	2	375	750	brown coal	Yallourn Energy
Southern Hydro	2	12	24	hydro	Southern Hydro
	1	150	150	hydro	Southern Hydro
	2	60	120	hydro	Southern Hydro
	6	18	108	hydro	Southern Hydro
	1	13.55	14	hydro	Southern Hydro
	4	15.5	62	hydro	Southern Hydro

There is an additional 1500 MWh available from the Snowy River scheme and 250 MWh which can be drawn from Sth Australia through the national grid if available.

REFINING DIFFUSION MODELS FOR STATE-DEPENDENT QUEUES

Toshikazu Kimura

Hokkaido University, Graduate School of Economics, Sapporo 060-0809, Japan

Abstract A diffusion model is developed for a class of queues with Poisson arrivals and general services, both of which are state-dependent. This class covers many practical queueing systems with multiple servers, finite waiting spaces, finite sources, discouraged arrivals and more. For each of these systems, a large number of diffusion models have been developed individually. Our diffusion model unifies and refines those previous models. The main focus is on the steady-state distribution of the number of customers in system. The discrete process of the number of customers is approximated by a continuous diffusion process with piecewise-constant infinitesimal parameters defined on a closed interval in the non-negative real line. A conservation law together with a consistent discretization generate an approximate formula for the steady-state distribution, which is more accurate than the previous models.

Keywords: queues; diffusion models; state dependency

1. INTRODUCTION

We consider a class of queues with Poisson arrivals and general services, both of which are *state-dependent*. This class covers many practical queueing systems with multiple servers, finite waiting spaces, finite sources, discouraged arrivals and more. The analysis of those queueing systems is useful for evaluating performance of computer, communication and manufacturing systems. For example, an $M/G/s$ queue with finite waiting spaces has been used as a simple model for buffer capacity design in these systems (see Kimura, 1996). A buffer design problem is to find the smallest buffer capacity such that the proportion of lost customers (*i.e.*, transactions, messages, parts and so on) is below an acceptance level. To obtain a concise expression for the optimal buffer size, we need an explicit formula for the loss probability. However, no exact solutions are available except for some special cases, for example, when service times are exponentially distributed. Under this restrictive condition, the number of customers in system can be analyzed by a birth-death (BD) process (see Cooper, 1981). In general, it is quite difficult to weaken the exponential condition and yield approximations that have been of practical interest. See Kimura (1987) for other possible applications of state-dependent queues, including performance evaluation of a multiaccess communication system with ALOHA packet-switching scheme, a switching system with overload control and so on.

A diffusion approximation is a technique in which a complicated stochastic process is replaced by an appropriate diffusion (Brownian motion) process. Diffusion processes have a great deal of mathematical structure, and hence they are more mathematically tractable than the original process. (Hereafter we use the word "diffusion model" instead of "diffusion approximation" to avoid confusion with heavy traffic limit theorems.) There have been a large number of diffusion models for particular queueing systems but fewer ones for a general class of state-dependent queues. To my best knowledge, there are only two diffusion models developed by Pujolle (1978) and Kimura (1987). Unfortunately, both models have a serious defect in that they are *not* consistent with any available exact results for particular cases. The inconsistency defect is commonly observed in other diffusion models; see Whitt (1982) where he discussed the necessity of refinement in a heuristic way. No doubt, the lack of consistency has quite lowered the value of diffusion models. The purpose of this paper is to develop a systematically refined diffusion model for the general class of state-dependent queues by extending my previous diffusion models for $GI/G/s$ queues (Kimura, 1995, 1999). Our diffusion model is consistent with the BD result under the exponential condition, unifying most diffusion models previously developed for particular cases.

2. A REFINED DIFFUSION MODEL

2.1 Basic Assumptions and Notations

For a generic state-dependent queue, let $A(t)$ ($D(t)$) denote the cumulative number of arrivals (departures including lost customers) in the time interval $(0, t]$, and let $L(t)$ denote the number of customers (either waiting or being served) in the system at time t (≥ 0). Then, there is the basic relation

$$L(t) = L(0) + A(t) - D(t), \qquad t \geq 0 \tag{1}$$

among these processes. The fundamental idea of diffusion models for queues is to approximate the *discrete* valued process $\{L(t); t \geq 0\}$ by a time homogeneous diffusion process $\{X(t); t \geq 0\}$ with a *continuous* path on an interval $\mathcal{S}_c \equiv [x_0, x_N]$ ($-\infty < x_0 < x_N \leq +\infty$), utilizing asymptotic properties of the arrival and service processes. To realize this idea in our diffusion model, we first define an interval $\mathcal{I}_k$ of $\mathcal{S}_c$ corresponding to the event $\{L(\cdot) = k\}$ ($k \in \mathcal{S}_d \equiv \{0, \ldots, N\}$). In this paper we use a general set of intervals defined by $\mathcal{I}_0 = \{0\}$ and $\mathcal{I}_k = (x_{k-1}, x_k]$ ($k = 1, \ldots, N$) for an increasing sequence $\{x_k; k \in \mathcal{S}_d\}$ with $x_0 = 0$. Clearly, $\mathcal{S}_c = \bigcup_{k \in \mathcal{S}_d} \mathcal{I}_k$. To regulate the process $X(\cdot)$ in $\mathcal{S}_c = [0, x_N]$, we simply assume that each of the boundaries is reflecting. Confer Pujolle (1978) and Kimura (1987) where the elementary return boundaries (Feller, 1954) were assumed.

Let $dX(\tau)$ be the increment in the diffusion process $X(\cdot)$ accrued over a time interval of length τ (> 0), *i.e.*, $dX(\tau) = X(\tau) - X(0)$. Then, apart from the boundary behavior, the diffusion process $X(\cdot)$ can be characterized by the limits

$$b(x) = \lim_{\tau \to 0} \frac{1}{\tau} E[dX(\tau) \mid X(0) = x],$$

$$a(x) = \lim_{\tau \to 0} \frac{1}{\tau} E[\{dX(\tau)\}^2 \mid X(0) = x],$$

for $x \in (0, x_N)$. These limits, if exist, are termed the *infinitesimal parameters* of the process. In particular, $b(\cdot)$ is called the *infinitesimal mean* and $a(\cdot)$ the *infinitesimal variance*. Assume that each of these parameters is *piecewise constant* (PC), *i.e.*, for $x \in \mathcal{I}_k$ ($k = 1, \ldots, N$),

$$\begin{aligned} b(x) &= b_k, \\ a(x) &= a_k, \end{aligned} \tag{2}$$

where $\{b_k;\ k \geq 1\}$ and $\{a_k;\ k \geq 1\}$ are bounded sequences and $a_k > 0$ for all k. Of course, the PC assumption is based on the natural correspondence between the events $\{L(\cdot) = k\}$ and $\{X(\cdot) \in \mathcal{I}_k\}$ ($k \in \mathcal{S}_d$). This idea was first introduced by Kimura (1983) in his diffusion model for the $M/G/s$ queue, and it was subsequently used in diffusion models for various queues. As noted in Kimura (1987), the process $X(\cdot)$ having the PC parameters (2) is well defined and still has a continuous path.

To obtain the parameters $\{b_k\}$ and $\{a_k\}$ of the diffusion process $X(\cdot)$ approximating the process $L(\cdot)$, we have to specify the counting processes $A(\cdot)$ and $D(\cdot)$ under the condition $L(\cdot) = k$ ($k = 1, \ldots, N$). Let $A_k(\cdot)$ ($D_k(\cdot)$) denote the conditional arrival (departure) process, and let $\lambda_k = \lim_{t \to \infty} E[A_k(t)]/t$ ($\mu_k = \lim_{t \to \infty} E[D_k(t)]/t$). Then, from the general theory of point processes (Franken et al., 1981) and the basic relation (1), we immediately obtain

$$b_k = \lambda_k - \mu_k, \qquad k = 1, \ldots, N. \tag{3}$$

The expression of $\{a_k\}$, however, depends on further information of $A(\cdot)$ and $D(\cdot)$.

2.2 Steady-State Distribution

Let $p(x,t\,|\,x^\circ)$ be the probability density function (pdf) of $X(t)$ starting from $X(0) = x^\circ$, *i.e.*, $p(x,t\,|\,x^\circ)dx = P\{x \le X(t) < x + dx\,|\,X(0) = x^\circ\}$. Then, it has been known that the pdf $p(x,t\,|\,x^\circ)$ satisfies the Kolmogorov forward (or Fokker-Planck) equation

$$\frac{\partial p}{\partial t} = \frac{1}{2}\frac{\partial^2}{\partial x^2}\{a(x)p(x,t\,|\,x^\circ)\} - \frac{\partial}{\partial x}\{b(x)p(x,t\,|\,x^\circ)\},$$

so that the steady-state pdf $p(x) = \lim_{t\to\infty} p(x,t\,|\,x^\circ)$, if existing, satisfies the ordinary differential equation

$$\frac{1}{2}\frac{d^2}{dx^2}\{a(x)p(x)\} - \frac{d}{dx}\{b(x)p(x)\} = 0, \qquad 0 < x < x_N. \tag{4}$$

At the origin $x = 0$ and $x = x_N$, the pdf $p(x)$ also satisfies the boundary conditions

$$\left.\frac{1}{2}\frac{d}{dx}\{a(x)p(x)\} - b(x)p(x)\right|_{x=0,\,x_N} = 0. \tag{5}$$

Solving (4) together with (5) and using the PC assumption, we obtain

$$p(x) = Cq(x), \qquad 0 \le x \le x_N,$$

where C is the normalization constant,

$$q(x) = \frac{1}{a_k}\exp\left\{\frac{2b_k}{a_k}(x - x_{k-1})\right\}\prod_{j=1}^{k-1}\gamma_j, \qquad x \in \mathcal{I}_k, \quad k = 1,\ldots,N, \tag{6}$$

and

$$\gamma_j = \exp\left\{\frac{2b_j}{a_j}(x_j - x_{j-1})\right\}, \qquad j = 1,\ldots,N. \tag{7}$$

Let L be the random variable representing the number of customers in the steady-state system at arbitrary time, and let $p_k = P(L = k)$ $(k \in S_d)$. To *extract* an approximation for the steady-state distribution $\{p_k\}$ from the steady-state pdf $p(x)$, we use a *pointwise* discretization defined by

$$p_k = p_0 q(x_k) = \frac{p_0}{a_k}\prod_{j=1}^{k}\gamma_j, \qquad k = 1,\ldots,N-1. \tag{8}$$

To determine the remaining probabilities p_0 and p_N, we use a conservation law: Since the rates of arrivals and departures are equal in the steady-state (Equation (4.3.45) in Franken et al., 1981), we have

$$\sum_{k=0}^{N-1} \lambda_k p_k = \sum_{k=1}^{N} \mu_k p_k. \tag{9}$$

This principle or (9) is called a *rate conservation law* that is true even for a non-renewal arrival process. Solving (9) and the normalizing condition $\sum_{k=0}^{N} p_k = 1$ for p_0 and p_N, we obtain

$$p_0 = \frac{\mu_N}{\sum_{j=0}^{N-1} (\lambda_j + \mu_N - \mu_j) \prod_{i=1}^{j} \gamma_i} \quad \text{and} \quad p_N = \frac{\sum_{j=0}^{N-1} (\lambda_j - \mu_j) \prod_{i=1}^{j} \gamma_i}{\sum_{j=0}^{N-1} (\lambda_j + \mu_N - \mu_j) \prod_{i=1}^{j} \gamma_i}, \tag{10}$$

where we set $\mu_0 \equiv 0$.

Let $\{p_k^*\}$ be the exact steady-state distribution for the case that service times are exponentially distributed. From the local balance conditions in the BD process, we have

$$\frac{p_k^*}{p_{k-1}^*} = \frac{\lambda_{k-1}}{\mu_k}, \qquad k = 1, \ldots, N, \tag{11}$$

while the approximation (8) gives

$$\frac{p_k}{p_{k-1}} = \frac{a_{k-1}}{a_k} \gamma_k, \qquad k = 1, \ldots, N-1, \tag{12}$$

where we set $a_0 \equiv 1$. We will determine the sequence $\{x_k\}$ in $\{\gamma_k\}$ so that the approximation $\{p_k\}$ in (8) for the BD case agrees with $\{p_k^*\}$: Equating (12) for the BD case with the exact result (11), we obtain

$$\gamma_k^* = \frac{a_k^*}{a_{k-1}^*} \frac{\lambda_{k-1}}{\mu_k}, \qquad k = 1, \ldots, N-1, \tag{13}$$

where we attach the asterisk $*$ to the quantities contained in the approximation $\{p_k\}$ for the BD case; note that $b_k^* = b_k$ because of (3). By (7) and (13), we obtain the recursive equation

$$x_k - x_{k-1} = \frac{a_k^*}{2b_k} \ln \left(\frac{a_k^*}{a_{k-1}^*} \frac{\lambda_{k-1}}{\mu_k} \right), \qquad k = 1, \ldots, N-1 \tag{14}$$

for the partitioning sequence $\{x_k\}$. Here we assume that $\{x_k\}$ is invariant with respect to the service-time distribution of the target queue. Then, by (14), (7) reduces to

$$\gamma_k = \left(\frac{a_k^*}{a_{k-1}^*}\frac{\lambda_{k-1}}{\mu_k}\right)^{\alpha_k} \quad \text{with} \quad \alpha_k = \frac{a_k^*}{a_k}, \qquad k = 1, \ldots, N-1, \tag{15}$$

and hence we obtain

$$p_k = \begin{cases} \dfrac{\mu_N \xi_k}{\sum_{j=0}^{N-1}(\lambda_j + \mu_N - \mu_j)\xi_j}, & k = 0, \ldots, N-1 \\[2ex] \dfrac{\sum_{j=0}^{N-1}(\lambda_j - \mu_j)\xi_j}{\sum_{j=0}^{N-1}(\lambda_j + \mu_N - \mu_j)\xi_j}, & k = N, \end{cases} \tag{16}$$

with

$$\xi_j = \begin{cases} 1, & j = 0 \\ \dfrac{1}{a_j}\prod_{i=1}^{j}\left(\dfrac{a_i^*}{a_{i-1}^*}\dfrac{\lambda_{i-1}}{\mu_i}\right)^{\alpha_i}, & j = 1, \ldots, N-1. \end{cases} \tag{17}$$

Due to the principle of consistent discretization, the approximation (16) should be consistent with the BD case. In fact, from (17), we have $\xi_j^* = \prod_{i=1}^{j}(\lambda_{i-1}/\mu_i)$ $(j = 1, \ldots, N-1)$, so that

$$\sum_{j=0}^{N-1}(\lambda_j + \mu_N - \mu_j)\xi_j^* = \mu_N \left(\sum_{j=0}^{N}\prod_{i=1}^{j}\frac{\lambda_{i-1}}{\mu_i}\right),$$

which certainly yields the well-known BD result; see Section 2.3 in Cooper (1981).

3. SOME EXAMPLES

3.1 A Finite-Capacity $M/G/s$ Queue

Consider the standard $M/G/s/s+r$ queueing system with s (≥ 1) identical servers in parallel, r (≥ 0) extra waiting spaces, and the first-come first-served discipline. Customers are arriving at the system according to a Poisson process with rate λ. If a customer arrives when all s servers are busy and all r waiting spaces are full, then he is *lost*, that is, he leaves without receiving service and without affecting future arrivals. Service times of customers allowed to enter the system are independent and identically distributed (iid) with a general distribution with mean μ^{-1} and the squared coefficient of variation (scv) c^2, being independent of the arrival process. For notational convenience, let $\rho \equiv \lambda/s\mu$ that means the traffic intensity. The process $L(\cdot)$ in the $M/G/s/s+r$ queue is analytically intractable except for a few special cases, so that approximations with heuristic nature have been developed so far by some researchers. See Kimura (2000) for their relationship.

For a more general $GI/G/s/s+r$ queue, three diffusion models have been developed by Sunaga et al. (1982), Yao and Buzacott (1985) and Kimura (1999). The first two diffusion models have the inconsistency defect, whereas Kimura's model is consistent with the exact results for the $M/G/s/s$ and $M/M/s/s+r$ queues. For the Poisson arrival case, his model coincides with our particular result for the $M/G/s/s+r$ queue.

From the description of the $M/G/s/s+r$ queue, we immediately obtain the parameters of our diffusion model, *i.e.*, $N = s + r$, $\lambda_k = \lambda$, $\mu_k = \min(k, s)\mu$, and hence $b_k = \lambda - \min(k, s)\mu$. For the conditional infinitesimal variance $\{a_k\}$, following a renewal-process approximation in Kimura (1999), we use

$$a_k = \lambda + \min(k, s)\mu\delta_k^2, \qquad k = 1, \ldots, s + r - 1 \tag{18}$$

where δ_k^2 is the scv of inter-departure times of $D_k(\cdot)$, which is given by

$$\delta_k^2 = \begin{cases} 1, & k = 1, \ldots, s - 1 \\ 1 + \min(\rho, 1)(c^2 - 1), & k = s, \ldots, s + r - 1. \end{cases} \tag{19}$$

From the forms of the infinitesimal parameters, we see that $\xi_k = \xi_s\hat{\rho}^{k-s}$ for $k = s, \ldots, s+r-1$, where $\hat{\rho} = \rho^{\alpha_s}$ with $\alpha_s = (\rho+1)/(\rho+\delta_s^2)$. Hence, the steady-state distribution in the $M/G/s/s+r$ queue is given by

$$p_k = \begin{cases} p_0\xi_k, & k = 1, \ldots, s - 1 \\ p_0\xi_s\hat{\rho}^{k-s}, & k = s, \ldots, s + r - 1 \\ 1 - \sum_{j=0}^{s+r-1} p_j, & k = s + r, \end{cases} \tag{20}$$

with

$$p_0^{-1} = \begin{cases} \displaystyle\sum_{j=0}^{s-1} \left(\rho + 1 - \frac{j}{s}\right) \xi_j + \frac{1 - \hat{\rho}^r}{1 - \hat{\rho}} \rho\xi_s, & \rho \neq 1 \\ \displaystyle\sum_{j=0}^{s-1} \left(2 - \frac{j}{s}\right) \xi_j + r\xi_s, & \rho = 1. \end{cases} \tag{21}$$

3.2 The Machine Interference Problem

Consider a situation where a group of K (≥ 1) machines are maintained by a team of s ($1 \leq s \leq K$) servers or operatives. Each machine has two states, either *up* (running) or *down* (requiring repair service). When a machine breaks down, it is repaired by one of the free servers. If all s servers are busy, the machine must wait for service. Running times for each of the machines are assumed to be iid with an exponential distribution with mean λ^{-1}. Repair times of failed machines are iid with a general distribution with mean μ^{-1} and the scv c^2, being independent of the breakdown process. In the queueing context, this system is classified into finite-source queues and is denoted by $M/G/s/\cdot/K$, and the arrival process is often called *quasirandom input*; see Section 3.5 of Cooper (1981). Benson and Cox (1951) obtained the steady-state distribution for the $M/M/s/\cdot/K$ case by using the BD formulation. Bunday and Scraton (1980) analyzed the $GI/M/s/\cdot/K$ case to find the insensitivity property that the steady-state distribution of the number of failed machines is the same in the $M/M/s/\cdot/K$ and $GI/M/s/\cdot/K$ cases. For other related work, see references in Bunday and Scraton (1980).

There have been a few diffusion models developed for solving the machine interference problem: Halachmi and Franta (1977) proposed a basic diffusion model for the $GI/G/s/\cdot/K$ queue; see also Biswas and Sunaga (1981) and Jain and Sharma (1986). All of these diffusion models have the inconsistency defect.

We can specify our diffusion model by using the fact that the conditional arrival process $A_k(\cdot)$ is the sum of $K-k$ independent and continuously running processes. Since the superposition of independent Poisson processes is still Poisson, we see that $A_k(\cdot)$ is the Poisson process with rate $\lambda_k = (K-k)\lambda$ ($k = 1, \ldots, K-1$). The conditional departure process $D_k(\cdot)$ has the same structure as in the $M/G/s/s{+}r$ queue with $r = K - s$. Hence, the infinitesimal parameters are given by

$$\begin{aligned} b_k &= (K-k)\lambda - \min(k,s)\mu, \\ a_k &= (K-k)\lambda + \min(k,s)\mu\delta_k^2, \end{aligned} \tag{22}$$

for $k = 1, \ldots, K-1$ and δ_k in (19).

It is worth noting that our diffusion model having the parameters (22) also can be applied to the analysis of the $M/G/s/s/K$ queue, for which it is well known that the steady-state distribution is insensitive to the service-time distribution except its mean and takes the form of a truncated binomial distribution; see Section 3.7 in Cooper (1981). In other words, the steady-state distribution is the same in the $M/M/s/s/K$ and $M/G/s/s/K$ cases. We can easily see that this insensitive property holds in our diffusion model with δ_k^2 in (19), so that the model is also consistent with the exact result for the $M/G/s/s/K$ queue.

3.3 An $M/G/s/s+r$ Queue with Balking

Consider a system where increase in the queue length discourages customers from joining the queue. Such a system has often been called a queue with ***discouraged arrivals*** or ***balking***; see Rao (1965). In this section, we apply our diffusion model to analyze the $M/G/s/s+r$ queue with balking, using the same assumptions and notations as in Section 3.1. Let ε_k be the probability that an arriving customer joins the queue when $L(\cdot) = k$ $(k = 0, \ldots, s+r)$, and assume that ε_k is nonincreasing in k with $\varepsilon_k = 1$ for $k = 0, \ldots, s-1$ and $\varepsilon_{s+r} = 0$. A customer decides whether to join the queue or not by himself, independent of any other customers. Then we see that the conditional arrival rate is $\lambda_k = \lambda\varepsilon_k$ and the infinitesimal parameters are given by

$$\begin{aligned} b_k &= \lambda\varepsilon_k - \min(k, s)\mu, \\ a_k &= \lambda\varepsilon_k + \min(k, s)\mu\delta_k^2, \end{aligned} \tag{23}$$

for $k = 1, \ldots, s+r-1$ and δ_k in (19).

4. CONCLUSION

For a general class of state-dependent queues with Poisson arrivals, we have developed a unifying approximation formula for the steady-state distribution $\{p_k\}$ by using the diffusion approximation. The formula is more accurate than the previously established diffusion approximations in the sense that it is consistent with the exact result when service times are exponentially distributed.

Note that $\{p_k\}$ given in (16) is the steady-state distribution of $L(\cdot)$ at arbitrary time. In some applications including the optimal buffer design, however, we often need the distribution of $L(\cdot)$ just before an arrival epoch. For queues with non-Poisson arrivals, these two probabilities are *not* equal. It would be an important subject of future studies to generalize the arrival process; see Kimura (1999) for a heuristic trial in this direction.

5. ACKNOWLEDGMENTS

This research was supported in part by the Grant-in-Aid for Scientific Research provided by the Ministry of Education, Science, Sports and Culture of Japan (Grant No. 10480084 in 1998–2001).

References

Benson, F. and Cox, D.R. The productivity of machines requiring attention at random intervals. Journal of the Royal Statistical Society, Series B 1951; 13: 65–82

Biswas, S.K. and Sunaga, T. Diffusion approximation method for the solution of two-stage cyclic queueing problems. Memoirs of the Faculty of Engineering, Kyushu University 1981; 41: 17–32

Bunday, B.D. and Scraton, R.E. The $G/M/r$ machine interference model. European Journal of Operational Research 1980; 4: 399–40

Cooper, R.B. Introduction to Queueing Theory. 2nd Ed., New York: North-Holland, 1981

Feller, W. Diffusion processes in one dimension. Transactions of the American Mathematical Society 1954; 77: 1–31

Franken, P., König, D., Arndt, U. and Schmidt, V. Queues and Point Processes. Berlin: Akademie-Verlag, 1981

Halachmi, B. and Franta, W.R. A diffusion approximate solution to the $G/G/k$ queueing system. Computers and Operations Research 1977; 4: 37–46

Jain, M. and Sharma, G.C. Diffusion approximation for the $GI/G/r$ machine interference problem with spare machines. Indian Journal of Technology 1986; 24: 568–572

Kimura, T. Diffusion approximation for an $M/G/m$ queue. Operations Research 1983; 31: 304–321

Kimura, T. A unifying diffusion model for state-dependent queues. Optimization 1987; 18: 265–283

Kimura, T. An $M/M/s$-consistent diffusion model for the $GI/G/s$ queue. Queueing Systems 1995; 19: 377–397

Kimura, T. Optimal buffer design of an $M/G/s$ queue with finite capacity. Stochastic Models 1996; 12: 165–180

Kimura. T. A consistent diffusion approximation for finite-capacity multi-server queues. In Stochastic Models in Engineering, Technology and Management, Proceedings of the First Western Pacific/Third Australia-Japan Workshop, R.J. Wilson, S. Osaki and M.J. Faddy (eds.), pp. 305–314, Christchurch, New Zealand, 1999

Kimura. T. Equivalence relations in the approximations for the $M/G/s/s+r$ queue. Mathematical and Computer Modelling 2000; 31: 215–224

Pujolle, G. Files d'attente avec dépendence de l'état et application à un système de communication multilignes. Research Report 1978; No. 315, IRIA Laboria

Rao, S.S. Queueing models with balking, reneging and interruptions. Operations Research 1965; 13: 596–608

Sunaga, T., Biswas, S.K. and Nishida, N. An approximation method using continuous models for queueing problems. II (multi-server finite queue). Journal of the Operations Research Society of Japan 1982; 25: 113–128

Whitt, W. Refining diffusion approximations for queues. Operations Research Letters 1982; 1: 165–169

A SYSTEM DYNAMICS MODEL FOR HELICOPTER OPERATIONS

Yi Yue

Land Operations Division
Defence Science and Technology Organisation
PO Box 1500, Salisbury SA 5108, Australia
(Tel: +61-8-8259-5687; Fax: +61-8-8259-5624; E-mail: yi.yue@dsto.defence.gov.au)

Abstract The introduction of armed reconnaissance helicopters into the Australian Army will add not merely a new platform, but a completely new capability to the Land Force. As such, many issues need to be sorted out before it can be effectively employed. For example, current organisation structures and doctrines will need to evolve in order to meet these new requirements. This calls for a better understanding of the dynamic behaviour of helicopter operations as a system, re-examining the way it operates and identifying deficits. Here the major concern is the total system behaviour, not a detailed study of individual entities. To address this issue, this paper develops a system dynamics model for helicopter operations. It first examines interrelationships of various elements of the system by constructing an influence diagram, so as to capture the essence of the system and understand the total system's dynamic behaviour. Then a quantitative model is built and computer simulations carried out to identify deficits and bottlenecks, where further resources, and possible changes of operational procedures or policies, are required to improve the system's performance. Finally the model's behaviour is optimised against an objective function and the significance of the results explained. This process shows that a system dynamics model can be used to determine the optimal platform and crewmember ratio, and reveals the potential problems that can be caused by an inadequate maintenance sector. Further work may involve a more detailed model of the maintenance sector, which takes into account the issues such as personnel training level and spare parts stocks. It also shows the usefulness of system dynamics modelling in determining pressure points and the role it can play as a test bed for decision makers with regard to military operations.

Keywords: system dynamics, helicopter operations, influence diagram, simulation.

1. INTRODUCTION

The definition of system in the Oxford Dictionary is: "Complex whole, set of connected things or parts, organised body of material or immaterial things." It is clear that system refers to interconnected parts, which act together through certain rules, or structures to serve predefined purposes. Dynamics means behaviour that changes with time. Here time is the most important element. Most real world systems are complex, non-linear and dynamic.

Usually, we are interested in a system because we want to understand it in order to manage it. Once human factors become integrated parts of a system, the system turns into a managed system. Managed systems have several distinctive characteristics. First, in a managed system, humans play a significant role by designing policies for the system. Second, the system behaves differently from the simple sum of its parts. Small changes in one part can produce big changes in overall behaviour. Third, the system's structure determines its behaviour, that is if different people are immersed into the same system, they are inclined to generate similar results. Fourth, causes and results can be spatially and temporally separated. This is where our conventional ways of thinking and the reality of complex systems' behaviours can differ significantly. Fifth, it contains feedback structures, which either reinforce (positive feedback) or stabilise (negative feedback) the system.

Since all management systems possess these characteristics, we would like to have a tool or methodology to help us understand overall performance of the system, predict consequences of a specific action, know long term impacts of policies that are meant to solve current problems, and design robust policies against uncertainty. However, traditional hard Operations Analysis (OA) methods fail to address these characteristics and are impotent in dealing with systems in which behaviour changes as a function of time are significant. To fill this gap, Professor J. W. Forrester pioneered a unique scientific discipline - system dynamics (Forrester 1961) - which takes a novel and completely different paradigm.

System dynamics is a rigorous method of studying the managed system's time-dependent behaviour in terms of its feedback structures and control policies. Its foundation rests in engineering science, not statistics. Usually, system dynamics modelling entails constructing a model based on mental models, qualitative knowledge, insights of the system and numerical information. Feedback control principles are applied to assess and improve the quality of the model. Then computer simulation is used to understand and improve the system's behaviour by varying its structures and policies. Finally, ways of improving system performance can be determined through optimisation and the scientific results are translated into implementation. Here a word of caution about optimisation. We must bear in mind that the purpose of system dynamics modelling is searching for useful solutions to real problems rather than a theoretically optimised solution.

System dynamics attempts to capture the highly complex, non-linear feedback relationships existing in the real world and incorporates the variable time delays that separate decisions from events. It recreates dynamic behaviour under a wide range of scenarios so as to allow the testing of possible alternative assumptions, decisions and policies using a model. This dynamic cyber environment provides a method to anticipate and plan for likely future events.

Since its founding in the 1960's, system dynamics has grown rapidly and been applied successfully in diverse fields such as defence, sociology, economics, biology, ecology and engineering. Its utility in the defence domain is well demonstrated by papers published in the open literature (Wang and Huang, 2000, Coyle et al 1999, Kearney 1998, Coyle 1996a, Coyle 1996b, Moffat 1996, Clark 1993, Coyle 1992a, Coyle 1992b, Wolstenholme et al 1992, Coyle and Gardiner 1991, Wolstenholme 1990, Coyle 1987, Wolstenholme 1986, Coyle 1985, Coyle 1981). It has been widely accepted that system dynamics integrates all elements of management processes to provide a larger picture and can fundamentally improve the effectiveness of decision-making and management.

This paper describes an application of system dynamics modelling to helicopter operations. It first examines the interrelationships of various elements of the system by constructing an influence diagram, to capture the essence of the system and understand the total system's dynamic behaviour. Then a quantitative model is built and computer simulations carried out to identify deficits, which need further resources, and possible changes of operation procedures or policies, which will improve the system's performance. Finally the model's behaviour is optimised against an objective function and the significance of the results is explained. It should be pointed out that although it attempts to model military helicopter operations, the data used in this paper are fictitious.

2. THE SYSTEM DYNAMICS MODEL FOR HELICOPTER OPERATIONS

2.1 Background

The introduction of armed reconnaissance helicopters into the Australian Army will add not merely a new platform, but a completely new capability. Many issues need to be resolved before it can be effectively employed in operations. New organisation structures and doctrine must be developed to meet the requirements. This calls for a better understanding of the dynamic behaviour of the system, re-examining the way it operates, and identifying deficits, which need further resources, to make it suitable for the operation of the new platform. It is also desirable to obtain guidance to limit the parameter space for subsequent detailed study. It is clear that to solve the problem we have to study the total system behaviour, not the details of individual entities. The purpose of the study is to provide guidelines to help to define the problem. This is exactly what the system dynamics' information/action/consequences paradigm is meant to do.

As the first attempt to untangle the enigma, the rest of this section presents a system dynamics model constructed to capture the essence of helicopter operations as a system.

2.2 Narrative Description

Interviewing aviation personnel produced the following narrative description of the system. Armed reconnaissance helicopters can be used to carry out missions of reconnaissance as well as attacks. They will be very high value assets and highly in demand during a land dominant military operation. It is assumed that during a period of intensive operations the requirement for an armed reconnaissance helicopter mission arrives randomly with a mean of one hour per mission. After a Command and Control (C2) and tasking delay, which takes a value of 0.5 to 1.5 hours, if there are enough helicopters (a pair) and crewmembers (two per helicopter) available, the mission will be flown, otherwise the mission will be abolished (That is the mission can not be carried out and means other than armed reconnaissance helicopters have to be used). There are a total of 12 armed reconnaissance helicopters and 24 crewmembers. Average mission duration is 2 hours. After each mission, the helicopters need to be refuelled and subjected to routine check and maintenance, which takes on average one hour, before it is available for the next mission. Also, on average, after 100 hours of flight time a helicopter undergoes a major maintenance, which takes an average of 48 hours. The crewmembers can work continually for 10 hours before requiring an 8-hour rest. One performance indicator of the system is the percentage of required missions that have actually been flown.

2.3 The Influence Diagram

An influence diagram illustrates cause-and-effect relationships, information flows and control actions within the system, and incorporates management policies. It serves as a vehicle for communication and discussion, allows qualitative study and provides a basis for simulation modelling. Based on the narrative description, an influence diagram is constructed, as shown in Figure 1, for the helicopter operation system.

There is a broad range of diagram conventions used in the system dynamics community and two of them (influence diagram or causal loop diagram and stock/flow diagram) are most popular. Lane (2000) has commented on their respective advantages and disadvantages. Here the influence diagram convention of Coyle (1996) is adopted. In the diagram, solid lines represent actual physical flows and dotted lines represent information flows and control actions. Arrows show influence direction. Positive sign indicates increment and negative sign indicates decrement. External inputs or parameters, which are not influenced by other elements in the system, are written in bold, italic characters and underlined. Character *D* near a line means a delay that is an interruption between an action and its consequences. Delays occur frequently in dynamic systems and often result in overshooting a desired outcome. The ‖ symbol marks a non-negativity constraint.

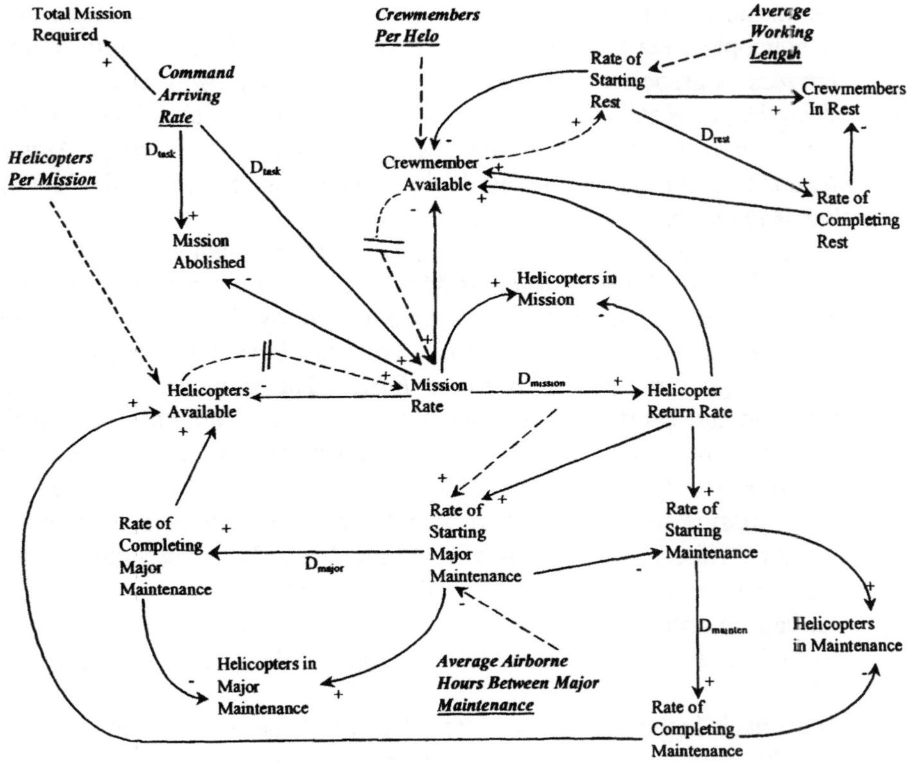

Figure 1. The influence diagram.

Mission Rate is the major driving factor in the system. It is determined by *Command Arriving Rate* as well as the availability of armed reconnaissance helicopters and crewmembers, which in turn are affected by the number of helicopters per mission (for safety reasons, it is assumed that helicopters always operate in pairs) and the predetermined minimum number of crew members for a mission. At any instant, the number of *Helicopters Available* for duty is compared with the required number of helicopters. If there are enough armed reconnaissance helicopters available, an auxiliary parameter called *Helicopter Trigger*, which isn't shown in the influence diagram, is turned on. Similarly the number of *Crewmember Available* is compared with the required number of crewmembers. If there are enough crewmembers available, *Crewmember Trigger* is turned on. Only when both *Helicopter Trigger* and *Crewmember Trigger* are on, that is both the required minimum number of helicopters and crew members are available for duty, can the commander actually dispatch helicopters for mission.

Helicopters Available is decreased by the *Mission Rate* and increased by the *Rate of Completing Maintenance* and the *Rate of Completing Major Maintenance*. *Rate of Starting Major Maintenance* is affected by *Helicopter Return Rate, Average Airborne Hours Between Major Maintenance* and average mission duration ($D_{mission}$). The more missions flown, the higher the rate of helicopters starting major maintenance. *Rate of Completing Major Maintenance* is associated with *Rate of Starting Major Maintenance* via a time *delay* D_{major}. *Rate of Starting Maintenance* (routine one) is increased by *Helicopter Return Rate* and decreased by *Rate of Starting Major Maintenance*. *Rate of Completing Maintenance* is associated with *Rate of Starting Maintenance* via a time delay $D_{mainten}$.

The crewmember section is similar. *Crewmember Available* is reduced by *Rate of Starting Rest* and *Mission Rate* with a proper multiplier *Crewmembers Per Helicopter* and supplemented by *Rate of Completing Rest*. *Crewmember Available* and *Average Working Length* of a crewmember govern *Rate of Starting Rest*. The more crew members available and the shorter the average time that a crewmember can work before a decent rest, the higher the rate that crewmembers start rest. *Rate of Completing Rest* is related to *Rate of Starting Rest* through a time delay D_{rest}.

The mission section consists of two delayed flow modules. *Command Arriving Rate* affects *Mission Rate* after a time delay D_{task}. The latter in turn affects *Helicopter Return Rate* after another time delay $D_{mission}$.

It can be noted that the influence diagram describes main operations and most of variables, however some fine details of the simulation model are omitted.

Feedback loops are fundamental structures that define complex real systems. In these systems, many variables, linked in complicated webs of interrelationships, affect each other in often surprising ways. We are sometimes surprised by dynamic behaviour of a system because we cannot anticipate how the myriad influences feedback on each other. There are two kinds of feedback loops: positive feedback loops and negative feedback loops. Positive feedback loops generate exponential growth and collapse - chain reaction effects. Usually, there exist limiting or balancing effects -

negative feedback - against these loops. Negative feedback loops generate resistance, maintain stability and help achieve equilibrium – goal seeking. Several feedback structures of the model can be identified by inspecting the influence diagram. For example, one positive feedback loop starts at *Crewmember Available*, goes through *Rate of Starting Rest* and *Rate of Completing Rest*, then back to *Crewmember Available*. The more crewmembers there are, the more begin resting, and after a delay for resting, the more will again become available. This is countered by a negative feedback loop from *Crewmember Available* to *Rate of Starting Rest* back to *Crewmember Available*, which helps to ensure that the total number of crewmembers will remain constant. The more crewmembers there are, the more start resting and the fewer crewmembers available. Similar feedback loops can be found in the helicopter section.

It can be seen that the influence diagram puts the whole helicopter operations as a system on a piece of paper and clearly shows all the relationships between various variables. It also shows at a glance the structure and assumptions of the model. It helps one to analyse a complex system qualitatively.

3. THE QUANTITATIVE RESULTS

To obtain a quantitative prediction of the system's behaviour, based on the influence diagram, an executable model was constructed using Powersim Constructor 2.51 software package. The parameters *Helicopters Per Mission* and $D_{mission}$ are chosen to be 2 helicopters and 2 hours respectively. The first one is based on the safety consideration that helicopters always go in pairs so that they can support each other in emergencies and cover each other's sensor shadows. The second one is determined by the helicopter's endurance, speed, and average transit distance. It is assumed that there are 12 helicopters and 24 crewmembers. On a mission, each helicopter needs two crewmembers. The commander will not dispatch any helicopters unless there are the minimum required number of helicopters and crewmembers available. The *Average Airborne Hours Between Major Maintenance* is 100 hours and *Average Working Length* of a crewmember is 10 hours. The other time delays used in the model are $D_{mainten} = 1$ hours, $D_{major} = 48$ hours, $D_{rest} = 8$ hours, and D_{task} takes a random value between 0.5 and 1.5 hours. Except for D_{task}, all delays are treated as third order material delays. The variation of the C2 and tasking delay is modelled by generating a random delay time; therefore a pipeline delay function is used. *Command Arriving Rate* is modelled using a PULSE function with a random interval between 0.5 and 2.

To ensure the validity of the model, several tests were conducted. A careful check was made to ensure that the influence diagram corresponded to the narrative description and the equations corresponded to the influence diagram. To ensure that the dynamics produced were due to the internal relationships of the system and not to faulty initial conditions, the PULSE function's first value was set to 5 to prevent anything happening until Time=5. All the model variables were clearly defined and the dimensional validity was thoroughly checked. A mass balance section was built into the model to check that both numbers of crewmembers and helicopters were conserved. Simulation time step was set as 0.0625, which was determined by the smallest delay $D_{mainten}$. A simulation of a 500-hour operation was carried out and every variable value was printed out and examined carefully. The system's behaviour shows that the model does produce reasonable results and behaves as the real system would for the same inputs and constraints.

A variety of insights about the helicopter operations as a system were derived from the simulation. For example, it showed that, during the 500 hour period, 475 missions were required. Of these, 42.7 missions were abolished, that is 91% of required missions were actually flown. Figure 2 shows the total number of required missions and the number of missions abolished, as a function of time.

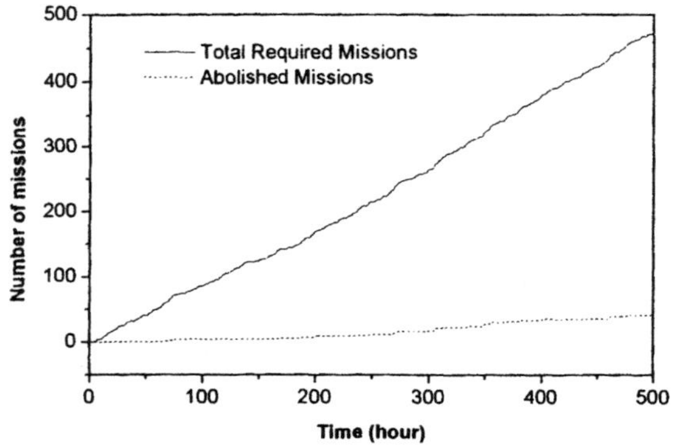

Figure 2. Number of missions abolished and number of missions required as a function of time.

The following four figures (Figure 3 - Figure 6) are bar graphs showing the number of helicopters at various states: in mission, in routine maintenance, in major maintenance and ready for a mission. These numbers should add up to a constant equal to the total number of helicopters. The data used in these figures have been rounded to the nearest integer.

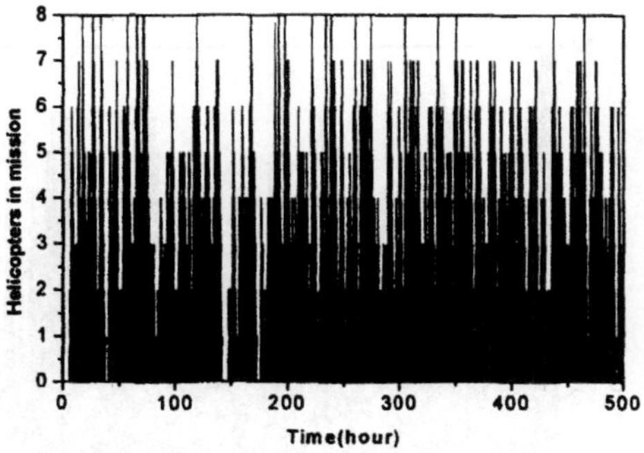

Figure 3. Number of helicopters in mission as a function of time.

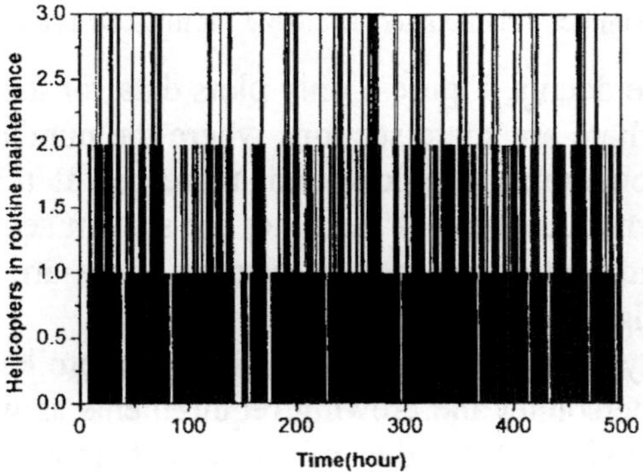

Figure 4. Number of helicopters in routine maintenance as a function of time.

Figure 3 shows that at any time there are up to 8 helicopters in missions with an average value of 4 helicopters. At any time, the maximum number of helicopters in routine maintenance (Figure 4) is 3, with an average value of 2 helicopters. Except during the initial stage, there are always one to two helicopters in major maintenance (Figure 5).

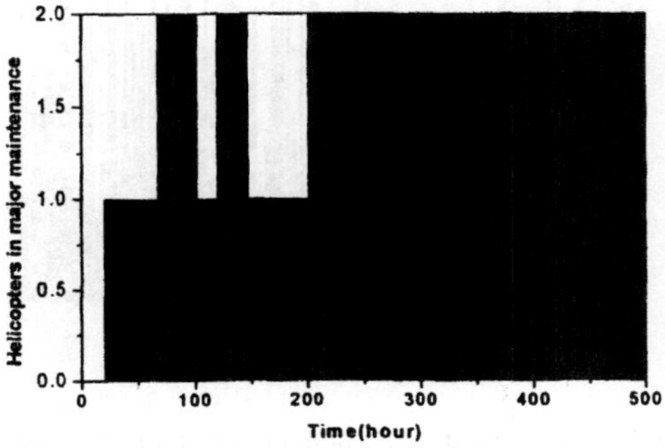

Figure 5. Number of helicopters in major maintenance as a function of time.

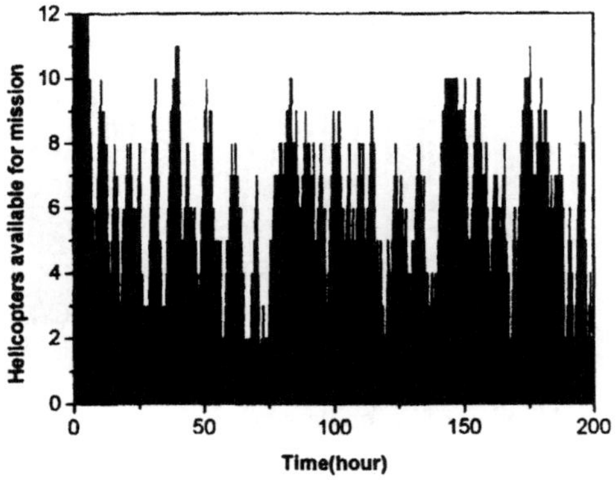

Figure 6. Number of helicopters available for mission as a function of time.

To show more details, Figure 6 only plots data for the first 200 hours. It can be seen that there are some sections where the number of helicopters available for mission drops to below two, therefore at that time no new mission can be flown. But most of the time, the system seems to cope with this level of demand quite well and on average, there are five helicopters available for new missions.

The system dynamics model provides insights into key aspects of the helicopter systems personnel and crewing requirements as well. Figures 7 - 9 are bar graphs showing the number of crewmembers at various states (in mission, resting, and available for a mission). The data used in the graph have been rounded to the nearest integers.

Figure 7 shows that at any time there are 0 to 17 crewmembers flying missions, with a mean of 7, and Figure 8 shows that on average there are 7 crewmembers in rest throughout the duration.

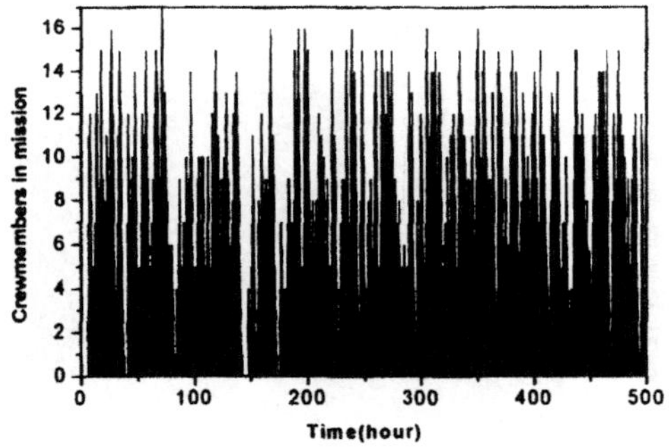

Figure 7. Number of crewmembers in mission as a function of time.

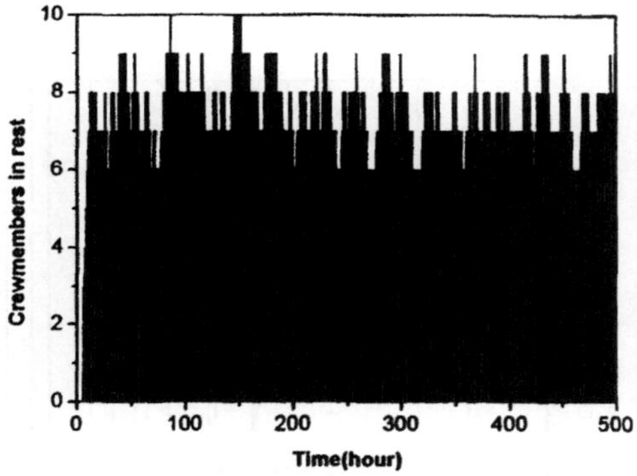

Figure 8. Number of crewmembers in rest.

Figure 9 shows that, although on average there are 10 crewmembers available, on several occasions this number drops below 4. This would result in armed reconnaissance helicopter missions being abolished due to a lack of aircrew, which reflects a key constraint on real military helicopter operations. Note that Figure 9 only plots results for the first 150 hours to show more details.

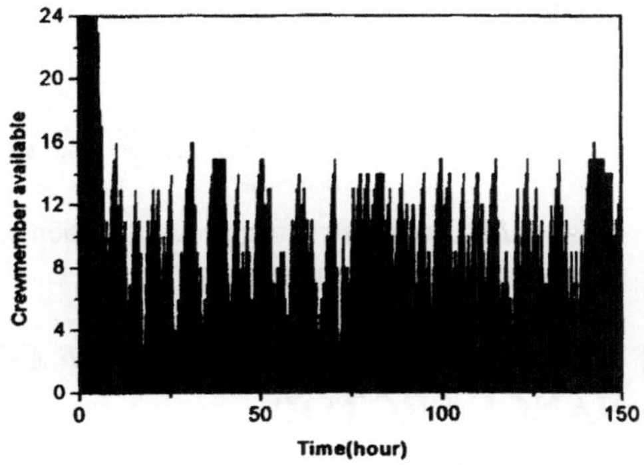

Figure 9. Number of crewmembers available as a function of time.

Figure 10 and Figure 11 show the Helicopter Trigger and the Crewmember Trigger respectively. The vertical lines in the graph represent when these triggers are off, that is take a value of 0 rather than 1. When either of them equals 0, indicating that the required corresponding resource is not available, the mission will be abolished.

It is clear that quite a few missions are abolished because of the shortage of crewmembers rather than helicopters. So, in this circumstance, 12 helicopters with 24 crewmembers is not an optimal combination. Simulation results show that, if the total number of crewmembers is increased to 29, the number of abolished missions is reduced to 40.5. This aspect of modelling of personnel requirements is a key objective of applying system dynamics modelling to armed reconnaissance helicopter capability.

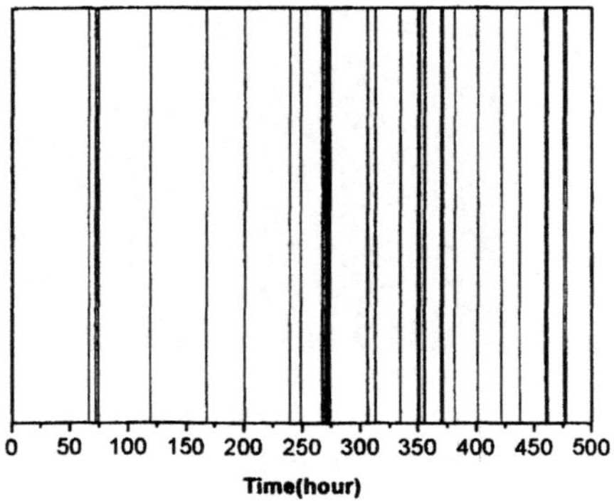

Figure 10. Helicopter trigger as a function of time.

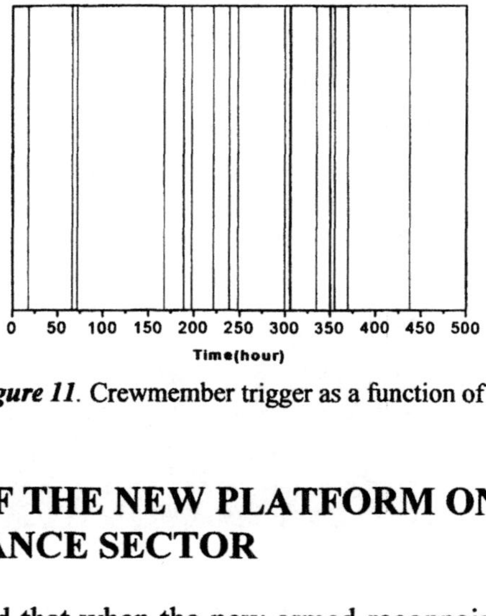

Figure 11. Crewmember trigger as a function of time.

4. IMPACT OF THE NEW PLATFORM ON THE MAINTENANCE SECTOR

It was perceived that when the new armed reconnaissance helicopters are introduced, the bottleneck will lie in the maintenance sector. Since the experience of the maintenance personnel is so little, if any parts are broken, it will take them a lot longer to fix them. To solve this problem, a sufficient amount of spare parts and additional intensive training of maintenance personnel will be needed. If the new platforms are introduced with the current maintenance system, they will not be operational. Perhaps more maintenance personnel with diverse expertise are needed, since it may be unrealistic to demand that the maintainers be expert in every aspect of the new platform, for example one person may not be an expert in both radar and thermal imaging systems. Because the new platform will be operating along with the old platforms, the demand on maintainers will be extremely high. The maintainer for one platform may not be able to maintain the other platform.

One of the objectives of building this system dynamic model is to test this hypothesis. Simulations were carried out doubling both maintenance times. The number of abolished missions sharply increased to 96.2, ie only 80% of the required missions can be flown. This clearly highlighted the importance of paying great attention to the maintenance sector while introducing the new platform into services.

5. OPTIMISATION

The major aim of a system dynamics study is to help decision makers design better policies. In the last few sections, it has been shown that the combination of an influence diagram and simulation can be used to identify bad or good policies and illuminate the way to improve things. Optimisation is another powerful technique provided by some system dynamics software packages. It searches for the best possible policy structures and parameters given the constraints of the situation. The above simulation results show that the ratio of the number of armed reconnaissance helicopters to the number of crewmembers greatly influences the system's performance. Can we find an optimum value of this ratio in order to guide the designing of the personnel aspect of the armed reconnaissance capability?

The measure of the system performance is the percentage of required missions that have actually been flown. Suppose that the commander has a desired performance level in mind, for example 95% of the missions should be accomplished, a reasonable objective function is the square of the difference between the actual percentage of missions flown and the desired performance level. As discussed in Section 3, for the base case, only 91% of the missions are flown, falling below the expected 95%. The purpose of optimisation is to find a suitable parameter value combination that lifts the performance level to 95%.

A STEP function is used in the objective function to make sure that its value is zero before the start of the operation and a scale factor sets its final value for the base case to 100. To optimise the model's behaviour is to minimise this objective function (see the equation below).

$$\text{Minimise}\left(Step\,(1,5) * \left(\frac{Mission\_Flown}{Total\_Mission} - 95\,\% \right)^2 \Big/ 62899.371 \right)$$

Optimisations were carried out using Powersim Solver 2.0 software. First the optimisation was run with the optimising parameter of initial number of crewmembers to determine the best level of human resources to run 12 armed reconnaissance helicopters. It is found that 28.2 are the optimal number of crewmembers. The objective function value is 75.7 and the abolished missions are reduced to 40.2.

The second optimisation was done with optimising parameters of both initial numbers of crewmembers and armed reconnaissance helicopters. It is found that with 13.4 armed reconnaissance helicopters and 30.3 crewmembers, the value of the objective function can be reduced to 0.001. As it is impossible to have 0.4 helicopters or 0.3 crewmembers, the results simply imply that with one more armed reconnaissance helicopter and six more crewmembers, the performance of the system can be remarkably enhanced and nearly reaches the preset goal.

Major maintenance is another area that can be improved. Thus the third optimisation was carried out with two optimising parameters: D_{major} and *Average Airborne Hours Between Major Maintenance* (*AAHBMM*). The results show that if *AAHBMM* can be extended to 110 hours and D_{major} reduced to 26.3 hours, then the value of objective function can be reduced from 100 to 31.9. This would represent a large improvement in the system's performance.

Finally, an optimisation was conducted with all four optimising parameters and the results are listed in Table 1 and shown in Figure 12. The optimisation results suggest that the best solution to the problem is to have one more armed reconnaissance helicopter and increase the number of crewmembers to 29. At the same time, extend AAHBMM for another 10 hours and reduce the length of major maintenance a little. It is quite obvious that adding more armed reconnaissance helicopters and crewmembers is the most effective way of enhancing the system's performance. However this is a very expensive solution. When the budget is limited and this is out of the question, with some effort in major maintenance sector the system's performance can be improved, though it will not reach the preset goal.

Table 1. Optimisation Results.

Parameter	Final value	Original value	Lower limit	Upper limit
Initial Helicopters	13.4	12	10	15
Initial Crewmembers	28.8	24	20	35
D_{major}	43.0	48	24	60
AAHBMM	109.7	100	80	120
Initial value of objective function			100	
Final value of objective function			0.001	

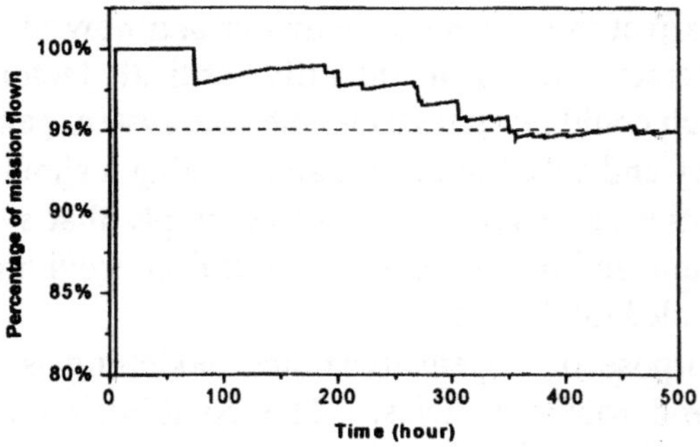

Figure 12. The optimisation results: the solid line is the percentage of mission flown and the dashed line is the 95% benchmark.

6. CONCLUSION

Defence problems are usually very complicated and unpredictable. It is extremely hard to solve this kind of problem with any strictly defined analytical methodologies, especially when human factors are involved. However, under most circumstances, any specific defence problem may be viewed as an integrated system managed by people and the real issue becomes whether the system behaviour can be improved and, if so, then how. System dynamics is a methodology that takes a systematic point of view and deals with overall system performance as time passes. Its information/action/consequences paradigm makes it ideal as an approach to the analysis of defence problems.

This paper presents the results of applying the system dynamics modelling technique to solve an armed reconnaissance helicopter operation problem. This case study has provided a greater understanding of the system, and it helps to lay a solid foundation for the future capability development of armed reconnaissance helicopters in the Australian Army. The model and simulations show that the system dynamics modelling technique can be used to determine the optimal platform and crewmember ratio, and reveal the potential problems that can be caused by an inadequate maintenance sector. Further work may involve a more detailed model of the maintenance sector, which takes into account issues such as personnel training level and spare parts stocks.

During the modelling exercise, several benefits of using system dynamics modelling technique become apparent.

- The process of constructing an influence diagram is flexible and helps both analysts and military officials to better define the problem, to determine the really critical issues and be able to ignore unimportant details.

- By closely examining feedback loops, the analyst is able to explain how various factors affect the system's behaviour and how the different parts of the system interact, so as to identify critical factors and operating procedures, which could be changed to enhance system performance.
- The simplicity and speed of coding and running a simulation model, once a satisfactory influence diagram is in place, imply that system dynamics is economical in time and resources and is useful in providing initial guidance before other detailed modelling.
- The main purpose of system dynamics modelling is to help managers design sound and robust policies. The combination of qualitative and quantitative analysis supplies very flexible experimental techniques.
- The iteration process of exploring various possibilities by studying the influence diagram, testing combinations of parameters and structures by repeated simulations, inspecting and elucidating the results carefully, and then thinking and developing other possibilities, is quite thought provoking. It helps one understand the system thoroughly and adopt operational policies that are highly adaptable in uncertain and ever changing situations.
- It can often show counter-intuitive behaviour of the system and demonstrates that obvious solutions are sometimes wrong, or even worse, measures that are intended to solve the system's problem, adopted without careful research, may well be the reason that the system performs poorly.
- The techniques of system dynamics optimisation are very powerful and facilitate multi-dimensional parameter searches to derive a better managed system. It enables the decision makers to gain insights into the performance of defence systems.

In summary, the work presented here illustrates the unique capability of the system dynamics model in representing the complex, changing and nonlinear defence helicopter operations. Its ability to model the feedback structures, which are inherent to the system, is very important. It enhances our understanding of the system, allows an easy evaluation of alternative policies and resolution of conflicting viewpoints, and helps to develop a consensus about future action. The work highlights the usefulness of the system dynamics modelling in determining pressure points, and the role it can play as a test bed for decision makers in the course of developing the critical new armed reconnaissance helicopter capability. System dynamics is a wonderful approach to analysing defence problems, if it is applied properly and its limitations are well recognised.

References

Clark, R.H. The Dynamics Of Force Reduction And Reconstitution. Defense Analysis 1993; 9: 51-68

Coyle J. M., Exelby, D. and Holt, J. System Dynamics In Defence Analysis: Some Case Studies. Journal of the Operational Research Society 1999; 50: 372-382

Coyle, R. G. The Dynamics Of The Third World War. Journal of the Operational Research Society 1981; 32: 755-765

Coyle, R. G. A System Description Of Counter-Insurgency Warfare. Policy Sciences 1985; 18: 55-78

Coyle, R. G. A Model For Assessing The Work-Processing Capacity Of Military Command And Control Systems. European Journal of Operational Research 1987; 28: 27-43

Coyle, R. G. A System Dynamics Model Of Aircraft Carrier Survivability. System Dynamics Review 1992a; 8: 193-212

Coyle, R. G. The Optimisation Of Defence Expenditure. European Journal of Operational Research 1992b; 56: 304-318

Coyle, R. G. System Dynamics Applied To Defense Analysis: A Literature Survey. Defense Analysis 1996a; 12: 141-160

Coyle, R. G. *System Dynamics Modelling: A Practical Approach.* London, UK: Chapman & Hall, 1996b

Coyle, R. G. and Gardiner, P. A. A System Dynamics Model Of Submarine Operations And Maintenance Schedules. Journal of the Operational Research Society 1991; 6: 453-462

Forrester, J. *Industrial Dynamics*. Portland Oregon, USA: Productivity Press,1961.

Kearney, J. Defence Applications Of System Dynamics Models. Journal of Battlefield Technology 1998; 1: 33-34

Lane, D. C. Diagramming Conventions In System Dynamics. Journal of the Operational Research Society 2000; 51: 241-245

Moffat., J. The System Dynamics of Future Warfare. European Journal of Operational Research 1996; 90: 609-618

Wang, K.-d. and Huang, Z.-l. The System Dynamics (SD) Is Used For Researching On Battle Simulation. Journal of Systems Science and Systems Engineering 2000; 9: 367-372

Wolstenholme, E. F. Defense Operational Analysis Using System Dynamics. European Journal of Operational Research 1986; 34: 10-18

Wolstenholme, E. F. *System Enquiry: A System Dynamics Approach.* Chichester, UK: John Wiley & Sons, 1990

Wolstenholme, E. F., Gavine, A., Watts, K. and Henderson, S. The Design, Application And Evaluation Of A System Dynamics Based Methodology For The Assessment Of Computerised Information Systems. European Journal of Information Systems 1992; 1: 341-350

---

* The views expressed in this paper are solely those of the author and have no official standing of any description.

INDEX

A
Agent 163
Airship control 205
Ant colony optimisation 41, 111
Ant system 41
Arc routing 1

B
Branch-and-cut 1

C
Collaborative learning 215
Combinatorial optimisation problems 41
Composite concave programming 299
Concept map 215
Constructive metaheuristics 111

D
Dairy industry 71
Data envelopment analysis 299, 333
Decision making 149
Decision support system 281, 317
Diffusion models 395
Discrete dynamic game 93
Division-of-labour problems 129
Duality 333

E
Efficiency 299
Electricity market 375

F
FISM 215
Flowshop 21
Forecasting 375

G
Genetic algorithm 93, 149
Global convergence property 227
Graph-partitioning problem 163

Greedy randomised search 111

H
Helicopter operations 405
Heuristics 177, 189
Hyper sensitivity analysis 299

I
Immune cell-cooperation 129
Implicit enumeration 351
Influence diagram 405
Inventory control 243

L
Linear fractional programming 333
Logistics 261

M
Maintenance 177
Management 363
Mateselection 71
Mathematical programming 281, 317
Max-min optimisation 351
MHC and immune network 129
Mixed-model assembly 21
Model management 281, 317
Modelling 281, 317
Multi-agent system 41, 129
Multi-commodity problem 53
Multiple knapsack problem 351
Multiple objective programming 299

N
Nurse scheduling problem 149

O
Open loop 93
Operations scheduling 243
Optimisation 129, 281, 317

P
Parallel algorithm 163
Pareto ranking scheme 149
Planning 363

Precision farming 363
Production planning 243
Productivity 299
Push/pull hybrid production system 189

Q
Queues 395

R
Ranking of fuzzy numbers optimisation 261
Real time recurrent learning 227
Recurrent neural network 227
Reinforcement learning 205
Routing 53
Rural postman problem 1

S
Scheduling 21, 177, 189
Sequencing 21
Simulated annealing 163, 189
Simulation 405
Stochastic modeling 243
Storage 53
System dynamics 405

T
Tabu search 1, 71, 163
Transmission 53
Travelling salesman problems 41
Triangular fuzzy number 261

V
Variable neighbourhood search 1

List of Contributors

Abbass Hussein
Brown Alan
Bulmer Michael
Burdett Robert
Churilov Leonid
Corry Paul
Driessen Theo
Endo Satoshi
Ferreira Luis
Fukushima Masao
Ghiani Gianpaolo
Hertz Alain
Jan Ahmad
Kaji Taichi
Kakazu Yukinori
Kawamura Hidenori
Keiko Motoyama
Kimura Toshikazu
Kozan Erhan
Lai K. K.
Lake Maree
Laporte Gilbert
Leung Stephen
Maeda Takashi
McAllister Ryan
Miyagi Hayao
Mukohara Tsuyoshi
Namekata Tsuneyuki
O'Brien G. C.
Ohuchi Azuma
Okuno Taku
Ouveysi Iradj
Preston Peter
Randall Marcus
Saito Hajime
Sekiguchi Yasuki
Sesana L.
Sniedovich Moshe
Suzuki Keiji
Takamatsu Kazuki
Takeo Yamada
Tanino Tetsuzo
Tatsumi Keiji
Tobin Patrick
Toma Naruaki
Towsey Michael
Van der Werf Julius
Wirth Andrew
Wu Lifen
Yamada Koji
Yamamoto Masahito
Yue Yi

GPSR Compliance

The European Union's (EU) General Product Safety Regulation (GPSR) is a set of rules that requires consumer products to be safe and our obligations to ensure this.

If you have any concerns about our products, you can contact us on ProductSafety@springernature.com

In case Publisher is established outside the EU, the EU authorized representative is:

Springer Nature Customer Service Center GmbH
Europaplatz 3
69115 Heidelberg, Germany

Batch number: 10755284

Printed by Printforce, the Netherlands